AERONOMY OF THE MIDDLE ATMOSPHERE

ATMOSPHERIC SCIENCES LIBRARY

Aeronomy of the Middle Atmosphere

Chemistry and Physics of the Stratosphere and Mesosphere

by

GUY BRASSEUR

Institut d'Aéronomie Spatiale and
Université Libre de Bruxelles, Brussels, Belgium

and

SUSAN SOLOMON

Aeronomy Laboratory, National Oceanic and Atmospheric Administration,
Boulder, Colorado, U.S.A.

Second edition

D. Reidel Publishing Company

A MEMBER OF THE KLUWER ACADEMIC PUBLISHERS GROUP

Dordrecht / Boston / Lancaster / Tokyo

Library of Congress Cataloging in Publication Data

Brasseur, Guy.
  Aeronomy of the middle atmosphere.

  (Atmospheric science library)
  Includes bibliographies and index.
  1. Stratosphere. 2. Mesosphere. 3. Middle atmosphere.
I. Solomon, Susan, 1956–   . II. Title. III. Series.
QC881.2.S8B73 1986   551.5′142   86–20358
ISBN 90–277–2343–5
ISBN 90–277–2344–3 (pbk.)

Published by D. Reidel Publishing Company,
P.O. Box 17, 3300 AA Dordrecht, Holland.

Sold and distributed in the U.S.A. and Canada
by Kluwer Academic Publishers,
101 Philip Drive, Assinippi Park, Norwell, MA 02061, U.S.A.

In all other countries, sold and distributed
by Kluwer Academic Publishers Group,
P.O. Box 322, 3300 AH Dordrecht, Holland.

First edition © 1984.
Second revised edition © 1986.
Reprinted 1995.
Reprinted 1996.
Reprinted 1998.

Contents

Preface

The reader may be surprised to learn that the word "aeronomy" is not found in many of the standard dictionaries of the English language (for example, Webster's International dictionary). Yet the term would appear to exist, as evidenced by the affiliations of the two authors of this volume (Institut d'Aeronomie Spatiale, Brussels, Belgium; Aeronomy Laboratory, National Oceanic and Atmospheric Administration, Boulder, CO, USA). Perhaps part of this obscurity arises because aeronomy is a relatively new and evolving field of endeavor, with a history dating back no farther than about 1940. The Chambers dictionary of science and technology provides the following definition:

> "aeronomy (Meteor.). The branch of science dealing with the atmosphere of the Earth and the other planets with reference to their chemical composition, physical properties, relative motion, and reactions to radiation from outer space"

This seems to us an appropriate description, and it is reflected throughout the content of this volume.

The study of the aeronomy of the middle atmosphere experienced rapid growth and development during the 1970's and 1980's, particularly due to concern over the possibility of anthropogenic perturbations to the state of the middle atmosphere and its protective ozone layer. As a result, much has been learned regarding both the natural behavior of the atmosphere and the impact of man's activities upon it. In this book we shall attempt to describe the current state of the art as we see it. Our particular viewpoints on aeronomy certainly reflect the influences of our teachers, who have done much to shape not only our personal understandings of the field, but have also been among the central figures in its development. We most gratefully acknowledge the people who gave us our first insights into what aeronomy is, Marcel Nicolet (GB), and Paul Crutzen (SS).

A word for which we have not found a simple definition, either in the dictionary or in the real world, is "aeronomer". For indeed, aeronomy is a highly interdisciplinary science, drawing on the fields of chemistry, physics, fluid dynamics, meteorology, statistics, etc., and aeronomers come from all of these fields as well as several more. This book is intended to provide an overview of the field in a manner understandable to persons familiar with college level chemistry and physics at about the senior undergraduate level, and also to review the field of aeronomy in such a way as to be a useful reference to researchers already working in the area. To this end we have written Chapter 2, entitled "Chemical concepts in the atmosphere", which might also be called "Aspects of aeronomy for non-chemists". Our goal in that section is to summarize and review the basic concepts from physical chemistry which are of relevance to the field of aeronomy. Similarly, Chapter 3, entitled "Structure and dynamics" might alternatively be titled "Dynamics for non-meteorologists", since it presents a highly simplified view of the basic elements of dynamics as they pertain to atmospheric transport of chemical species. Chapter 4 outlines the role of radiation in the chemical and thermal budgets of the middle atmosphere. Chapter 5 describes the chemical composition of the atmosphere, and includes of necessity a detailed account of all the chemical processes of importance. If this seems rather boring to the reader who is neither accustomed to nor interested in the list of the several hundred reactions which must be noted, we emphasize the more fundamental concepts which are also to be found there, namely the descriptions of the importance of transport processes on each species, and the discussion of what may be learned from available measurements. We have constructed subsections in that chapter which should help the reader to sort through the detailed chemistry and identify the roles played by each reaction more easily. Chapter 6 presents a similar analysis for ionic species of importance to the middle atmosphere. Finally, Chapter 7 examines the importance of coupling and the roles of both natural and anthropogenic perturbations.

Although the middle atmosphere may be roughly defined as the region between the tropopause (about 12 km) and the homopause (about 100 km; see Chapter 3 for definitions), we have sometimes found it necessary to discuss the aeronomy of the troposphere (surface to about 12 km) and thermosphere (region above about 100 km) in order to explain the behavior of the middle atmosphere as a result of its neighbors' influences. We emphasize, however, that our decisions regarding where these other atmospheric layers should be mentioned are somewhat arbitrary, and that no attempt at a complete discussion of the neighboring layers has been made in this volume.

Due to the large number of topics discussed here, we have found it impossible to provide a complete review of the available literature. In many cases, we have opted to reference more recent work to illustrate our ideas; we anticipate that the interested reader can trace the detailed historical development of the indicated concepts by using these as starting points.

Critical reviews of portions of the manuscript by our colleagues have certainly improved the quality of this work. We particularly thank Rolando R. Garcia, whose recommendations and suggestions are reflected throughout Chapter 3. He contributed greatly to the discussion of the dynamical equations, and to much of the material on transport processes. The comments provided by James R. Holton also were important to our development of Chapter 3, as were those of Paul J. Crutzen to Chapter 5. We gratefully acknowledge as well the suggestions of M. Ackerman, D. Albritton, E. Arijs, L. Bossy, M. Carroll, D. Fahey, F. Fehsenfeld, E. Ferguson, M. Geller, D. Hartmann, A. Henderson-Sellers, C. Howard, J. Kiehl, D. Kley, J. Lenoble, S. Liu, J. Mahlman, R. Meier, G. Mount, C. Nicolis, J. Noxon, J. Olivero, G. Reid, B. Ridley, J. Roberts, A. Schmeltekopf, W. Sedlacek, P. Simon, C. Tricot, R. Turco, and D. Wuebbles.

We appreciate the assistance and honesty of Dr. David J. Larner, who made our interaction with Reidel Publishing Company a pleasant one.

We also acknowledge permission to publish uncopyrighted figures, graciously extended by Dr. J. London and Dr. A. O'Neill.

We deeply appreciate important help with computer problems provided by R. Winkler and M. Lightner. Finally, we gratefully thank M. Neary, B. Sloan and D. Hoge for important assistance in text preparation, J. Schmitz and A. Simon for their expert figure drafting, and clerical assistance by M. DeClercq, M. Sauvage, and E. Rigo.

G. B.

S. S.

February, 1984
Boulder, Colorado

"Concern for man himself and his fate must always form the chief interest of all technical endeavors ..... in order that the creations of our mind shall be a blessing and not a curse to mankind. Never forget this in the midst of your diagrams and equations" - *Albert Einstein*

List of Principal Symbols

The following symbols are used throughout this volume. Some symbols which are of local use only are not included here. In a few cases, the same symbols have different meanings in different chapters; these are indicated.

a	Radius of the Earth
a_e	Electron-neutral attachment rate
b	Collision cross section
c	Speed of light
	and gravity wave phase speed (Chapter 3 only)
c_p	Specific heat at constant pressure (per gram)
c_1	First radiation constant
c_2	Second radiation constant
d	Mean line spacing
d_p	Photodetachment rate
e	Electron charge
e_s	Saturation vapor pressure of H_2O
f	Coriolis parameter
f_d	Negative ion collisional detachment rate
f_i	Mixing ratio of species i
$f(r_a)$	Aerosol size distribution function
f	Wave frequency
g	Acceleration of gravity
h	Planck's constant
$j(z,E)$	Flux of ionizing particles
k	Boltmann's constant
k_a	Absorption coefficient
k_f	Wave propagation factor
k_s	Scattering coefficient
k_ν	Total extinction coefficient at frequency ν
$k_{act, uni, d, r, 1,2,3..}$	Reaction rate constants
$m, m_{A,B}$	Mass of atom or molecule A, B
m_e	Mass of electron
n, n_i	Number density of ith constituent
n_e	Electron density
n^-	Negative ion density
n^+	Positive ion density
n	Index of refraction
p	Pressure

$p(\nu); p(\Theta)$	Phase function for frequency ν or angle Θ
q	Solar flux
q_m	Solar flux at frequency ν
r	Distance from center of the earth
r_a	Aerosol radius
r_s	Sun-earth distance
$\vec{r}$	Position vector
r	Energy deposition rate
s	Length
t	Time
u	Zonal wind speed
$\bar{u}$	Mean zonal wind speed
u'	Eddy deviation from mean zonal wind speed
u	Dimensionless optical depth $= Sw/A_o$
v	Meridional wind speed
v_r	Relative particle speed
$\vec{v}$	Vector speed of an air parcel
$\bar{v}$	Eulerian mean meridional wind speed
$\bar{v}^*$	Residual Eulerian mean meridional wind speed
v'	Eddy deviation from mean meridional wind speed
w	Vertical wind speed
$\bar{w}$	Eulerian mean vertical wind speed
$\bar{w}^*$	Residual Eulerian mean vertical wind speed
w'	Eddy deviation from mean vertical wind speed
w	Optical path
z	Altitude
A	Albedo
A_D	Absorptance for a Doppler line
A_L	Absorptance for a Lorentz line
A_o	Effective band width
$A_{\delta\bar{\nu}}$	Absorptance
B	Rotational constant
B_ν, B_i	Planck function
$\vec{B}$	Induction
C_p	Heat capacity (per mole of substance)
D	Depolarization factor
$\vec{D}$	Displacement vector
E	Energy
$\vec{E}$	Electric field vector
E_{act}	Energy of activation
$\vec{F}, F_\lambda, F_\phi, F_z$	Frictional force, in the λ, ϕ, z directions (Chapter 3)
$F, (F(z), F(\vec{r}, \vec{w}), F\uparrow, F\downarrow)$	Radiation flux density (Chapter 4)
F_{ij}	Rate of CO_2 transfer from reservoir i to j

$F_{10.7}$	10.7 cm radio wave flux
ΔG	Gibbs free energy of formation
H	Scale height of neutral atmosphere
H_i	Scale height of ith constituent
$\vec{H}$	Magnetic field vector
ΔH	Enthalpy of formation
H	Hour angle
I	Ionization frequency
$I(\lambda)$	Radiation intensity at wavelength λ
J	Current density
$J_i; J(x)$	Photolysis frequency for species i
J_m	Source function
K	Absorption coefficient
K_r	Rayleigh friction coefficient
K_z	One-dimensional eddy diffusion coefficient
$K_{zz}, K_{yz}, K_{yy}, K_{zy}$	Two-dimensional eddy diffusion coefficients
$L, L_\nu, L\uparrow, L\downarrow$	Radiance
L_i	Photochemical loss frequency for species i
M	Third body, any atmospheric species
M_E	Emitted energy flux (exitance)
M_λ, M_ν	Number of particles per unit volume
N	Brunt-Vaisala frequency (Chapter 3) and total oxygen column (Chapter 4)
$N(i); N(i,z,\chi)$	Column abundance of species i, i=O_2, O_3, N_2, O
P, P_i	Rate of photochemical production for species i
P_H	Gross radiative heating rate
Q	Net radiative heating rate
Q	Ionization rate
R	Gas constant
R_o	Rossby number of the flow
R_s	Solar radius
RH	Relative humidity
S	Static stability parameter (Chapter 3) and integrated line intensity (Chapter 4)
S_c	Solar constant
ΔS	Entropy of formation
S	Unit surface for radiative flux
T	Temperature, and effective transmission (Chapter 4 only)
$T(\lambda), T_\nu$	Transmission at wavelength λ or frequency ν
$\vec{V}$	Electron speed
W	Energy per ion pair produced
α	Radiative relaxation rate
α_a	Mie parameter

α_d	Electron-ion recombination rate
α_{eff}	Effective recombination rate
α_i	Ion-ion recombination rate
β	Mean line width parameter
β_s	Dilution factor for solar radiation
γ_D	Half-width at $1/e$ maximum intensity
γ_L	Half-width at half maximum for emitting line
δ	Declination angle
ϵ	Quantum yield
ϵ_0	Permittivity of free space
ϵ_p	Potential energy
θ	Angle relative to normal of S
κ	R/c_p
λ	Longitude (Chapter 3), wavelength (Chapters 2, 4)
	Negative ion to electron ratio (Chapter 6)
μ	Angle of incidence for radiation flux
μ_0	Magnetic permeability
ν	Frequency
ν_c	Collision frequency
$\bar{\nu}$	Wavenumber
ρ	Density
σ	Stefan-Boltzmann constant
$\sigma_a, \sigma_\nu, \sigma_{\bar{\nu}}$	Cross section
τ_a	Optical depth for absorption
τ_i	Chemical time constant for species i
τ_s	Optical depth for scattering
$\tau_\nu, \tau(\lambda)$	Total optical depth
$\tau_{chem,dyn}$	Chemical or dynamical time constant
ϕ	Latitude (Chapter 3),
	Zenith direction for radiation (Chapter 4)
$\phi(i)$	Flux of chemical species i
$\phi(r_a, n)$	Extinction efficiency factor for particle of radius r_a
χ	Solar zenith angle
ω	Solid angle for radiative flux
ω_r	Radio wave frequency
Γ	Temperature lapse rate
Γ_d	Dry adiabatic lapse rate
Θ	Potential temperature
Φ	Geopotential (Chapter 3), and irradiance (Chapter 4)
$\Omega, \vec{\Omega}$	Angular rotation rate of the Earth
Ω_ν	Albedo for single scattering

Chapter 1

The Middle Atmosphere
and Its Evolution

1.1 Introduction

Our atmosphere is the medium for life on the surface of the planet, and the transition zone between Earth and space. Man has always been interested in the characteristics, manifestations and perturbations of the atmosphere around him - in its changing weather patterns, and its brilliant sunsets, rainbows, and aurorae.

The lower part of the atmosphere has been continuously studied over many years through meteorological programs. The development of rocket and satellite technology during the past 25 years has also led to the investigation of the upper atmosphere, establishing a new field of research.

The intermediate region which extends from about 10 to 100 km altitude is generally called the middle atmosphere. This region is somewhat less accessible to observation and has only been systematically studied for the past 10 to 15 years. The purpose of this volume is to outline some of the factors which control the behavior of this layer of the atmosphere, a layer which is particularly vulnerable to external perturbations such as solar variability or the emission of anthropogenic material, either at the surface or at altitude.

One of the most important chemical constituents in the middle atmosphere is ozone, because it is the only atmospheric species which effectively absorbs ultraviolet solar radiation from about 250-300 nm, protecting plant and animal life from exposure to harmful radiation. Therefore, the stability of the ozone layer (located near 20 to 25 km) is a central part of the study of the middle atmosphere.

In the study of planetary atmospheres it has been customary to distinguish between thermodynamic and dynamic aspects, which constitute a portion of the field of meteorology, and the chemical and photochemical aspects, which are part of the domain of aeronomy. However, we are now beginning to realize that the interactions between these different disciplines play an important role, particularly in the altitude region which we shall address here, so that an effort to examine the coupled dynamical, chemical and radiative problem must be made in order to understand the observed variations in atmospheric chemical constituents.

1.2 Evolution of the Earth's atmosphere

The terrestrial atmosphere has certainly evolved in the course of time. The details of this evolution, however, are not well established, and not all scholars of the process agree on how it occurred. In the following, we briefly summarize the viewpoint presented by Walker (1977), which represents one possible interpretation of this poorly known history.

Evidence indicates that the Earth has a secondary atmosphere (one produced by gradual release of gases from its interior), rather than a primordial atmosphere retained by the planet at the time of its original formation. Part of the support for this idea stems from the depletion of inert gases such as He, Ne, Ar, and Kr on Earth, relative to their observed abundance in the solar system. Since there is no known process which could remove these very unreactive species from the atmosphere, their low abundance suggests that they never represented as large a fraction of our atmosphere as they do elsewhere in our solar system (e. g., Moulton, 1905).

The composition of gases released from the solid earth, in particular their oxidation state, is dependent on the free iron present in the crust and upper mantle. Because the contemporary upper mantle contains no free iron, volcanic gases are mostly oxidized, containing much more water vapor than hydrogen or hydrocarbons, and more carbon dioxide than monoxide (Walker, 1977). But the atmosphere may have formed so long ago that the composition of the mantle was different from that found today. Indeed, geologic evidence on the growth of the Earth's crust and the oceans, and the deposition of sedimentary rocks, indicates that most of the degassing occurred early in the planet's history (see, e. g. Walker, 1977). However, evidence from core samples, particularly their observed high nickel content, suggests that the mantle never contained much free iron (e. g. Ringwood, 1959), so that we assume that the oxidation state of released gases was probably similar to what is found today in volcanic emissions.

Contemporary volcanoes are believed to release large amounts of CO_2, SO_2, Cl, F, H_2O, and N_2, and trace amounts of other substances. The fate of these constituents upon release depends in part on the thermal conditions found at the planet surface, which can be estimated rather simply, at least to a first approximation. The rate of emission of energy from the surface can be equated with the incoming net flux of solar energy, which is equal to $\pi a^2 S_c (1-A)$, where a is the Earth's radius, A is the albedo (ratio of reflected to incident light), and S_c is the flux of incident solar radiation:

$$\pi a^2 S_c (1-A) = 4\pi a^2 \sigma T^4 \qquad (1.1)$$

where σ is the Stefan-Boltzmann constant (see Appendix A), and T is the mean surface temperature. Assuming that the Earth's albedo prior to the establishment of its atmosphere was equal to the present albedo of Mars and assuming that S_c was the same as its contemporary value, we estimate an original surface temperature of about 260 K (Walker, 1977). Under these conditions, the water vapor released by volcanoes would remain gaseous, and begin to warm the Earth through absorption of infrared radiation (the "greenhouse" effect) until its vapor pressure reached about 10 mb, when supersaturation would begin, resulting in condensation, and the formation of the oceans. CO_2 would then begin to dissolve in the oceans and establish an equilibrium with the atmosphere.

This scenario, however, assumes that S_c has not changed; it is far more likely that the flux of solar radiation has increased since the earth was formed. Sagan and Mullen (1973) have shown that assuming more probable values of S_c, the surface temperature of this ancient Earth could have been as low as 238 K. Under these circumstances, the degassing water would be frozen, which does not agree with geologic evidence. It is therefore likely that additional greenhouse gases were present in the early stages of the Earth's evolution. Sagan and Mullen (1973) suggest that NH_3 may have been responsible for the initial greenhouse effect which warmed the planet enough for water to remain in the liquid and gaseous phases. Other studies imply that it is likely that CO_2 provided the necessary warming (see e.g., Owen et al., 1979), and that a cirrus cloud layer may also have contributed significant greenhouse effects (Rossow et al., 1982). In any case, the evolution of life requires liquid water, so that the temperature of the primitive Earth must somehow have been nearer to 260 K.

The gaseous water vapor present in the atmosphere would photolyze, producing some molecular oxygen. This source, however, is insufficient to account for the large percentage of oxygen found in the present atmosphere. Further, the evolution of life requires that a reducing atmosphere be present initially (Miller and Orgel, 1974).

The evolution of oxygen was probably closely tied to the evolution of life. Some of the first living creatures survived by fermentation. These were followed by chemoautotrophs which obtained energy from chemical reactions, and finally photosynthetic organisms began to produce oxygen in much larger amounts, eventually increasing the oxygen content to contemporary levels.

Present photosynthetic life, however, is protected from harmful solar radiation by oxygen and ozone. Thus we must ask how these original creatures survived and evolved to the present state even as they formed the protective shield which their descendants would enjoy. It is possible that they were a form of algae, protected by liquid water from the sun's rays. It has also been suggested that primitive microbes could be protected by layers of purine and pyrimidine bases, which absorb in the ultraviolet (Sagan, 1973).

Thus, it is clear that the evolution of the Earth's atmosphere depended on many factors, including its albedo, the biosphere (plant and animal life), the oceans, the sun, and the composition of the solid earth. How stable is the current state of the atmosphere, considering all the complex interactions which were involved in its evolution? In the rest of this chapter and throughout this volume, we will discuss the natural processes which control the present atmosphere, and assess the possible impact which man's activities may have upon it.

1.3 Possible perturbations

The possible modification of the chemical composition of the atmosphere, and the attendant climatic effects, is a problem which attracts the attention of the international scientific community. Some of the processes which may produce these modifications will be briefly identified in this section. A more detailed analysis will be presented in Chapter 7.

We first consider the effects of the agricultural revolution wrought by man and ask whether the modification of the vegetation on the Earth's surface may influence the atmosphere. In tropical regions, for example, (Brazil, Central Africa, Southeast Asia) large regions of forest and savannah are burned for agricultural reasons. It has been estimated that 160000 km^2 of wooded land are destroyed each year (e.g., Kandel, 1980). The combustion produces numerous chemical species, such as CO_2, CO, H_2, N_2O, NO, NO_2, COS, and CH_3Cl (Crutzen et al., 1979) which eventually reach the middle atmosphere, where they can influence the budgets of several minor species, particularly ozone. Further, the introduction of modern agricultural techniques, especially the intensive use of nitrogen fertilizers, has altered the natural nitrogen cycle by increasing the fixation of this element in the form of ammonia, amino acids, and nitrates. During nitrification and denitrification, part of the nitrogen is emitted to the atmosphere in the form of N_2O rather than N_2. Since N_2O provides the principal source of NO in the middle atmosphere, the use of nitrogen fertilizers ultimately leads to an acceleration of ozone destruction determined by the following reactions:

$$NO + O_3 \rightarrow NO_2 + O_2$$

$$NO_2 + O \rightarrow NO + O_2$$

$$\boxed{Net: \quad O + O_3 \rightarrow 2O_2}$$

Note that this process is catalytic (NO initiates the ozone destruction process, but is regenerated, so that no net consumption of nitrogen oxides occurs). Indeed, each stratospheric NO molecule can catalytically destroy about 1×10^{12} to 1×10^{13} ozone molecules during its lifetime in the stratosphere. Therefore, possible perturbations in the nitrogen oxide content of the atmosphere could have significant effects on the ozone layer.

The rate of nitrogen fixation on the Earth's surface is not known with great accuracy. The contributions of natural biological fixation, as well as the effects of lightning, and combustion must be considered. It is, however, estimated (Crutzen, 1976, McElroy et al., 1976, Liu et al., 1977) that the annual amount of nitrogen fixed biologically is between 180 and 260 MT (megaton), while fixation by combustion represents about 20 to 40 MT, and that the contribution from nitrogen fertilizers was near 40 MT in 1975. Since the latter increases at a mean annual rate of about 10 percent, the effect of this artificial source on the ozone layer (pointed out by Crutzen, 1974, and McElroy et al., 1976) could become significant in the future.

The possible effects of nitrogen oxides on the middle atmosphere have been very actively studied since Crutzen (1970) and Johnston (1971) indicated that the injection of large quantities of these species in aircraft exhaust could profoundly alter the protective ozone layer. Later, calculations showed that the effect introduced by such aircraft depends directly on the flight altitude: an injection of nitrogen oxide could contribute to ozone production at altitudes below about 15 km, but to ozone destruction if the craft flew at higher altitudes. Another source of nitrogen oxides which could be significant is provided by intense nuclear explosions, which were particularly numerous in the 1950's and 1960's. For example, the production of nitrogen oxides by the Soviet tests in the fall of 1962 (equivalent power of 180 MT of TNT) totaled between 0.7 and 2.7×10^{34} molecules of NO. This figure should be compared to the total number of NO molecules present between 10 and 50 km ($4-16 \times 10^{34}$; Bauer, 1978), or the global annual rate of natural production of NO from N_2O (4.5×10^{34}; Johnston et al., 1979).

The effects of industrial activities, especially the emission of gaseous effluents and their possible effects on climate, have also attracted considerable attention. The most well known problem is certainly that of CO_2, but the cycling of this gas between the atmosphere, biosphere, hydrosphere (clouds and oceans), and the lithosphere (solid earth) remains uncertain. The emission of CO_2 as a result of combustion of fossil fuels (coal, oil) adds to the natural biospheric production, and appears to be responsible for the increase in the CO_2 concentration which has been observed in the atmosphere. Thus, it seems (e.g., Bach, 1976) that between 1960 and 1975 the total amount of CO_2 in the atmosphere increased from 2.23 to 2.47×10^{12} tons. Such an increase influences the thermal budget of the atmosphere, both by cooling of the upper layers due to an increase in the emission of infrared radiation to space and through heating in the lower layers due to absorption of the infrared emission by the Earth's surface as well as by atmospheric gases. Global models indicate that a doubling of the contemporary amount of CO_2 would increase the surface temperature of the Earth by 1.5 to 4.5K (NAS, 1983), which could greatly influence the polar ice caps and global precipitation processes. This in turn can affect the surface albedo, further raising the temperature and introducing a possible positive feedback whose eventual consequences could be quite significant. The exact magnitude of the effect of increased CO_2 is difficult to establish because of the complicated feedbacks which exist between the atmosphere, oceans, and biosphere. Observations indicate that a systematic increase in the Earth's mean surface temperature occurred from 1900 to about 1940, but did not continue after that year. Since the secular temperature variation is also dependent on the aerosols emitted by major volcanic eruptions (see Chapter 7), this trend cannot be unambiguously attributed to CO_2.

Changes in the atmospheric temperature alter the rates of many photochemical processes and therefore the concentrations of minor species. For example, the increase in CO_2 between 30 and 50 km should lead to a small increase in the ozone content, according to current models. Thermal perturbations can also influence atmospheric dynamics (see Chapter 3).

The emission of chlorine compounds also is a matter of interest, because ozone can be destroyed by the following catalytic cycle (Stolarski and Cicerone, 1974; Molina and Rowland, 1974):

$$Cl + O_3 \rightarrow ClO + O_2$$

$$ClO + O \rightarrow Cl + O_2$$

$$\boxed{\text{Net:} \quad O + O_3 \rightarrow 2O_2}$$

The chlorine atom is present in many industrial products dispersed in the atmosphere, but many of these are rapidly destroyed at low altitudes and therefore do not reach the middle atmosphere where the bulk of the atmospheric ozone layer resides. Some industrial compounds do, however, have sufficiently long lifetimes to present a potential perturbation: primarily carbon tetrachloride (CCl_4 or F-10), trichlorofluoromethane ($CFCl_3$ or F-11), dichlorodifluoromethane (CF_2Cl_2 or F-12) and tricloroethane (CH_3CCl_3), all of which can be transported to the middle atmosphere and emit chlorine atoms. The industrial halocarbons are used for a number of purposes, especially as propellants in aerosol cans, solvents, and as cooling elements in refrigeration. The amount of halocarbon emitted to the atmosphere can be determined from its industrial production. It can be estimated that in 1976, for example, the rate of emission for the entire world was as follows (Bauer, 1978): 308 kilotons of $CFCl_3$, 380 kilotons of CF_2Cl_2, 51 kilotons of CCl_4, and 401 kilotons of CH_3CCl_3. Only for carbon tetrachloride has the rate of production not increased in the past thirty years. The rate of atmospheric emission of the majority of the halocarbons has increased along with industrial development, but has stabilized or even slightly decreased since 1974 due to control measures. CH_3Cl (methyl chloride) was present before the industrial era, and must also be considered in evaluating the impact of the halocarbons on the atmosphere. The quantitative model estimates of the effect of the industrial chlorine perturbation have changed considerably as the values of the reaction rate constants of several chemical processes became known with greater precision. All the models indicate that the time required for the middle atmosphere to respond to surface emission of industrial halocarbons is very long (several decades) and that the reduction of the ozone concentration is greatest near 40 km altitude. We will later discuss these effects in more detail. It should also be noted that the halocarbons are infrared active and can affect the atmospheric radiative balance.

The analysis of perturbations to the middle atmosphere must also include natural processes, such as the effects of volcanic eruptions, which produce large quantities of fine particles as well as water vapor and SO_2, which eventually produces H_2SO_4. The amount of gas injected, the composition and the maximum altitude of injection vary with the intensity of the eruption. Such events can alter the budgets of some atmospheric constituents and are clearly reflected in the middle atmospheric aerosol content. We have also noted the important role of the sun in establishing the thermal and photochemical conditions on the planet surface. Variations in the solar output are caused by its 27-day rotation period, the 11-year solar cycle, and by solar flares. Each of these processes may account for part of the natural variability of the atmosphere.

The effect of human activities on the structure and composition of the atmosphere is probably far from negligible even if the effects are not at present clearly evident. Agricultural development and industrial activities are likely to influence the ecosystem and the terrestrial atmosphere. It is therefore important to understand the behavior of the middle atmosphere, which is so vulnerable to external perturbations and so essential to the climatic and photochemical environment on Earth.

References

Bach, W., Global air pollution and climatic change, Rev. Geophys. Space Phys., 14 , 429-474, 1976.

Bauer, E., A catalog of perturbing influences on stratospheric ozone, Federal Aviation Admin. Report, FAA-EQ-78-20, 1978.

Crutzen, P. J., The influence of nitrogen oxide on the atmospheric ozone content, Quart. J. Roy. Met. Soc., 96 , 320, 1970.

Crutzen, P. J., Estimates of possible variations in total ozone due to natural causes and human activity, Ambio, 3 , 201, 1974.

Crutzen, P. J., Upper limits on atmospheric ozone reductions following increased application of fixed nitrogen to the soil, Geophys. Res. Lett., 3 , 169-172, 1976.

Crutzen, P. J., L. E. Heidt, J. P. Krasnec, W. H. Pollock, and W. Seiler, Biomass burning as a source of the atmospheric gases CO, H_2, N_2O, NO,

CH_3Cl, and COS, Nature, 282 , 253, l979.

Johnston, H. S., Reduction of stratospheric ozone by nitrogen oxide catalysts from supersonic transport exhaust, Science, 173 , 5l7, l97l.

Johnston, H. S., Serang, O., and J. Podolske, Instantaneous global nitrous oxide photochemical rates, J. Geophys. Res., 84 , 5077, 1979.

Kandel, R. S., *Earth and Cosmos,* Pergamon Press, (Oxford), l980.

Liu, S. C., R. J. Cicerone, T. M. Donahue, and W. L. Chameides, Sources and sinks of atmospheric N_2O and the possible ozone reduction due to industrial fixed nitrogen fertilizers, Tellus, 29 , 251, l977.

McElroy, M. B., J. W. Elkins, S. C. Wofsy, and Y. L. Yung, Sources and sinks for atmospheric N_2O, Rev. Geophys and Space Phys., 14 , 143, l976.

Miller, S. L., and L. E. Orgel, *The origins of life on Earth,* Prentice Hall, (Englewood Cliffs, N. J.), l974.

Molina, M. J., and F. S. Rowland, Stratospheric sink for chlorofluoromethanes: Chlorine atom catalyzed destruction of ozone, Nature, 249 , 810, l974.

Moulton, F. R., On the evolution of the solar system, Astrophys. J., 22 , 165, l905.

NAS (National Academy of Sciences), *Changing climate,* Report of the carbon dioxide assessment committee, National Academy Press, (Washington, D. C.), l983.

Owen, T., Cess, R. D., and V. Ramanathan, Enhanced CO_2 greenhouse to compensate for reduced solar luminosity on early Earth, Nature, 277 , 640, l979.

Ringwood, A. E., On the chemical evolution and densities of the planets, Geochim. Cosmochim. Acta, 15 , 257, l959.

Rossow, W. B., A. Henderson-Sellers, and S. K. Weinreich, Cloud feedback: a stabilizing effect for the early Earth?, Science, 217, 1245, l982.

Sagan, C., Ultraviolet selection pressure on the earliest organisms, J. Theo. Bio., 39 , 195, 1973.

Sagan, C., and G. Mullen, Earth and Mars: Evolution of atmospheres and surface temperatures, Science, 177 , 52, l973.

Stolarski, R. S., and R. J. Cicerone, Stratospheric chlorine: a possible sink for ozone, Can. J. Chem., 52 , 1610, l974.

Walker, J. C. G., *Evolution of the atmosphere,* McMillan Pub., (New York, N. Y.), l977.

Chapter 2

Chemical Concepts in the Atmosphere

2.1 Introduction

Aeronomy is a highly interdisciplinary field, drawing its participants from areas as diverse as chemistry, physics, electrical engineering, mathematics, meteorology, and many others. It therefore seems appropriate to briefly review some of the chemical and physical concepts which form the basis for the chemistry of importance in atmospheric processes.

Almost all of the constituents which are present in the middle atmosphere undergo chemical and photochemical processes which greatly affect their distributions. The energy provided by the sun can break the chemical bonds of many species, producing reactive fragments which interact with other compounds. Most of these fragments are *free radicals* (particles containing one or more unpaired electrons). Much of the chemistry of importance to the middle atmosphere involves free radical reactions, as we shall discuss in Chapter 5. The absorption of solar energy also plays an important role in the thermal budget and dynamical properties of the middle atmosphere. To understand the aeronomy of the middle atmosphere, therefore, we must first understand some of the basic principles of photochemistry, kinetics and thermodynamics. In this chapter, we attempt to outline how these enter into the chemistry of the middle atmosphere, and present examples of their use. The discussion presented here will be very elementary; the reader may wish to examine the bibliography for detailed treatments.

2.2 Thermodynamic considerations

The basic principles governing the reactivity of chemical substances lie in the thermodynamic properties of these species. In order to evaluate the feasibility of chemical processes, the enthalpy of reaction must be examined. The enthalpy of formation of chemical constituents or "heat content" represents the energy required to make and break the chemical bonds which compose that substance, starting only from the elements in their most stable form (e. g., O_2, N_2, S, etc.), which are conventionally assigned an enthalpy of formation of zero. Consider, for example, the formation of two ground state oxygen atoms, starting from an oxygen molecule. The energy required to break the O_2 bond is about 120 kcal/mole at room temperature, so that the enthalpy of formation of each atom is about 60 kcal/mole (See Table 2.1). Let us examine a general reaction:

$$A + B \rightarrow C + D \qquad (2.1)$$

The enthalpy change associated with such a reaction is given by the difference between the enthalpy of formation of the products and the enthalpy of formation of the reactants:

$$\Delta H_R^0 = \Delta H_f^0(C) + \Delta H_f^0(D) - \Delta H_f^0(A) - \Delta H_f^0(B) \qquad (2.2)$$

where the superscript 0 refers to the standard state, which is 298.15K by convention. If ΔH_R^0 is negative, heat is liberated by the reaction, and it is called *exothermic*. If ΔH_R^0 is positive, heat must be supplied in order for the reaction to proceed, and it is then called *endothermic*. Exothermic reactions can proceed spontaneously in the atmosphere, while endothermic processes require additional energy, and this necessary external energy may not be available.

It must be noted that ΔH is a function of temperature. Tabulations of ΔH generally refer to $\Delta H_{298.15K}$. The temperature dependence of ΔH is given by

$$\Delta H_T = \Delta H_{298.15K}^0 + \int_{298.15}^{T} C_p dT \qquad (2.3)$$

where T is the temperature and C_p is the heat capacity at constant pressure (cal/deg mol). This correction is generally small over the range of atmospheric temperatures so that the tabulated values can be applied as a first approximation.

Entropy may also influence the spontaneity of chemical reaction. In general terms the entropy represents the disorder or randomness associated with a particular process. The Gibbs free energy of reaction includes both the enthalpy and entropy associated with chemical processes, and is defined as

$$\Delta G^0 = \Delta H^0 - T\Delta S^0 \qquad (2.4)$$

and the ΔG_R^0 is defined similarly to ΔH_R^0:

$$\Delta G_R^0 = \Delta G_f^0(C) + \Delta G_f^0(D) - \Delta G_f^0(A) - \Delta G_f^0(B) \qquad (2.5)$$

If ΔG_R^0 is negative, then the reaction can proceed spontaneously at the temperature under consideration. Table 2.1 presents values of ΔG^0, ΔH^0, S^0 and C_p^0 for a number of atmospheric species.

One of the most important uses of thermodynamic data is the evaluation of the feasibility of proposed chemical reactions, which have perhaps never been studied in the laboratory. We imagine, for example, that the reaction of NO_3 with H_2O might be an important source of HNO_3:

$$NO_3 + H_2O \rightarrow HNO_3 + OH \tag{2.6}$$

Is this reaction possible? This question can be answered immediately by examining the Gibbs free energy of reaction:

$$\Delta G_R^0 = \Delta G_{HNO_3}^0 + \Delta G_{OH}^0 - \Delta G_{NO_3}^0 - \Delta G_{H_2O}^0 \tag{2.7}$$

which equals +17.8 kcal/mole. Therefore, the reaction is thermodynamically impossible without an external source of energy; we need not await laboratory studies to rule out the possibility of its occurrence in the atmosphere.

Another important use of thermodynamic data in atmospheric applications is the evaluation of the exothermicity of chemical reactions which may play a role in the thermal budget of the atmosphere. Consider, for example, the recombination of atomic oxygen:

$$O + O_2 + M \rightarrow O_3 + M \tag{2.8}$$

where M can be any arbitrary atom or molecule (see the discussion of termolecular reactions below). We may write

$$\Delta H_R^0 = \Delta H_{O_3}^0 + \Delta H_M^0 - \Delta H_O^0 - \Delta H_{O_2}^0 - \Delta H_M^0 \tag{2.9a}$$

$$\Delta H_R^0 = \Delta H_{O_3}^0 - \Delta H_O^0$$

From Table 2.1, $\Delta H_R^0 = 34.1 - 59.5 = -25.4$. Thus this reaction is exothermic, liberating 25.4 kcal per mole. We shall see later in Chapter 4 that this process can serve as an important source of heat in the upper part of the middle atmosphere.

Table 2.1 Thermodynamic data for atmospheric species.
From Hampson and Garvin, 1978 (and Shum and Benson, 1983, for HO_2)

Species (gaseous)	ΔH_f^0 kcal/mole	ΔG_f^0 kcal/mole	S^0 cal/deg mol	C_p^0 cal/deg mol
$O(^3P)$	59.553	55.389	38.467	5.237
$O(^1D)$	104.78			
$O(^1S)$	156.40			
$O_2(^3\Sigma_g^-)$	0	0	49.003	7.016
$O_2(^1\Delta_g)$	22.54			
$O_2(^1\Sigma_g)$	37.51			
O_3	34.1	39.0	57.08	9.37
H	52.103	48.588	27.391	4.9679

H_2	0	0	31.208	6.889
OH	9.31	8.18	43.890	7.143
HO_2†	$3.5\,^{+1.0}_{-0.5}$	4.4	54.73	8.34
H_2O	-57.796	-54.634	45.104	8.025
H_2O_2	-32.58	-25.24	55.6	10.3
N	112.979	108.883	36.622	4.968
N_2	0	0	45.77	6.961
NO	21.57	20.69	50.347	7.133
NO_2	7.93	12.26	57.35	8.89
NO_3	17.0	27.7	60.4	11.2
N_2O	19.61	24.90	52.52	9.19
N_2O_4	2.19	23.38	72.70	18.47
N_2O_5	2.7	27.5	85.0	20.2
NH	82.0	80.6	43.29	6.97
NH_2	45.5	47.8	46.51	8.02
NH_3	-10.98	-3.93	46.05	8.38
HNO	23.8	26.859	52.729	8.279
HNO_2	-19.0	-11.0	60.7	10.9
HNO_3	-32.28	-17.87	63.64	12.75
S	65.7	56.52	40.085	5.658
SO	1.2	-5.0	53.02	7.21
SO_2	-70.94	-71.74	59.30	9.53
SO_3	-94.58	-88.69	61.34	12.11
HS	33.3	26.3	46.73	7.76
H_2S	-4.93	-8.02	49.16	8.18
CO	-26.416	-32.780	47.219	6.959
CO_2	-94.051	-94.254	51.06	8.87
CH_3	34.8	35.3	46.38	9.25
CH_4	-17.88	-12.13	44.492	8.439
HCO	-25.95	-24.51	53.68	8.26
H_2CO	-25.95	-24.51	52.26	8.46
CH_3O	3.9	6.4	64.2	8.9
CH_3O_2	6.7			
CH_3OOH	-30.8	-17.4	67.9	13.70
CH_3NO_2	-17.86	-1.65	65.69	13.70
CH_3ONO	-16.5	-1.5	69.7	16.5
CH_3NO_3	-29.8	-9.4	76.1	16.1
Cl	28.992	25.173	39.457	5.220
ClO	24.47	23.68	53.78	8.23
OClO	24.5	28.8	61.36	10.03
ClO_2	21.3	25.1	63.0	11.0
ClO_3	37			

HCl	-22.062	-22.777	44.646	6.96
HOCl	-18.7	-15.7	56.54	8.88
NOCl	12.36	15.77	62.52	10.68
$ClNO_2$	3.0	13.0	65.02	12.71
$ClNO_3$	6.28			
Cl_2O	19.20	23.3	64.07	11.48
CF_4	-223.04	-212.37	62.45	14.59
$CClF_3$	-169.2	-159.5	68.17	15.99
CF_2Cl_2	-117.5	-108.2	71.91	17.31
$CFCl_3$	-69.0	-59.6	74.00	18.65
CCl_4	-22.94	-12.83	74.02	19.93
CH_3Cl	-19.59	-13.98	55.99	9.73
CH_2Cl_2	-22.83	-16.49	64.57	12.16
C_2Cl_4	-2.97	5.15	82.05	22.84
CH_3CCl_3	-34.01	-18.21	76.49	22.07

† Shum and Benson (1983)

An endothermic reaction can proceed if the required energy is provided by some external source, such as a photon. The energy of a photon of wavelength λ is given by

$$E = \frac{hc}{\lambda} \qquad (2.10)$$

where $h = 6.62 \times 10^{-27}$ erg sec is Planck's constant, and $c = 2.998 \times 10^{10}$ cm s^{-1} is the speed of light in a vacuum. Other commonly used units are the frequency, ν, which is related to λ by

$$\nu = \frac{c}{\lambda} \qquad (2.11)$$

and the wavenumber $\bar{\nu}$, given by

$$\bar{\nu} = \frac{1}{\lambda} = \frac{\nu}{c} \qquad (2.12)$$

Appendices A and B present some conversion factors and useful physical constants for related units of energy. Thus consider, for example, the photolysis of molecular oxygen to form two O^3P atoms:

$$O_2 + h\nu \rightarrow 2O(^3P) \qquad (2.13)$$

$$\Delta H_R = 2(59.55) - \frac{hc}{\lambda} = 119.10 \text{kcal/mole} - \frac{hc}{\lambda} \qquad (2.14)$$

This process requires that at least 119.10 kcal/mole, or 8.27×10^{-15} ergs/molecule, be provided by the photon. This corresponds to $\lambda \leqslant 240$ nm.

2.3 Elementary chemical kinetics

The preceding section describes the thermodynamics associated with determining whether a reaction can proceed spontaneously at a given temperature. The next question one might wish to pose is how fast such a reaction will occur. That is the concern of chemical kinetics. The concentrations of almost all atmospheric chemical species depend critically on reaction kinetics, which determine the rates at which these constituents are produced or destroyed. The observed density of most chemical species in the atmosphere is dependent on the balance between the rates of photochemical production and destruction, and the rate of atmospheric transport processes. Reactions involving decomposition of a single particle, $A \rightarrow B + C$, are called first order or unimolecular; reactions involving two particles are second order, two body, or bimolecular; processes involving three particles are referred to as third order, three body, or termolecular. The rate of chemical reaction describes the rate of formation of the products or the rate of disappearance of the reactants.

For the special case of a first order process:

$$(k_u); \quad A \rightarrow C + D \tag{2.15}$$

the rate is given by

$$\text{Rate} = -\left(\frac{d(A)}{dt}\right) = \frac{d(C)}{dt} = \frac{d(D)}{dt} = k_u(A) \tag{2.16}$$

where k_u is the reaction rate constant $\left(\text{sec}^{-1}\right.$ for a unimolecular process$\left.\right)$ and (X) denotes the concentration of species X $\left(\text{molec cm}^{-3}\right)$. Rearranging this equation, we can write

$$\frac{d(A)}{(A)} = -k_u dt \tag{2.17}$$

whence

$$\ln(A) = -k_u t + C \tag{2.18}$$

where C is the constant of integration. Assuming that initially $(A) = (A)_0$ at time $t = 0$, we then have

$$(A) = (A)_0 e^{-k_u t} \tag{2.19}$$

The time required for the concentration of A to decrease to $1/e$ of its initial value if this were the only reaction process occurring is therefore equal to $1/k_u$. This defines the chemical lifetime of species A. The concept of lifetime is an important one in atmospheric applications. If, for example, we wish to know whether an atmospheric species is likely to be directly affected by transport processes, we can compare its photochemical lifetime to the time scale appropriate to transport. If photochemistry is much faster than transport for that particular species then the direct effects of transport can be neglected to a first approximation. This will be discussed in more detail in Chapters 3 and 5.

Referring to the general bimolecular equation $A + B \rightarrow C + D$, the rate at which such a reaction proceeds is described by

$$\text{Rate of reaction (molec cm}^{-3}\text{s}^{-1}) = \frac{d(C)}{dt} = \frac{d(D)}{dt} \qquad (2.20)$$

$$= - \left(\frac{d(A)}{dt} \right) = - \left(\frac{d(B)}{dt} \right)$$

$$= k_r(A)(B)$$

where k_r represents the reaction rate constant $(\text{cm}^3 \text{ molec}^{-1} \text{ s}^{-1}$ for a two body reaction). Second and third order reactions $(A + B \rightarrow C + D$ and $A + B + C \rightarrow D + E)$ are commonly reduced to a pseudo first order form to examine the chemical lifetime, for example

$$(A) \approx (A)_o e^{-k_r(B)t} \text{ or } (A) \approx (A)_o e^{-k_r(B)(C)t} \qquad (2.21)$$

and the lifetime is:

$$\tau_A = \frac{1}{k_r(B)} \text{ or } \frac{1}{k_r(B)(C)} \qquad (2.22)$$

These expressions can be used only if the concentrations of species B and C can be considered constant over the time τ_A. This is not always the case, since B and C may also be undergoing production and destruction reactions which introduce a time dependence in their densities over the time τ_A.

In a system like the atmosphere, in which numerous gases are interacting simultaneously, examination of the lifetime and the rate of change of a species must involve the sum of all of its possible production and loss processes, i. e., assume the following reactions

$$(k_1); \quad A + B \rightarrow \text{Products}$$

$$(k_2); \quad A + C + M \rightarrow \text{Products}$$

$$(k_3); \quad A + F \rightarrow \text{Products}$$

$$(k_4); \quad G + H \rightarrow A + \text{Products}$$

$$\frac{d(A)}{dt} = -k_1(A)(B) - k_2(A)(C)(M) - k_3(A)(F) + k_4(G)(H)$$

$$\tau_A = \frac{1}{k_1(B) + k_2(C)(M) + k_3(F)}$$

If this lifetime is short compared to transport times and if the densities of B, C, F, G, and H are not changing over the time scale considered, then photochemical stationary state (also called *steady state)* can be assumed; this implies that

A is in instantaneous equilibrium as determined by its sources $(\sum_i P_i)$ and sinks $(\sum_i L_i)$:

$$\frac{d(A)}{dt} = 0 = \sum_i P_i - \sum_i L_i(A) \qquad (2.23)$$

$$(A) = \frac{\sum_i P_i}{\sum_i L_i} \qquad (2.24)$$

$$(A) = \frac{k_4(G)(H)}{k_1(B) + k_2(C)(M) + k_3(F)} \qquad (2.25)$$

2.3.1 Collision theory of bimolecular reactions

Simple collision theory provides a useful conceptual framework to understand two body reactions. It cannot be applied to first and third order reactions without considerable modifications.

Let us again examine the reaction of A and B. Clearly the reaction cannot occur more rapidly than the rate at which A and B collide. If we assume that the particles behave as hard spheres, then the collision rate constant between these two particles is defined by

$$k_{AB} = bv_r \qquad (2.26)$$

where b is the collision cross section and v_r is the relative particle speed. In the hard sphere approximation, the collision cross section is given by

$$b = \pi\left(r_A + r_B\right)^2 \qquad (2.27)$$

where r_A and r_B are the radii of particles A and B. For typical molecular radii, the value of b is about 10^{-15} cm^2. The mean relative speed, v_r is

$$v_r = \left(\frac{8kT(m_A + m_B)}{\pi\, m_A m_B}\right)^{1/2} \qquad (2.28)$$

where m_A and m_B are the respective molecular masses of A and B, and k is Boltzmann's constant. For typical masses, the mean relative speed is about 4×10^4cm s^{-1}, and the value of k_{AB} is about 4×10^{-11} cm^3 molec^{-1} s^{-1}. We can also see that the temperature dependence of the rate constant thus defined is $T^{1/2}$.

Unfortunately, the actual interactions between reacting gases are considerably more complex than this and not all two-body rate constants are equal to 4×10^{-11} (in fact few of them do conform to this simple description). The reasons for this lie in the differences between the actual interactions between reacting molecules and the "hard sphere" assumption made above. In general, particles form an intermediate species during reaction, and this intermediate is often called an activated complex. Indeed, several detailed theories of chemical interactions achieve successful descriptions of certain processes simply by considering the energy modes available to the activated complex (e. g. RRK theory, see Rice and Ramsperger, 1927, and Weston and Schwartz, 1972). Here we simply note that formation of such a complex generally involves an enthalpy of reaction, such that the overall process can be crudely represented as shown in Figure 2.1. In general, a certain "energy of activation" is associated with forming the activated complex. If the reactants have sufficient energy to surmount the barrier involved in complex formation, then products will be formed.

The energy distribution of particles is often expressed as a Maxwell-Boltzmann distribution:

$$\frac{dn^*}{n_o} = 4\pi \left(\frac{m}{2\pi kT} \right)^{3/2} c^2 e^{-(m\frac{c^2}{2} + \epsilon_p)/kT} dc \qquad (2.29)$$

(see, e. g., Castellan, 1971). where ϵ_p is the potential energy of the particle at point x,y,z, and where dn^* is the number of particles per cubic centimeter at the position x,y,z which have speeds from c to $c + dc$, and m is the particle mass. The exponential term accounts for both the kinetic and potential energies of the particle. For a given potential energy, the kinetic energy distribution depends only on T. Given the form of this expression and the reaction barrier discussed above, it is not surprising that the experimentally derived kinetic rate expression for many processes is:

$$k_{reaction} = A e^{(-E_{act}/RT)} \qquad (2.30)$$

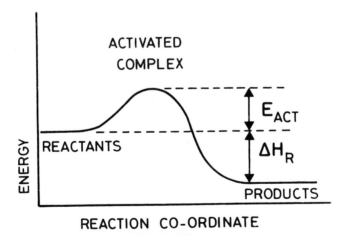

Fig. 2.1. Enthalpy variation during a reaction.

where E_{act} represents the energy of activation, R is the gas constant, and A is the "pre-exponential" factor. In this case, we would then expect the exponential term to denote the fraction of colliding molecules which have the energy required to surmount the barrier, while A represents roughly the collision rate constant, k_{AB} as presented in equation (2.26). Once again, we find experimentally that this is not always the case. One simple explanation for this is that only certain orientations are conducive to reactions, especially for complex molecules. Thus we can introduce a "steric factor" to represent the probability that the molecules will be in the proper geometric arrangement when they collide. Finally, another area of possible failure in the hard sphere model lies in the true nature of the approaching particles: they are not hard spheres, rather, they exert attractive forces on one another. In the case of ion-molecule reactions, the attractive forces are coulombic, and exert an influence even at relatively large distances. Therefore these reactions often proceed very much faster than values derived from simple hard sphere collision theory. Even for neutral unpolarized molecules, London forces produce an interaction potential. Thus the interactions between particles become multi-dimensional and quite complex. Molecular scattering experiments, in which reaction kinetics are examined as functions of energy and orientation, have led to considerable progress in our study of these processes. See, for example, Johnston (1966), Weston and Schwarz (1972) and Smith (1980).

2.3.2 Unimolecular reactions

The special case of a unimolecular reaction deserves a brief discussion. Under what circumstances does a molecule decompose into its constituent parts? Clearly this type of process cannot be understood in the context of collision theory as discussed above for bimolecular processes.

The first successful theory of unimolecular reactions was proposed by Lindemann in 1922. He introduced the idea that apparently unimolecular reactions were really the result of the following processes:

$$(k_{act}); \quad A + M \rightarrow A^* + M \text{ activation} \tag{2.31}$$

$$(k_d); \quad A^* + M \rightarrow A + M \text{ deactivation} \tag{2.32}$$

$$(k_r); \quad A^* \rightarrow B + C \text{ reaction of energized molecule} \tag{2.33}$$

The rate of decomposition of A is

$$-\left(\frac{d(A)}{dt}\right) = k_r(A^*) \tag{2.34}$$

The steady state concentration of A^* is derived as follows:

$$\frac{dA^*}{dt} = 0 = k_{act}(A)(M) - [k_d(M) + k_r](A^*) \tag{2.35}$$

$$A^* = \frac{k_{act}(A)(M)}{k_r + k_d(M)} \tag{2.36}$$

so that

$$\frac{d(A)}{dt} = -\left(k_r \frac{k_{act}(A)(M)}{k_r + k_d(M)} \right) \tag{2.37}$$

Thus A is conceived to be transformed to an "energized" state, A^*, by collision. A^* can then be deactivated by collision, or it may decompose. At sufficiently high pressures (high pressure limit), the rate of deactivation is much greater than the rate of decomposition; the rate limiting step is then equation (2.33) and the reaction will be found to be independent of pressure.

$$k_d(M) >> k_r; \quad \frac{d(A)}{dt} = -k_r \frac{k_{act}(A)}{k_d} \tag{2.38}$$

At very low pressures (low pressure limit), the observed rate will be proportional to the pressure, since the rate limiting step will be the production of the energized molecules by equation (2.31).

$$k_d(M) << k_r; \quad \frac{d(A)}{dt} = -k_{act}(A)(M) \tag{2.39}$$

Between these two extremes, the reaction is neither first nor second order, and the pressure dependence must be carefully parameterized. The thermal decomposition of N_2O_5,

$$N_2O_5 \rightarrow NO_2 + NO_3$$

is an important stratospheric reaction which is of intermediate order at stratospheric pressures.

2.3.3 Termolecular reactions

Reactions involving three particles do not generally occur via a single collision involving all three particles at once; indeed, the probability of such an event is quite small. In general a termolecular reaction can be understood in the context of activated complex theory as discussed earlier. The initial step involves an interaction of two particles:

$$(k_{act}); \quad A + B \rightarrow C^* \tag{2.40}$$

which must be followed by deactivation

$$(k_d); \quad C^* + M \rightarrow C + M \tag{2.41}$$

in order that the energized molecule not decompose back to reactants:

$$(k_{uni}); \quad C^* \rightarrow A + B \tag{2.42}$$

Thus this type of process is the reverse of unimolecular decay. The rate of formation of the product, C, is given by

$$\frac{d(C)}{dt} = k_{act}(M)(C^*) = k_{act}(M)\frac{k(A)(B)}{k_d(M) + k_{uni}} \qquad (2.43)$$

(in which steady state is assumed for C^* as shown above). If $k_{uni} \gg k_d(M)$, then the apparent kinetic rate will depend linearly on pressure. Most of the termolecular processes in the middle atmosphere behave in this manner. However, as in the case of unimolecular processes, it is certainly possible for a reaction to be neither second nor third order. Then a more detailed model of the pressure dependence must be used.

Since k_{uni} is generally larger than $k_d(M)$, termolecular processes often are found to exhibit negative energies of activation. Physically, this may be thought of as related to the degree of energization of C^*. At higher temperatures, the C^* intermediate will be formed with higher mean internal energy, and fewer of them will live long enough for the rate limiting stabilizing reaction, k_d, to occur.

2.4 Term symbols and their use

Atoms and molecules can possess energy in a variety of forms. One is the energy associated with the particular electronic configuration of the particle, describing the arrangement of electrons in the available atomic or molecular orbitals. The energy difference between electronic states is relatively large, of the order of electron volts (eV), and the photons capable of inducing transitions between electronic states are generally in the ultraviolet and visible regions of the spectrum. Molecules also possess vibrational and rotational energy, but the

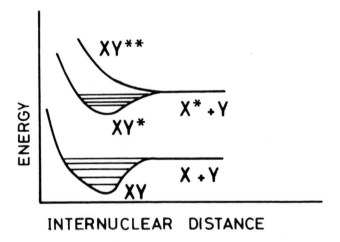

Fig. 2.2. Typical potential energy diagram for a diatomic molecule XY.

energy differences associated with vibrational and rotational transitions are much smaller, so that infrared and far-infrared photons are involved in purely vibrational or rotational transitions. Finally, atoms and molecules also possess a certain amount of kinetic energy, and for our purposes we may assume that this energy is not quantized.

The arrangements of electrons in the available quantized energy levels are conventionally denoted with shorthand term symbols; these will be the subject of this section.

Figure 2.2 presents a typical potential energy diagram for a diatomic molecule. The abscissa represents the distance between atoms. At very large distances, the particles are separate atoms exerting no significant attractive or repulsive forces on one another. As the particles approach one another in the ground state of the molecule, they begin to exert attractive forces, as indicated by the decrease in potential energy. The ground electronic state of the molecule, XY, exhibits maximum stability at the minimum in the potential curve, so that this represents a stable, bound configuration for the molecule. As the distance becomes even smaller, the atoms begin to repel one another, and the potential energy increases. Electronically excited states, such as XY^* and XY^{**}, may be repulsive at any distance, or they may have a stable bound configuration like the ground state, as shown. In the bound states, vibrational energy levels are depicted by the horizontal lines in the potential well.

Transitions between rotational, vibrational, and electronic states are quantized, and obey certain selection rules which can be derived from quantum mechanical principles (e. g., Steinfeld, 1978). It is beyond the scope of the

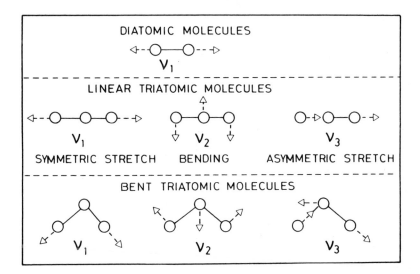

Fig. 2.3. Vibrational modes of diatomic and triatomic molecules.

present work to derive the selection rules. In the following, we will only sum-marize the basic meaning of the term symbols which are commonly used to denote the electronic and vibrational states of molecules and note the most important selection principles without offering proofs; the reader is directed to the bibliography where several excellent treatises on these subjects may be found. Our intent here is merely to briefly review the terminology and most important rules which appear in aeronomy.

Vibrational energy transitions play a particularly important role in the thermal balance of the middle atmosphere, because they are the fundamental processes responsible for most of the infrared cooling occurring there (see Chapter 4). Therefore, a brief description of vibrational modes and term sym-bols will be given here.

Vibrational quanta are distributed among the available vibrational modes of the molecule, which correspond physically to stretching and bending the chemical bonds of the molecule. These are conventionally referred to as ν_1, ν_2, etc. Figure 2.3 presents a schematic diagram of the possible vibrational modes for diatomic, linear triatomic, and bent triatomic molecules. The vibrational term symbol provides an ordered list of quantum numbers for each mode. For example, (010) means that the quantum number is zero for ν_1, one for ν_2 and zero for ν_3. As we shall see in Chapter 4, the 15 μm ν_2 band of CO_2 plays a dominant role in cooling throughout the middle atmosphere.

The electronic configurations are also denoted by term symbols which pro-vide a description of the electronic state. The significance of these terms to atmospheric chemistry is twofold: 1)Different electronic configurations of the same particle generally have very different reactive properties; for example, the excited O^1D atom behaves very differently upon collision with a water molecule than does the ground O^3P atom and 2) Photochemical processes involving absorption and emission of photons induce transitions from one state to another, and these occur more readily between certain electronic states than others. The transitions between electronic states which result from absorption of relatively high energy photons serve two important purposes in the middle atmosphere. They represent the major source of heating, as will be discussed in Chapter 4, and they initiate photochemical processes (e. g., photodecomposition or *photolysis* and photoionization) which play a major role in establishing the distributions of minor constituents (this is described in detail in Chapters 5 and 6). There are particular selection rules which describe the ease with which such processes occur. Note that it is common to refer to transitions as "forbidden" and "allowed". In practice, even highly "forbidden" transitions occur under certain circumstances, although much more slowly than "allowed" transitions. Therefore, we will refer to transitions here as "unfavorable" or "allowed". Much of the following discussion is similar to that presented in the excellent book by Campbell (1977).

2.4.1 General

In general, term symbols indicate both spin angular momentum (S for atoms, Σ for diatomic molecules) and orbital angular momentum (L for atoms, Λ for diatomic molecules). The term symbol is written We will not concern ourselves here with the derivation of L and Λ. Part of the reason for this is the fact that the selection rules involving orbital angular momentum are often not strictly followed in actual transitions; the restrictions regarding spin are the ones which tend to be most rigidly true, and we shall concentrate only on these. For increasing L from 0 to 6, the term takes on the values S, P, D, F, G, H. For increasing Λ, the Greek symbols Σ, Π, Δ, and Φ are used. See, for example, Karplus and Porter (1970).

The spin terms can be readily evaluated given knowledge of basic atomic and molecular orbital theory (see Karplus and Porter, 1970; Moore, 1962). The ground state of the oxygen atom is conventionally described as

$$(1s)^2(2s)^2(2p_1)^2(2p_2)^1(2p_3)^1$$

The superscripts denote the number of electrons in each atomic orbital. Electrons must be paired (i. e., of equal and opposite spin) in atomic orbitals containing two electrons. Paired electrons do not contribute to the atomic or molecular spin angular momentum. Hund's rule of maximum multiplicity states that in the lowest energy configuration, the electrons must be spread over as many available orbitals of equal energy as possible, in order to maximize the spin multiplicity. Since three 2p orbitals are available in which 4 electrons must be distributed, the lowest electronic state therefore has two unpaired electrons in each of two 2p orbitals. Each unpaired electron contributes 1/2 to the spin angular momentum. Thus $2S+1 = 3$ for the oxygen atom, and the term symbol is ^3P.

The oxygen molecule has the ground configuration

$$(\sigma_g 1s)^2(\sigma_u 1s)^2(\sigma_g 2s)^2(\sigma_u 2s)^2(\sigma_g 2p)^2(\pi_u 2p)^4(\pi_g 2p)^2$$

The two $\pi_g 2p$ electrons are unpaired, so that $2\Sigma+1 = 3$, and the ground state is $X^3\Sigma_g^-$. The X is used to denote that this is the ground state of the molecule; first and second excited states are indicated by A, B, or a, b, etc. More details regarding the significance of the term symbols can be obtained, for example, in Herzberg (1950).

Polyatomic molecules have different term symbols, but again the spin multiplicity is given by the left hand superscript. For example, the ground state of the ozone molecule is denoted by the term $^1A_1$. Note that these terms should not be confused with the excited states of diatomic molecules, which can also contain A, B, etc. in their term symbols.

2.4.2. Selection rules for electronic radiative processes

The importance of the term symbols lies in their use in predicting the ease with which photon absorption

$$A + h\nu \rightarrow A^* \tag{2.44}$$

and radiative decay can occur.

$$A^* \rightarrow A + h\nu \tag{2.45}$$

The selection rules for radiative processes permit us to evaluate the ease or difficulty with which a particular transition can proceed. The basic rules are as follows:

• Spin is unchanged in an allowed transition, for both atoms and molecules. This is generally the most important consideration in establishing the transition probability.

• L or Λ changes of 1 or zero are allowed, except in the case of atoms if $0 \rightarrow 0$ ($S \rightarrow S$). Changes of L or Λ of more than one (i. e., $D \rightarrow S$) are unfavorable.

• For molecules with g and u subscripts, $g \rightarrow u$ transitions are allowed, but $u \rightarrow u$ or $g \rightarrow g$ transitions are difficult.

Examples of such transitions which are of importance in the middle atmosphere are, for example,

$$O_2(^1\Delta_g) \rightarrow O_2(^3\Sigma_g^-) + h\nu \tag{2.46}$$

This process is unfavorable because it involves a change in spin, a change in Λ of two, and a $g \rightarrow g$ transition. Thus its probability is relatively low, which is reflected in a time constant for radiative decay of almost an hour. Similarly, the cross section for production of $O_2(^1\Delta_g)$ by photon absorption from the ground state

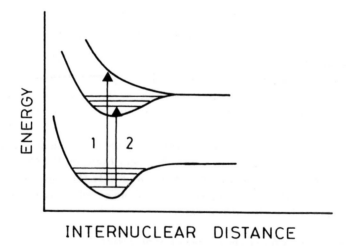

Fig. 2.4. Typical potential energy curves for a diatomic molecule

$$O_2(^3\Sigma_g^-) + h\nu \rightarrow O_2(^1\Delta_g) \qquad (2.47)$$

is small.

2.5 Photolysis processes

As has already been indicated, absorption of a photon of appropriate energy can lead to promotion of a molecule from one electronic state to another:

$$XY + h\nu \rightarrow XY^* \qquad (2.48)$$

If the excited state is unstable, it may decompose into its constituent atoms:

$$XY^* \rightarrow X + Y \qquad (2.49)$$

Photoionization may also occur:

$$XY^* \rightarrow XY^+ + e \qquad (2.50)$$

Figure 2.4 presents another typical potential energy diagram for a diatomic molecule, but in this case possible photon-induced transitions are also indicated. At atmospheric temperatures, most molecules are in their ground vibrational state. Possible transitions are represented by the vertical lines labeled 1 and 2, for example. Transition 1 leads from the ground state to an unbound state, and thus a continuum of wavelengths near the energy of 1 are possible. The

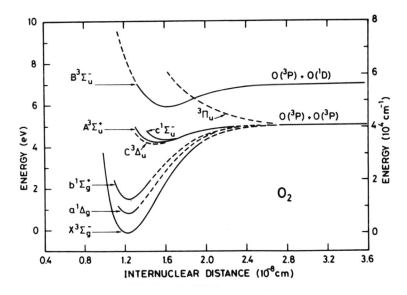

Fig. 2.5. Potential energy diagram for the principal states of the oxygen molecule. Adapted from Gilmore (1964).

spectrum should therefore consist of a smooth continuum at short wavelengths above the dissociation limit. Transition 2, on the other hand requires that absorption into a specific vibrational level of the upper electronic state occur. It is therefore quantized at nearly the energy defined by 2 (differences in rotational energy levels, which are too small to show in the figure, will lead to a band of possible energies about 2). The spectrum corresponding to these transitions will be banded at discrete wavelengths. Such spectra are exhibited by oxygen in the Herzberg continuum and Schumann-Runge bands and continuum. An approximate potential energy diagram for O_2 is shown in Figure 2.5, where these transitions are indicated. The Schumann-Runge bands correspond to transitions from the ground $^3\Sigma_g^-$ to the bound portion of the $B^3\Sigma_u^-$ state, and the Schumann-Runge continuum involves the same electronic states, but at shorter wavelengths above the dissociation limit. The Herzberg continuum involves excitation from the ground state to the repulsive $A^3\Sigma_u^+$ state.

If the upper state is bound, as it is for transition 2 in Figure 2.4, can such a transition lead to dissociation? Dissociation can often occur in such a case through curve crossing onto a predissociating state. Figure 2.6 presents a corresponding potential energy curve showing this phenomenon. The excited state, XY^*, which is initially produced, can cross into XY^{**} by a non-radiative transition or by collision induced crossing. From the predissociating state, ground state atoms are eventually produced. The indicated transition for the Schumann-Runge bands as depicted in Figure 2.5 involves crossing into the repulsive $^3\Pi_u$ state leading to dissociation. Nitric oxide photolysis in the middle atmosphere occurs largely through predissociation. Polyatomic molecules have more complex potential surfaces in three or more degrees of freedom, and

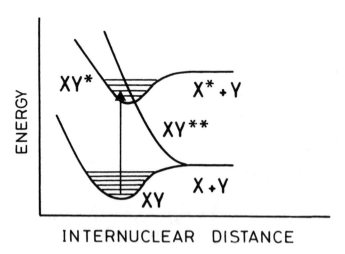

Fig. 2.6. Potential energy curves for a predissociative process in a diatomic molecule.

an increased probability of possible predissociating surfaces.

2.6 Excited species in the middle atmosphere

Photolysis reactions produce a number of electronically excited species which play an important role in the middle atmosphere. Some of these constituents are important because they are very reactive and can initiate chains of important free radical chemistry. Examples of such processes include

$$O_3 + h\nu \rightarrow O_3^* \rightarrow O(^1D) + O_2(^1\Delta_g)$$

$$O(^1D) + H_2O \rightarrow 2OH$$

which provides the major source of reactive hydrogen containing radicals in the stratosphere and troposphere. This will be discussed in much more detail in Chapter 5. Another excited species is $N(^2D)$, which produces large amounts of NO in the lower thermosphere (see Chapters 5 and 6) via

$$N(^2D) + O_2 \rightarrow NO + O$$

Another significant aspect of excited state chemistry, especially in the mesosphere and lower thermosphere, is the importance of fluorescence and chemiluminescence. Fluorescence is the rapid reemission of absorbed photons:

$$A + h\nu \rightarrow A^*$$

$$A^* \rightarrow A + h\nu$$

In chemiluminescence the emitting particle is excited by chemical reactions rather than direct photon absorption:

$$A + B \rightarrow C^* + D$$

$$C^* \rightarrow C + h\nu$$

Excited species can lose their energy by collision (or *quenching)* with other molecules. However, in the upper atmosphere, where the density of air molecules is relatively small and more high energy photons are available to produce excited species, many constituents fluoresce or chemiluminesce. Thus, photons of specific wavelengths are released, producing characteristic airglow emissions. Many of these emissions can also be used to deduce information about the distributions of atmospheric species. Table 2.2 presents a summary of some of the more important emitting species in the middle atmosphere.

Note the very long radiative lifetimes of O^1D, $O_2(^1\Delta_g)$, and N^2D. Transitions of these species to their respective ground states involve violation of the spin conservation rule, and are therefore relatively slow. The $O(^1S) \rightarrow O(^1D)$ transition, which produces the green line, is spin allowed, but corresponds to ΔL of two, and is therefore somewhat unfavorable. In contrast, the

Table 2.2 Emissions of some excited species in the middle atmosphere

Lower state	Excited state	Radiative lifetime (s)	λ (Å)	Name
$O(^3P)$	$O(^1D)$	110	6300	Red line
$O(^1D)$	$O(^1S)$	0.74	5577	Green line
$O_2(X^3\Sigma_g^-)$	$O_2(a^1\Delta_g)$	2.7(3)	12700+	Infrared atmospheric bands
$O_2(X^3\Sigma_g^-)$	$O_2(b^1\Sigma_g^+)$	12	7619+	Atmospheric bands
$O_2(X^3\Sigma_g^-)$	$O_2(A^3\Sigma_u^+)$	1	2600-3800	Herzberg bands
$OH(X^2\Pi)_{v=0,1,..}$	$OH(X^2\Pi)_{v=9,8,..}$	6(-2)	< 28007	Meinel bands
$N(^4S)$	$N(^2D)$	9.36(4)	5200	
$N(^4S)$	$N(^2P)$	12	3466	
$N_2(X^1\Sigma_g^+)$	$N_2(A^3\Sigma_u^+)$	2	2000-4000	Vegard-Kaplan bands
$NO(X^2\Pi)$	$NO(A^2\Sigma^+)$	2(-7)	2000-3000	γ bands

$NO(A^2\Sigma^+) \rightarrow NO(X^2\Pi)$ transition is fully allowed and the radiative lifetime of this state is correspondingly short.

The green line of atomic oxygen is now believed to arise mainly through the process

$$O + O + M \rightarrow O_2^* + M$$

followed by

$$O_2^* + O \rightarrow O(^1S) + O_2$$

and finally

$$O(^1S) \rightarrow O(^1D) + h\nu(5577\text{Å})$$

commonly referred to as the Barth mechanism (Barth and Hildebrand, 1961). Earlier, it had been proposed that this excitation occurred via

$$O + O + O \rightarrow O_2 + O(^1S)$$

first proposed by Chapman in 1931. For a number of years, considerable controversy existed as to the correct mechanism (see, for example, the excellent review by Bates, 1981), but it now appears that the Barth mechanism is more consistent with the observations. This emission provides a useful means of deducing the atomic oxygen density near 100 km, as shown for example by

Slanger and Black (1977).

$O_2(^1\Delta_g)$ is produced in the sunlit mesosphere principally via

$$O_3 + h\nu \rightarrow O(^1D) + O_2(^1\Delta_g)$$

and to a lesser extent from

$$O(^1D) + O_2 \rightarrow O_2(^1\Sigma_g^+) + O$$

or

$$O_2 + h\nu \rightarrow O_2(^1\Sigma_g^+)$$

followed by

$$O_2(^1\Sigma_g^+) + M \rightarrow O_2(^1\Delta_g) + M$$

$O_2(^1\Delta_g)$ is in turn destroyed either by radiative decay, or by quenching:

$$O_2(^1\Delta_g) + O_2 \rightarrow 2O_2$$

If the rates of all of these processes are known, then ozone densities can be derived from measurements of the infrared atmospheric volume emission rate (e. g. Evans et al., 1968; Thomas et al., 1983). This provides a useful means of measuring mesospheric ozone abundances.

The OH Meinel bands are produced primarily by

$$H + O_3 \rightarrow OH^\dagger(v=9,8,7,6) + O_2$$

The vibrationally excited OH can either chemiluminesce, or be quenched:

$$OH^\dagger + M \rightarrow OH + M$$

At present, the radiative and quenching lifetimes are not well known, making interpretation of this very interesting nightglow feature difficult (see, e. g. Llewellyn et al., 1978).

Emissions from $N(^2D)$ and $N(^2P)$ are particularly important in the aurora, where they are produced primarily through photoelectron impact on N_2 (see Chapter 5). As mentioned above, the reactions of N^2D or N^2P provide an important source of NO in the lower thermosphere, and this emission feature can indicate the yield of excited nitrogen production in auroral processes (see, for example, Frederick and Rusch, 1977; Zipf et al., 1980).

The fluorescence of NO in the γ bands represents a useful method of NO measurement in the lower thermosphere, as first pointed out by Barth (1966).

Excited state chemistry is one of the most important aspects of classical chemistry as applied to the middle atmosphere. Indeed, as shown in Figures 2.2 and 2.3, all photolysis reactions involve at least temporary production of electronically excited species; thus the photochemistry of the atmosphere is closely tied to the excited states of a number of important species. More detail regarding the chemistry of some of these states will be presented in Chapter 5.

References and bibliography

Barth, C. A., and A. F. Hildebrand, The 5577 Å airglow emission mechanism, J. Geophys. Res., 66, 985, 1961.

Barth, C. A., Nitric oxide in the upper atmosphere, Ann. Geophys., 22, 198, 1966.

Bates, D. R., The green light of the might sky, Planet. Space Sci., 29, 1061, 1981.

Calvert, J. G., and J. N. Pitts, *Photochemistry*, J. Wiley, (New York), 1966.

Campbell, I. M., *Energy and the atmosphere*, John Wiley and Sons (Chichester, G. B.), 1977.

Castellan, G. W., *Physical chemistry*, Addison-Wesley, (Reading, Mass.), 1971.

Chapman, S., Some phenomena of the upper atmosphere, Proc. R. Soc. Lond., Ser. A., 132, 353, 1931.

Evans, W. F. J., D. M. Hunten, E. J. Llewellyn and A. Vallance-Jones, Altitude profile of the infrared atmospheric system of oxygen in the dayglow, J. Geophys. Res., 73, 2885, 1968.

Frederick, J. E., and D. W. Rusch, On the chemistry of metastable atomic nitrogen in the F-region deduced from simultaneous satellite measurements of the 5200 Å airglow and atmospheric composition, J. Geophys. Res., 82, 3509, 1977.

Gilmore, F. R., Potential energy curves for N_2, NO, O_2, and corresponding ions, RAND corporation memorandum R-4034-PR, June, 1964.

Hampson, R. F., and D. Garvin, Reaction rate and photochemical data for atmospheric chemistry - 1977, U. S. Dept. of Commerce, NBS special publication 513, 1978.

Herzberg, G., *Spectra of diatomic molecules*, D. Van Nostrand Co., (New York), 1950.

Johnston, H. S., *Gas phase reaction rate theory*, Ronald Press, (New York), 1966.

Karplus, M., and R. N. Porter, *Atoms and molecules: an introduction for students of physical chemistry*, W. A. Benjamin, Inc., (Menlo Park, Cal.), 1970.

Lindemann, F. A., Discussion on radiation theory of chemical action, Trans. Far. Soc., 17, 598, 1922.

Llewellyn, E. J., B. H. Long, and B. H. Solheim, The quenching of OH^* in the atmosphere, Planet. Space Sci., 26, 525, 1978.

McEwan, M. J. and L. F. Phillips, *Chemistry of the atmosphere*, Edwards Arnold Ltd., (London), 1975.

Moore, W. J., *Physical chemistry*, Prentice Hall, (Englewood Cliffs, N. J.), 1962.

Rice, O. K., and H. C. Ramsperger, J. Am. Chem. Soc., 49, 1617, 1927.

Shum, L. G. S., and S. W. Benson, Review of the heat of formation of the hydroperoxyl radical, J. Phys. Chem., 87, 3479, 1983.

Slanger, T. G. and G. Black, O^1S in the lower thermosphere - Chapman vs. Barth, Planet. Space Sci., 25, 79, 1977.

Smith, I. W. M., *Kinetics and dynamics of elementary gas reactions,* Butterworths, (London), 1980.

Steinfeld, J. I., *Molecules and radiation: An introduction to modern molecular spectroscopy,* MIT press, (Cambridge, Mass.), 1978.

Thomas, R. J., C. A. Barth, G. J. Rottman, D. W. Rusch, G. H. Mount, G. M. Lawrence, R. W. Sanders, G. E. Thomas, and L. E. Clemens, Ozone density in the mesosphere (50-90 km) measured by the SME limb scanning near infrared spectrometer, Geophys. Res. Lett., 10, 245, 1983.

Weston, R. E. and H. A. Schwarz, *Chemical kinetics,* Prentice Hall, (Englewood Cliffs, N. J.), 1972.

Zipf, E. C., P. J. Espy, and C. F. Boyle, The excitation and collisional deactivation of metastable N^2P atoms in auroras, J. Geophys. Res., 85, 687, 1980.

Chapter 3

Structure and Dynamics

3.1 Introduction

The distributions of most chemical species in the middle atmosphere result from the influences of both dynamical and chemical processes. In particular, when the rates of formation and destruction of a chemical species are comparable to the rate at which it is transported by physical processes, then transport can play a major role in determining the constituent distribution (a detailed discussion of these concepts is provided in §3.4). This transport can be produced both by the prevailing winds (also called advection) and by turbulent mixing (diffusion). These will each be discussed in more detail below. In turn, the distributions of certain photochemical species, particularly ozone, can influence the radiative budget, affecting temperatures and the dynamic flow patterns. Therefore the study of aeronomy intersects greatly with those of fluid dynamics and meteorology. If we wish to understand why photochemical species behave as observed, certain concepts from these disciplines must be explored. In this chapter, the general structure of the middle atmosphere will be discussed, and then the transport processes of the stratosphere and mesosphere will be described. In our description of atmospheric motions, we have not sought to provide even a partial account of the dynamic meteorology of the middle atmosphere. The reader is referred to the monograph by Holton (1975) for a detailed treatment of observations and theory. Instead, the present chapter is directed towards an understanding of the effects of dynamics on chemical constituents. We will focus on the influence of longitudinally averaged motions on the distributions of photochemical constituents in the latitude-height plane.

The sections of this chapter deal with the following elements of atmospheric dynamics: observational evidence for transport of trace species by advection and by diffusive processes (§3.2), fundamental equations of atmospheric motion (§3.3), time constants appropriate to transport of chemical constituents and the relative importance of dynamical and chemical effects on photochemical species (§3.4), three-dimensional models of the middle atmosphere (§3.5), the mean meridional circulation and the use of the transformed

Eulerian formalism to illustrate the roles of mean meridional and eddy transports (§3.6), the important role of wave transience and dissipation (§3.7), and finally, the use of one-dimensional descriptions of atmospheric structure and transport (§3.8).

3.2 Vertical structure and some observed dynamical characteristics

The earth's atmosphere is commonly described as a series of layers defined by their thermal characteristics (Fig. 3.1). Specifically, each layer is a region where the change in temperature with respect to altitude has a constant sign. The layers are called "spheres" and the boundary between

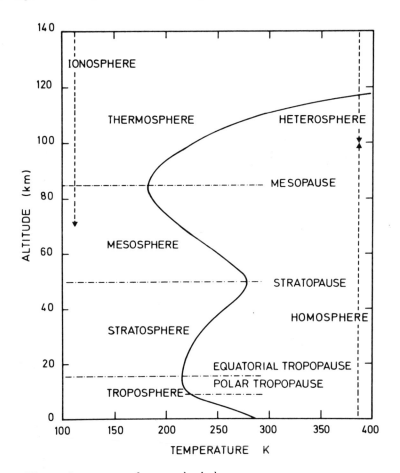

Fig. 3.1. Thermal structure of atmospheric layers.

connecting layers is the "pause". The lowest layer, called the *troposphere,* exhibits generally decreasing temperatures with increasing altitudes up to a minimum called the *tropopause.* The temperature and location of the tropopause vary with latitude and season. At the equator, its mean altitude is located near 18 km, and the corresponding temperature is about 190K, while in polar regions its elevation is only about 8 km, and the temperature roughly 220K (see Fig. 3.2). Above the tropopause, the *stratosphere* begins, exhibiting increasing temperatures with altitude up to a maximum of about 270K at the level of the *stratopause* located near 50 km. At still higher altitude, the temperature again decreases up to 85 km, where another temperature minimum is found. This layer is called the *mesosphere* and its upper boundary is the *mesopause.* In these layers, the major constituents, N_2 and O_2, make up about 80 and 20%, respectively, of the total number density, so that the mean

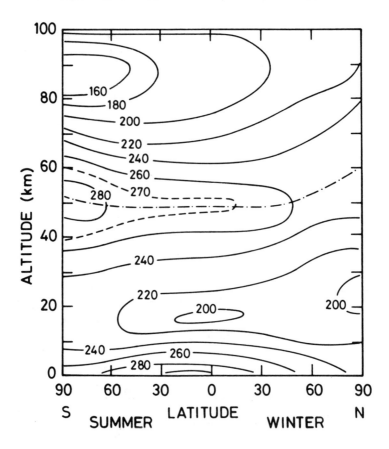

Fig. 3.2. Temperature structure of the middle atmosphere at solstice. From London (1980).

molecular weight of air varies little with altitude. Because of this common feature, the three layers are collectively referred to as the *homosphere*. Figure 3.2 presents a contour plot of the observed temperature structure of the homosphere at solstice, showing some of these features in more detail.

The region located above the mesopause is called the *thermosphere*. The temperatures there increase very rapidly with altitude and can reach 500 to 2000K, depending on the level of solar activity. The composition at these altitudes is very different from that of the lower regions due to an increasing proportion of atomic oxygen, whose density becomes comparable to and even greater than those of O_2 and N_2 above about 130 km. The abundances of O_2 and N_2 decrease, primarily as a result of rapid photodissociation. In contrast to the homosphere, the mean molecular weight of air in this region therefore varies with altitude; for this reason the region above 100 km is also called the *heterosphere*.

As was mentioned in Chapter 2, the major sources of heat in the middle atmosphere are provided by absorption of ultraviolet radiation, particularly by ozone and to a lesser extent by molecular oxygen. Radiative cooling occurs

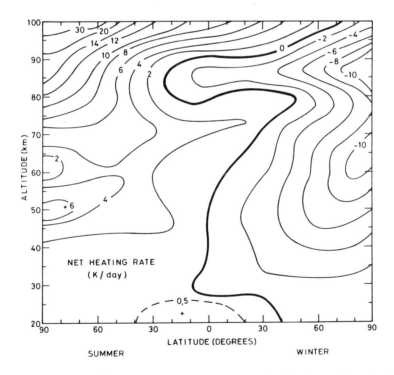

Fig. 3.3. Total net radiative heating rate associated with absorption of ultraviolet radiation by ozone and molecular oxygen, and cooling by CO_2 (°/day). From London (1980).

through infrared emissions associated with the vibrational relaxation of CO_2, H_2O, and O_3. Figure 3.3 presents a theoretical calculation of the rate of net radiative heating in the middle atmosphere. The large radiative heating rates found in the stratosphere are due primarily to the large amounts of ozone found there. The observed increase in temperature with altitude in the stratosphere is a result of heating by ozone, illustrating the important relationship of atmospheric chemical composition to the radiation budget and the thermal structure of the atmosphere.

Comparison of Figures 3.2 and 3.3 reveals several interesting features. For example, although the radiative heating rate at the summer mesopause is large, the temperatures observed there are much lower than those found in the winter hemisphere. The tropical tropopause, as we have already noted, is much colder than its counterpart at middle and high latitudes, although no dramatic variation in the radiative heating rate occurs there. Both cases illustrate the importance of dynamical effects in establishing the temperature structure of the middle atmosphere. Specifically, an air parcel displaced adiabatically upward undergoes expansion and net cooling; the inverse is true for adiabatic compression. These principles will be described in more detail below in connection with the thermodynamic equation, but we note at this point

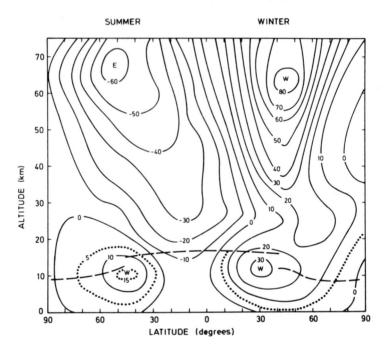

Fig. 3.4. Mean zonal wind distribution in the middle atmosphere. From Murgatroyd (1969), copyrighted by the Royal Meteorological Society.

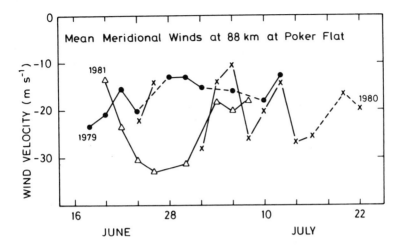

Fig. 3.5. Meridional winds in the mesosphere. From Nastrom et al. (1982). (Copyright by the American Geophysical Union).

that mean upward motion must be characteristic of the summer mesopause and the tropical tropopause based on these simple thermodynamic considerations. It should also be noted that the upward motion at the tropical tropopause is forced from below, and therefore represents one of the ways in which the middle atmosphere is coupled to the lower atmosphere. In particular, the formation of clouds and precipitation in tropical regions is accompanied by release of the latent heat of evaporation, providing the energy needed to drive much of the upward motion at the tropopause.

More direct observations of dynamics in the middle atmosphere are also available. One of the most easily measured atmospheric dynamical quantities is the *zonal* wind (wind speed in the *longitudinal* direction). Figure 3.4 presents a typical latitude height distribution of measured mean zonal winds, averaged over longitude. In the mesosphere, the *meridional* wind (wind speed in the *latitudinal* direction) can also be measured at a particular point, usually with radar techniques. Figure 3.5 shows measured long-term averaged meridional winds at Poker Flat, Alaska near 90 km from Nastrom et al. (1982). These are comparable to measurements reported at the same altitude at other locations as summarized by Nastrom et al. (1982), so that these values can probably be considered to be roughly representative of the mean meridional wind at 90 km.

Mean vertical and meridional winds in the stratosphere are very difficult to derive from observations because their magnitudes are very small (e. g., less than a few centimeters per second for vertical winds) compared to those of the zonal winds. They can, however, be estimated from theoretical studies, which are usually based on solutions to the equations of atmospheric dynamics, as will be described below.

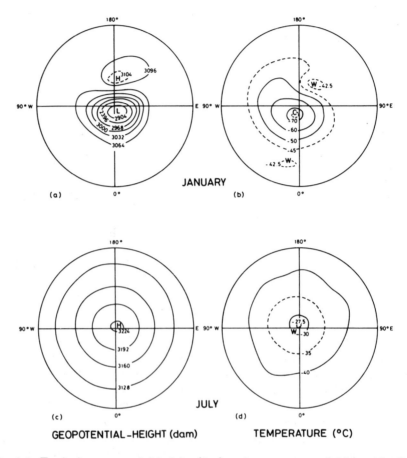

Fig. 3.6. Typical geopotential height (dm) and temperature fields at 10 mb in the northern hemisphere. From O'Neill (1980).

It should be emphasized that Figure 3.4 presents *the mean zonal wind,* that is, the wind speed averaged over longitude. In addition, large local variations in the zonal wind speed and direction are often observed. Particularly in the winter stratosphere, these variations are related to the presence of quasi-stationary planetary waves which are forced from the troposphere (see, e. g., Holton, 1979). Figure 3.6 presents typical observed temperature and geopotential height fields for summer and winter (the geopotential height will be defined in eqn. 3.21). Generally the local vector wind is directed approximately parallel to the contours of the geopotential height field. The summer temperature and geopotential height fields are nearly parallel to the latitude circles, and no substantial wave structure is apparent. Therefore, the local wind is almost uniformly in the zonal direction, and the flow pattern can be referred to as *zonally symmetric.* On the other hand, in winter, and particularly at high latitudes, the temperature, geopotential height, and associated flow patterns exhibit regions of relative highs and lows as a function of position in longitude, so that flow across latitude lines (e. g., in the meridional direction) occurs at certain longitudes. This flow pattern is *zonally asymmetric*. The wave structure in Figure 3.6 can be described as predominantly a wavenumber one pattern, exhibiting only one major ridge (high) and one trough (low) in the longitudinal direction. Observations (e. g. Hare and Boville, 1965) have established that large scale planetary waves of wavenumbers less than three account for most of the wave structure in the stratosphere. As we shall see later, the presence of waves plays an important role in the transport of some chemical species and in the theoretical description of atmospheric dynamics, particularly in the zonally averaged representation.

In addition to the winds, another possible mechanism for transport of chemical species is that of diffusion. The only true diffusion in the atmosphere occurs at the molecular level, via the random motion of atoms and molecules. This becomes important as a transport process only in the thermosphere. At lower altitudes, all transport results from motions of parcels of air rather than individual particles. However, the concept of *eddy diffusion* is a useful way to

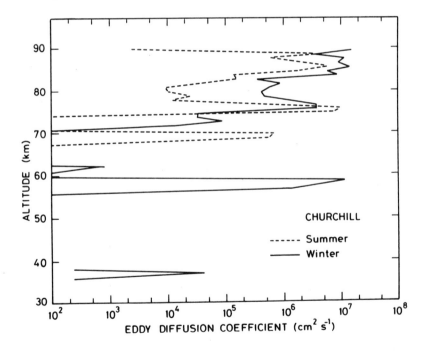

Fig. 3.7. Turbulent diffusion coefficients derived from tracer experiments. Adapted from Zimmerman and Murphy (1977). (Copyright by Reidel Publishing Company).

describe the mixing effects of fluid motions occurring on time and spatial scales much smaller than those of primary interest. This process is formally represented as a turbulent mixing, modeled after the behavior of turbulent diffusion. In this chapter we are primarily interested in gradients of photochemical constituents which occur over spatial scales of the order of kilometers in the vertical, and thousands of kilometers in the horizontal directions. Motions which occur over spatial scales much smaller than these may be understood for our purposes as diffusive, resulting in mixing rather than organized motion over the relevant distance. Evidence for the importance of diffusion in the mesosphere can be obtained by chemical release experiments, in which detectable substances are released from rockets, and their dispersion is monitored as a function of time. Figure 3.7 shows the diffusion profile derived from such an experiment by Zimmerman and Murphy (1977). In the stratosphere, the rate of turbulent diffusion is probably small compared to advection by the zonal winds, making direct observations of its effects relatively difficult. As we shall see below, identification of transports which are truly diffusive in nature and the evaluation of their magnitudes compared to the advective transports provided by the winds is a difficult and important problem in aeronomy.

3.3. Fundamental description of atmospheric dynamics

3.3.1 The primitive equations

To describe the theoretical dynamic and thermal behavior of the atmosphere, the fundamental equations of fluid mechanics must be employed. In this section these equations are presented in a relatively simple form. A more conceptual view will be presented in §3.6. The circulation of the Earth's atmosphere is governed by three basic principles: these are Newton's laws of motion, and conservation of energy and of mass. Newton's second law describes the response of a fluid to external forces. In a frame of reference which rotates with the Earth, it is given by:

$$\frac{d\vec{v}}{dt} + \frac{1}{\rho}\nabla p + 2\vec{\Omega} \times \vec{v} = \vec{g} + \vec{F} \qquad (3.1)$$

where $\vec{v}$ represents the vector speed of an air parcel, p is pressure, ρ is the mass per unit volume, $\vec{\Omega}$ is the angular rotation rate of the Earth, $\vec{g}$ is the gravitational acceleration, $\vec{F}$ is the frictional force due to viscosity, and t is time. The terms in the above equation each correspond to a different force acting on the fluid; these are the pressure gradient force $(1/\rho\nabla p)$, the Coriolis force $(2\vec{\Omega} \times \vec{v})$, the gravitational (g) and frictional (F) forces (these conventionally represent the force per unit mass in meteorological studies). Note that the Coriolis force is an apparent force which arises because the frame of reference used to describe the motion is rotating. See, for example, Holton (1979) for a detailed physical explanation of each of these terms.

The second equation is conservation of energy (first law of thermodynamics):

$$c_p \frac{dT}{dt} - \frac{1}{\rho} \frac{dp}{dt} = Q \qquad (3.2)$$

where T is the temperature, c_p represents the specific heat of air at constant pressure and Q is the net heating rate per unit mass (due to physical processes such as radiative effects or release of latent heat). The first two terms express the inverse relationship between pressure and temperature which describes expansion cooling or compression heating associated with adiabatic (energy conserving) processes taking place in a compressible fluid. The remaining term on the right hand side of the equation represents diabatic processes involving the rate of net heating and cooling of the parcel. In the stratosphere and mesosphere, this rate is principally due to the difference between heating by absorption of ultraviolet radiation by ozone and cooling by infrared emission by ozone, CO_2 and water vapor, and thus depends on the distributions of several important minor constituents, as has already been mentioned. A detailed account of the radiation and thermal budgets is presented in Chapter 4.

The third fundamental equation is conservation of mass (also called continuity):

$$\frac{d\rho}{dt} + \rho \nabla \cdot \vec{v} = 0 \qquad (3.3)$$

In equations (3.1), (3.2) and (3.3), the total time derivative (also called the *material derivative*) can be written:

$$\frac{d}{dt} = \frac{\partial}{\partial t} + \vec{v} \cdot \nabla \qquad (3.4)$$

The material derivative is the total time rate of change of a fluid property at a fixed point in space. The first term on the right hand side of this equation represents the local time rate of change and the second term describes the advection of a property of the fluid by the wind field. We emphasize that the material time derivative appears in this form because the equations are formulated in an Eulerian frame of reference, i.e., a frame of reference wherein the properties of the fluid are evaluated at fixed points in space. As will be shown in more detail in §3.6, this formalism influences our understanding of the importance of the various terms in the dynamical equations, particularly in their two-dimensional forms.

The net radiative heating (Q) is the only source of external forcing in the energy conservation equation. If we assume that Q = 0 and we consider a frame of reference which moves with a moving air parcel, we can examine the thermodynamics of adiabatic air parcel displacements. Equation (3.2) then becomes

$$c_p \frac{dT}{dt} = \frac{1}{\rho} \frac{dp}{dt} \qquad (3.5)$$

The ideal gas law (also called the equation of state) can be used to relate p and T:

$$p = \rho RT \tag{3.6}$$

where R is the gas constant for air. Substituting for ρ using (3.6), we obtain

$$\frac{dT}{T} = \frac{R}{c_p} \frac{dp}{p} \tag{3.7}$$

which can be integrated to yield

$$T = Ap^\kappa \tag{3.8}$$

where A is a constant and $\kappa = R/c_p = 0.286$ for dry air. This can be rewritten by adopting a reference pressure of 1000 mb,

$$\frac{T}{\Theta} = \left[\frac{p}{1000} \right]^\kappa \tag{3.9}$$

Θ is called the *potential temperature*, and it represents the temperature which an air parcel would attain if it were adiabatically compressed or expanded starting from a temperature T and pressure p to a pressure of 1000 mb. Θ is therefore a conservative property in any adiabatic air parcel displacement, and is sometimes used to evaluate air parcel trajectories (see, e. g., Danielsen, 1961).

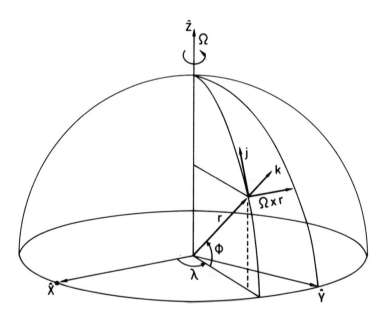

Fig. 3.8. Representation of the adopted coordinate system.

Introducing the potential temperature (as defined in eqn. 3.9) in the thermodynamic equation (3.2) provides a convenient simplification:

$$\frac{d\Theta}{dt} = \left[\frac{P_0}{P}\right]^{\kappa} \frac{Q}{c_p} \tag{3.10}$$

and allows for ready evaluation of the influence of external heating (Q) on Θ.

Newton's second law, relation (3.1) is generally written as three scalar equations, often expressed in spherical coordinates. Given the independent variables:

  λ the longitude in the easterly direction
  ϕ the latitude in the northerly direction
  z the altitude

also

  r = a + z is the distance from the center of the Earth (a is the Earth's radius).
  $u = r\cos\phi\, d\lambda/dt$ is the zonal component of $\vec{v}$ measured towards the East.
  $v = r\, d\phi/dt$ is the meridional component of $\vec{v}$ measured towards the North.
  $w = dz/dt$ is the vertical component of $\vec{v}$
  These components of motion are the scalar products of $\vec{v}$ with the unit vectors (see Fig. 3.8).

$$\vec{i} = \vec{\Omega} \times \vec{r} \tag{3.11a}$$

$$\vec{j} = \vec{k} \times \vec{i} \tag{3.11b}$$

$$\vec{k} = \vec{r}/r \tag{3.11c}$$

In the spherical coordinate system, the components of the acceleration $d\vec{v}/dt$ are not simply given by the derivatives of the components of $\vec{v}$ because the derivatives of the Cartesian unit vectors i, j, and k must be considered. Figure 3.8 shows a schematic representation of the adopted coordinate system.

The equations of motion then become:

$$\frac{du}{dt} = \frac{\tan\phi}{r}uv - \frac{1}{r}uw + 2\Omega v \sin\phi - 2\Omega w \cos\phi - \frac{1}{\rho r \cos\phi}\frac{\partial p}{\partial \lambda} + F_\lambda \tag{3.12a}$$

$$\frac{dv}{dt} = -\frac{\tan\phi}{r}u^2 - \frac{1}{r}uw - 2\Omega u \sin\phi - \frac{1}{\rho r}\frac{\partial p}{\partial \phi} + F_\phi \tag{3.12b}$$

$$\frac{dw}{dt} = \frac{1}{r}u^2 + \frac{1}{r}v^2 + 2\Omega u \cos\phi - g - \frac{1}{\rho}\frac{\partial p}{\partial z} + F_z \tag{3.12c}$$

The last of these three equations can be greatly simplified by retaining only the two dominant terms:

$$\frac{1}{\rho}\frac{\partial p}{\partial z} + g = 0 \qquad (3.13)$$

This simplification of the third momentum equation is commonly called the *hydrostatic approximation*. The approximation states that, in the vertical direction, the most important forces acting on a parcel of air are the vertical pressure gradient and gravity. This approximation is common in dynamical models, although it neglects some phenomena such as sound waves. The ideal gas law can be used to substitute for ρ in the above equation to obtain

$$\frac{dp}{p} = -\frac{dz}{H} \qquad (3.14)$$

where H is called the scale height. H is given by

$$H = \frac{RT}{g} = \frac{kT}{mg} \qquad (3.15)$$

where m is the molecular weight and k is Boltzmann's constant. H represents twice the distance through which atoms or molecules having the equipartition of translational energy, $1/2 \, kT$, can rise in the vertical direction against the force of gravity (Campbell, 1977).

This simple relationship can be used to characterize the vertical structure of the atmosphere in a convenient form. For example, the vertical distributions of the pressure and total number density can be represented by, respectively

$$p(z) = p_o \exp - \int_{z_o}^{z} \frac{dh}{H(h)} \qquad (3.16a)$$

and

$$n(z) = \frac{n_o T_o}{T(z)} \exp - \int_{z_o}^{z} \frac{dh}{H(h)} \qquad (3.16b)$$

where the subscript o denotes that the corresponding value is defined at some reference altitude z_o. If the scale height varies little with altitude, one can write

$$p(z) \approx p_o \exp \left[-\frac{z - z_o}{H} \right] \qquad (3.17a)$$

and

$$n(z) \approx n_o \exp \left[-\frac{z - z_o}{H} \right] \qquad (3.17b)$$

The atmospheric scale height is roughly 7 ± 1 km in the homosphere. This represents the altitude which corresponds to a reduction by $1/e$ in the pressure or the concentration. The distribution of any atmospheric constituent can also be characterized by its concentration, $n_i(z)$, or its scale height, $H_i(z)$. When the scale height is equal to the atmospheric scale height, $H(z)$, the constituent is well mixed and its mole fraction $f_i = n_i/n_{total}$ is constant with altitude. Table 3.1 presents a possible reference atmosphere for mid-latitudes, illustrating the location of the various layers and the variation of scale height with altitude.

The decrease of temperature with increasing altitude is often called the *lapse rate*. Combining the hydrostatic equation and the expression for potential temperature, we can evaluate the dry adiabatic lapse rate (Γ_d), i. e., the temperature decrease which would be associated with a vertical adiabatic displacement. Logarithmic differentiation of equation (3.9) holding Θ constant yields:

$$\frac{1}{T}\frac{dT}{dz} = \frac{\kappa}{p}\frac{dp}{dz} \qquad (3.18)$$

substituting for dp/dz from equation (3.13),

$$\Gamma_d = -\frac{dT}{dz} = -\frac{g}{c_p} = -10\,°K/km \qquad (3.19)$$

If potential temperature varies with height, then the actual lapse rate (Γ) will differ from the dry adiabatic lapse rate. This difference indicates the importance of non-adiabatic processes, such as the absorption of ultraviolet radiation by ozone, for example. The difference between the actual and the dry adiabatic lapse rates is also related to the tendency of a displaced air parcel to return to its original position. This quantity is called the static stability parameter (S) and is related to $\Gamma - \Gamma_d$. If $\Gamma = \Gamma_d$, then an air parcel adiabatically displaced from its position will tend to remain at its new location, since its temperature will be the same as that of its surroundings. On the other hand, if $\Gamma < \Gamma_d$, then an air parcel adiabatically lifted (lowered) from its equilibrium position will tend to sink (rise) back to its original position. The atmosphere is then referred to as statically stable or stably stratified. As can be derived by examining the vertical temperature gradients of the standard atmosphere given in Table 3.1, the middle atmosphere exhibits much greater static stability than does the troposphere, suggesting that it should be expected to be more stable with respect to vertical motions, and this is indeed found to be the case.

In order to facilitate the solution of the equations of motion, it is convenient to write them using log pressure as the vertical coordinate rather than the geometric coordinate system employed above (which is probably more familiar to aeronomers). As we shall see, the thermodynamic and continuity equations also assume much simpler forms in this coordinate system. Log pressure altitude is an isobaric (constant pressure) coordinate system defined by:

$$z_p = -H_z \ln(p/p_0) \qquad (3.20)$$

where p_0 is a reference pressure, and H_z is a scale factor, often chosen to be 7 km. H_z may be thought of as representative of the scale height of the atmosphere, which is generally not too far from 7 km as we have already seen. For an isothermal atmosphere where T = 239K, the log pressure and geometric

Table 3.1 Example of typical values of physical parameters below 120 km
at middle latitudes. From the U. S. Standard Atmosphere (1976).

Altitude (km)	Temperature (K)	Scale height (km)	Pressure (mb)	Concentration (cm^{-3})
0	288	8.4	1013.3	2.55(19)[*]
5	256	7.5	540.5	1.53(19)
10	223	6.5	265.0	8.61(18)
15	217	6.4	121.1	4.04(18)
20	217	6.4	55.3	1.85(18)
25	222	6.5	25.5	8.33(17)
30	227	6.6	12.0	3.83(17)
35	237	6.9	5.7	1.74(17)
40	250	7.3	2.3	6.67(16)
45	264	7.7	1.5	4.12(16)
50	271	7.9	8.0(-1)	2.14(16)
55	261	7.6	4.3(-1)	1.19(16)
60	247	7.2	2.2(-1)	6.45(15)
65	233	6.8	1.1(-1)	3.42(15)
70	220	6.4	5.2(-2)	1.71(15)
75	208	6.1	2.4(-2)	8.36(14)
80	198	5.8	1.1(-2)	4.03(14)
85	189	5.5	4.5(-3)	1.72(14)
90	187	5.5	1.8(-3)	6.98(13)
95	188	5.5	7.6(-4)	2.93(13)
100	195	5.7	3.2(-4)	1.19(13)
105	209	6.1	1.5(-4)	5.20(12)
110	240	7.0	7.1(-5)	2.14(12)
115	300	8.8	4.0(-5)	9.66(11)
120	360	10.5	2.5(-5)	5.03(11)

[*]read 2.55(19) for example, as 2.55×10^{19}.

altitudes are identical. Further, for the range of temperatures found in the
middle atmosphere, z_p is roughly equal to the actual geometric altitude. In
the following, we shall simply use z rather than z_p to refer to the log pressure
altitude.

It is also convenient to introduce the *geopotential*, Φ, defined as the work
required to raise a unit mass to height z from mean sea level:

$$\Phi = \int_0^z g\,dz \tag{3.21}$$

or, using the hydrostatic approximation,

$$d\Phi = g\,dz = -RT\,d\ln p = \frac{RT}{H}dz \tag{3.22}$$

We also assume that $r \approx a$. In spherical log pressure coordinates, the equations of motion then become:

$$\frac{du}{dt} - \frac{uv}{a}\tan\phi - 2\Omega\,\sin\phi\,v + \frac{1}{a\cos\phi}\frac{\partial\Phi}{\partial\lambda} = F_k \tag{3.23a}$$

$$\frac{dv}{dt} + \frac{u^2}{a}\tan\phi + 2\Omega\,\sin\phi\,u + \frac{1}{a}\frac{\partial\Phi}{\partial\phi} = F_u \tag{3.23b}$$

$$\frac{\partial\Phi}{\partial z} - \frac{RT}{H} = 0 \tag{3.23c}$$

The last equation is the hydrostatic equation in terms of the geopotential, and in log pressure coordinates, and is equivalent to equation (3.13). The laws of conservation of energy and mass (continuity) are given by

$$\frac{dT}{dt} + w\frac{\kappa T}{H} = \frac{Q}{c_p} \tag{3.24}$$

$$\frac{1}{a\cos\phi}\frac{\partial u}{\partial\lambda} + \frac{1}{a\cos\phi}\frac{\partial(v\cos\phi)}{\partial\phi} + \frac{1}{\rho_s}\frac{\partial\rho_s w}{\partial z} = 0 \tag{3.25}$$

where $\rho_s = \rho_0 \exp(-z/H)$, and ρ_0 is a reference density. The operator $d(\ )/dt$ is defined as

$$\frac{d}{dt} = \frac{\partial}{\partial t} + \frac{u}{a\cos\phi}\frac{\partial}{\partial\lambda} + \frac{v}{a}\frac{\partial}{\partial\phi} + w\frac{\partial}{\partial z} \tag{3.26}$$

Equations (3.23) through (3.25) are called the *primitive equations* because they contain almost no approximations, except for those of hydrostatic balance, the neglect of Coriolis acceleration terms due to the horizontal component of the Earth's rotation vector, and the substitution of a for r.

Equations (3.23a) and (3.23b) are statements of Newton's second law of motion applied to a fluid continuum. They are often referred to as the momentum equations because they describe the change of momentum per unit mass along the two horizontal coordinate directions.

Equations (3.23)-(3.25) define the basis for a theoretical treatment of the atmospheric circulation. Together with suitable boundary and initial conditions they are used in primitive equation models of the atmosphere (see §3.5). Although the primitive equations have the advantage that they can be used to represent all types of atmospheric motions (except sound waves, which are filtered out by the hydrostatic approximation), they are rather expensive to solve numerically. However, for a large class of scales of motion, the primitive equations can be replaced by an approximate equation for the *potential vorticity.* Vorticity can be thought of as a measure of the rotation of the fluid. The equation of vorticity conservation is essentially a statement of angular momentum conservation. The assumption involved in the derivation of this equation is that the horizontal wind field is very nearly in balance with the horizontal pressure gradient. This is a good approximation for motions of the spatial scales of greatest interest to us.

3.3.2 The quasi-geostrophic potential vorticity equation

Scale analysis of the primitive momentum equations, (3.23a,b), reveals that, for typical wind speeds and spatial scales observed in the atmosphere, the time derivative and curvature terms in these equations are small compared to the pressure gradient and Coriolis accelerations (see Holton, 1979, for details). In particular, if we scale the time derivative term in equation (3.23a) as follows:

$$\frac{du}{dt} \approx \frac{U}{(L/U)} \approx \frac{U^2}{L} \qquad (3.27)$$

where U is a typical zonal wind speed (order of 10-100 m/s for the stratosphere) and L is a typical length scale (order of 1000 km). The Coriolis acceleration is

$$2\Omega \sin\phi u \approx fU \qquad (3.28)$$

where f is the Coriolis parameter, roughly 10^{-4} s^{-1} except near the equator. The ratio of the inertial (du/dt) to Coriolis accelerations defines the *Rossby number* (R_o) *of the flow,* given by

$$R_o = \frac{U^2/L}{fU} = \frac{U}{fL} \qquad (3.29)$$

Substituting the numbers given above, we obtain $R_o \approx 10^{-1}$, indicating that the inertial term (du/dt) is much smaller than the Coriolis acceleration in the momentum equation for the case considered here. It can be shown that for this case, the only term in equation (3.23a) which is of comparable magnitude to the Coriolis term is the one representing the horizontal pressure gradient force, $1/a \cos\phi \, \partial\Phi/\partial\lambda$, and indeed these two terms are nearly in balance with one another to a very good approximation. It can also be shown that $R_o \leqslant 0.1$ is a good assumption for all scales of motion larger than about 1000 km, except very near the equator, where $f \to 0$.

Retaining only the Coriolis and pressure gradient terms in (3.23a) and (3.23b), we obtain the *geostrophic* wind equations:

$$fu = -\frac{1}{a}\frac{\partial\Phi}{\partial\phi} \qquad (3.30a)$$

$$fv = \frac{1}{a\cos\phi}\frac{\partial\Phi}{\partial\lambda} \tag{3.30b}$$

These equations provide the justification for the assertion made in §3.2, namely that stratospheric winds blow approximately parallel to the contours of the geopotential field. An alternative version of the geostrophic balance can be obtained by substituting the hydrostatic approximation as given in (3.23c) into (3.30a,b), after differentiating with respect to z:

$$f\frac{\partial u}{\partial z} = -\frac{R}{H}\frac{1}{a}\frac{\partial T}{\partial\phi} \tag{3.31a}$$

$$f\frac{\partial v}{\partial z} = \frac{R}{H}\frac{1}{a\cos\phi}\frac{\partial T}{\partial\lambda} \tag{3.31b}$$

These are the *thermal wind equations*, which state that the vertical shear of the horizontal wind field is proportional to the horizontal temperature gradient. These equations illustrate that the wind structure is strongly coupled to the temperature gradient, underscoring the close relationship between thermodynamic conditions and advective motion. Note, for example, that the zonal wind, u, as shown in Figure 3.4, increases (decreases) with altitude in the presence of an equatorward (poleward) temperature gradient (Fig. 3.2) in both the northern and southern hemispheres. Thus the region of maximum zonal wind (the "jet") near the polar night region can be understood to be a manifestation of geostrophic balance as expressed by the thermal wind equation.

Although the geostrophic wind equations are an excellent approximation of the horizontal wind field in atmospheric motion systems of scales greater than about 1000 km, they cannot be used to predict the evolution of the flow since they do not contain time derivatives, and are thus diagnostic, not prognostic (predictive) equations. However, by performing a more detailed scale analysis, it is possible to derive a system of prognostic equations which still embody the geostrophic balance as the lowest order approximation, but include in certain terms the possibility of departures from geostrophy. These are known as the *quasi-geostrophic* equations. There is no single quasi-geostrophic system, since the form of the equations depends, among other things, on the horizontal scale of the motions of interest compared to the Earth's radius. A common feature of these equations is that they can be used to construct prognostic equations for the *quasi-geostrophic potential vorticity*. For the purpose of illustration, we next consider briefly the case of mid-latitude motions whose scale is small compared to the Earth's radius (for more details, see, for example, Pedlosky, 1979; Gill, 1982; and the review article by Phillips, 1963). Defining the geostrophic vorticity as

$$\varsigma_g = \frac{1}{a\cos\phi}\frac{\partial v_g}{\partial\lambda} - \frac{1}{a}\frac{\partial u_g}{\partial\phi} \tag{3.32}$$

where u_g and v_g are the geostrophic velocities given by (3.30a,b), and letting

$$\beta = \frac{1}{a}\left[\frac{\partial f}{\partial \phi}\right]_0 \tag{3.33}$$

where the subscript o indicates that the derivative is evaluated at some middle latitude, ϕ_0, then the quasi-geostrophic potential vorticity equation is (Pedlosky, 1979):

$$\frac{d_g}{dt}\left[\varsigma_g + \beta y + \frac{1}{\rho_s}\frac{\partial}{\partial z}\rho_s \frac{\Theta}{S}\right] = \frac{1}{\rho_s}\frac{\partial}{\partial z}\left(\frac{\rho_s Q}{Sc_p}\right) \tag{3.34}$$

In equation (3.34),

$$\frac{d_g}{dt} = \frac{\partial}{\partial t} + \frac{u_g}{a\cos\phi}\frac{\partial}{\partial \lambda} + \frac{v_g}{a}\frac{\partial}{\partial \phi} \tag{3.35}$$

Θ is the potential temperature, S is the static stability parameter $(\Gamma - \Gamma_D)$, and $y = a\int d\phi$. The quantity in square brackets is the quasi-geostrophic potential vorticity, which is made up of contributions from the geostrophic vorticity of the flow, the planetary vorticity due to the Earth's rotation, and the vorticity due to stratification of the fluid, respectively. Equation (3.34) states that the potential vorticity is conserved following the motion for adiabatic conditions $(Q = 0)$.

The principle of conservation of potential vorticity is of central importance in dynamic meteorology. Not only is an equation like (3.34) much easier to solve than the full set of primitive equations, but the physical interpretation of atmospheric motions becomes more accessible when the powerful constraint of vorticity conservation can be applied. The fact that potential vorticity is conserved under adiabatic conditions also makes this quantity very useful as a tracer of atmospheric motions (see, Danielsen, 1961). McIntyre and Palmer (1983), for example, used a more general form of potential vorticity to study the mixing effect of large amplitude planetary waves in the stratosphere.

3.4 Effects of dynamics on chemical species

Now that the fundamental physical principles governing the atmospheric circulation have been delineated, we turn to a brief discussion of when and why atmospheric transport processes influence chemical species. The particular effects of various kinds of dynamical processes on many chemical species will be described in Chapter 5 when each individual species is discussed. It is useful, however, to establish a rough guide to evaluating the relative importance of dynamical and chemical effects, and this can be obtained from an analysis of the relevant time constants. In Chapter 2, the time constant appropriate to photochemical processes (τ_{chem}) was discussed, and it was shown that this time constant can be readily evaluated from knowledge of the rate of loss of chemical species. The time constants for dynamical effects on chemical species are somewhat more difficult to evaluate.

If we consider only the effects of advection and chemistry, the continuity equation for a chemical species i can be written as:

$$\frac{\partial n_i}{\partial t} = \sum_i P_i - \sum_i L_i n_i - \frac{\partial}{\partial z}(n_i w) - \frac{\partial}{\partial y}(n_i v) - \frac{\partial}{\partial x}(n_i u) \tag{3.36}$$

where n_i is the number density of species i, and where, for simplicity, we have used Cartesian coordinates. To derive the time constant for dynamical processes, some assumption must be made regarding the gradient of i. If we assume, for example, that the vertical distribution of i can be described by a scale height, H_i:

$$n_i = n_{i,o} \exp(-z/H_i) \tag{3.37}$$

where z is the altitude and $n_{i,o}$ is the density of i at a reference altitude, then

$$\frac{\partial n_i}{\partial z} = -\frac{1}{H_i} n_{i,o} \exp(-z/H_i) = -\frac{1}{H_i} n_i \tag{3.38}$$

Considering for the purpose of illustration, transport only by vertical winds:

$$\frac{\partial n_i}{\partial t} = -\frac{\partial}{\partial z}(n_i w) = \frac{w}{H_i} n_i \tag{3.39}$$

(assuming that w is constant over the spatial distance H).

$$n_{i,t} = n_{i,t_o} \exp(\frac{w}{H_i} t) \tag{3.40}$$

so that the time required for i to change by $1/e$ relative to its initial value, n_{i,t_o} is given by H_i/w. This defines a time constant for transport by the vertical winds. Typically, the scale height for chemically active species is smaller than that for the neutral atmosphere due to the importance of chemical processes. Assuming a typical scale height of about 5 km for chemical constituents and approximate values for the zonally averaged mean vertical winds as derived from models, the time constant is of the order of months in the stratosphere and days in the upper mesosphere.

Time constants for transport by the meridional and zonal winds can be derived in similar fashion, but in these directions it is somewhat more difficult to characterize the gradients of chemical species. Vertical gradients are generally better known, at least from measurements, than are meridional and longitudinal gradients. Clearly, a major difficulty with the use of dynamical time constants in general is that prior knowledge of the gradients must be available. If the real gradient is much different from that assumed for the derivation of these time constants, then the estimated effects of dynamics can be in error. Assuming a typical length scale of about 1000 km in the zonal and meridional directions, the time constant for transport by the zonal winds is of the order of days throughout the middle atmosphere, while that for mean meridional transport is of the order of months in the stratosphere and days in

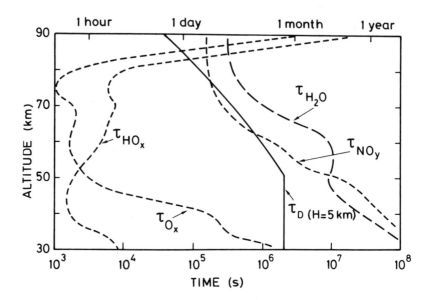

Fig. 3.9. Vertical profiles of the lifetimes of the HO_x, O_x, and NO_y families, as well as H_2O, and a typical one-dimensional diffusive lifetime.

the upper mesosphere. The time constant for transport by vertical eddy diffusion as used in one-dimensional models is given by H^2/K_z (where K_z is the one-dimensional diffusion coefficient, see § 3.8), and is of the order of months to years in the lower stratosphere, and days in the mesosphere. Figure 3.9 presents an altitude profile of the approximate time constant for one-dimensional eddy diffusion, as well as the chemical time constants of the O_x $(O+O_3)$, HO_x $(H+OH+HO_2)$, and NO_y $(N+NO+NO_2+NO_3+HNO_4+2\times N_2O_5+HNO_3)$ chemical families, and water vapor. As a general rule, wherever the time constant for the dynamical process under consideration is comparable to the chemical time constant, then the effects of transport and chemistry will both be important in determining the constituent's density (i.e., all the terms in eqn. 3.36 will be competitive). For example, in the one-dimensional framework, we can see that vertical eddy transport would be important for NO_y at all the altitudes shown, but not for HO_x below about 85 km. This is discussed more explicitly below.

Three different cases can be identified to characterize the effects of the competition between dynamics and chemistry in determining constituent distributions:

- $\tau_{chem} << \tau_{dyn}$. Under these circumstances, the species in question will be in photochemical equilibrium (see eqn. 2.23), and the effects of dynamical processes will not be directly important (note that the dynamical terms in eqn. 3.36 will be much smaller than the chemical terms). Dynamics may still be important in an indirect fashion through the effects of temperature or coupling between chemical species. For example, a very short lived species, such as HO_x can be produced or destroyed by reaction with a longer lived species which does depend on transport (for example, H_2O, see Fig. 3.9). Thus an indirect coupling can be important even for short lived species.

- $\tau_{chem} >> \tau_{dyn}$. In this case the effect of dynamics is to make the distribution of the constituent well mixed, eliminating gradients. For example, the chemical lifetime of N_2O is of the order of years in the lower stratosphere, while that of the zonal wind is of the order of days. Thus the effect of the zonal wind is to make the N_2O mixing ratio uniform in longitude. Moderate perturbations to the zonal wind speed will also not influence the constituent density. The major species, O_2 and N_2, are so long lived in the middle atmosphere that they are thoroughly mixed by atmospheric motions, and typical spatial and temporal variations in transport do not influence them. This is not the case in the thermosphere, where their chemical lifetimes are much shorter.

- $\tau_{chem} \approx \tau_{dyn}$. Under these circumstances, the effect of dynamics can be quite large, and the species distribution depends critically on both dynamics and chemistry. The time constants for meridional and vertical transport, are, for example, comparable to the photochemical lifetime of N_2O in the stratosphere, so that transport in the meridional plane is expected to be quite important in determining its density, in contrast to zonal transports, as discussed above.

The special case of zonal asymmetries should also be mentioned. The vector wind is generally aligned along latitude circles in summer, but in winter the influence of planetary waves can cause the vector wind to flow across latitude lines (see Fig. 3.6). Under these zonally asymmetric conditions, the local wind speed in the meridional direction is much greater than the *mean* meridional wind. Indeed, winds of zonal speeds (tens of meters per second) can flow across latitude lines, and any species exhibiting a latitude gradient will be affected. For example, the importance of planetary waves in determining the distribution of ozone has long been recognized (e. g., Berggren and Labitzke, 1968). Schmidt (1982) has demonstrated the importance of this effect for N_2O, and Solomon and Garcia (1983) have illustrated its role in the distributions of N_2O_5 and NO_2.

Dynamics will thus have important consequences for the distributions of almost every chemical species, either in a direct or indirect way, through transport in the meridional or zonal directions. In the remaining sections of this chapter we examine ways of describing atmospheric transport processes so that we may quantitatively examine their impact in aeronomy.

3.5 General circulation models

Models which provide solutions to some form of the general equations outlined in §3.3 are called general circulation models. General circulation models for the middle atmosphere have been developed by several groups (Kasahara et al., 1973; Hunt and Manabe, 1968; Manabe and Mahlman, 1976; Newson, 1974; Cunnold et al. 1975, 1980; Grose and Haggard, 1981; Rind et al., 1984; etc.). These models require very powerful computers, and at present a significant obstacle to their development and use in aeronomic applications is technological: contemporary computing capabilities are barely large enough to allow for detailed computations of both dynamics and chemistry. We briefly discuss below some of the aeronomic studies which have been presented using three-dimensional models. Essentially two classes of models are presently used: primitive equation and quasi-geostrophic. The governing equations for these two classes have been discussed briefly in §3.3. The choice of vertical coordinate also varies from model to model. Many dynamical models are formulated in pressure or log pressure coordinates. Several groups have adopted what is called a sigma coordinate system: since a pressure of 1000 mb does not necessarily correspond to the surface, it is sometimes convenient to account for the effects of orography by normalizing with respect to the surface pressure, which varies in space and time. This system has been used in the models developed by the Geophysical Fluid Dynamics Laboratory at Princeton (Smagorinsky, 1963) and the British Meteorological Office (Newson, 1974). Fels et al. (1980) have used a hybrid model combining the advantages of the sigma coordinate system in the troposphere with the simplicity of a pressure coordinate system in the stratosphere and mesosphere.

• *Solution of the equations*

The three-dimensional equations of atmospheric dynamics are generally solved using two types of techniques:

1. In the grid point method, the variables are defined by their values at particular, regularly spaced points in longitude, latitude and altitude. The spatial derivatives are replaced by finite differences. The stability and accuracy of the solution generally depends on the size of the time and spatial steps chosen.

2. In spectral models, the horizontal behavior of the meteorological fields (T, p, etc.) is expressed in terms of truncated spherical harmonic expansions (which can represent the effects of waves). In the vertical, the variables are evaluated at discrete grid points. In practice, the horizontal expansion is

usually somewhat truncated and therefore resolves only large scale motions. Cunnold et al., (1975), for example, consider only planetary waves of wavenumber less than or equal to 6 and 6 degrees of freedom in latitude. This method has the advantage of allowing exact calculation of the horizontal derivatives.

- *The use of three-dimensional models in aeronomic applications*

Certain limitations exist even in three-dimensional models. Not all of the relevant physical and chemical processes can be modeled in complete detail.

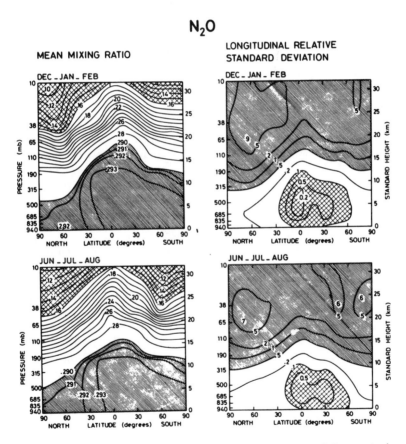

Fig. 3.10. Computed cross sections of the zonal mean N_2O mixing ratio (ppmv, left) and the longitudinal relative standard deviation (right) as calculated in a three-dimensional model study. From Levy et al. (1979). (Copyright by the American Geophysical Union).

For example, the exchange of heat between the atmosphere and ocean must be parameterized in some way. Similarly, very small scale processes such as cumulus convection may be quite important to the heat and moisture budgets, but are too small to be modeled in detail and a parameterization must be adopted to describe them in an approximate fashion. As previously mentioned, these processes play an important role in determining the forcing between the troposphere and the stratosphere, especially in the tropics. In spite of these limitations, however, three-dimensional models provide the only means of obtaining a complete representation of atmospheric transport processes in all dimensions. The processes which drive the chemistry of the atmosphere are photochemical (c. f., Chapter 2), and as such involve the influence of the sun. This varies according to the local solar irradiance, which is a function of latitude, longitude, altitude and time. Thus only a three-dimensional model can provide a completely realistic representation of both transport processes and photochemical effects (see, e. g., Tuck, 1979).

Levy et al. (1979) studied the distribution and variability of N_2O in a three-dimensional model. Figure 3.10 presents some of the results of that study, showing the computed zonal mean N_2O mixing ratios and the zonal mean cross sections of the longitudinal standard deviation of the mixing ratios. These will be discussed in more detail later. Mahlman et al. (1980) performed an "ozone precursor experiment" which revealed important information about the dynamic processes affecting total ozone variations, as well as a derivation of the flux of ozone across the tropopause. Levy et al. (1980) adapted this study to obtain the flux of nitrogen species across the tropopause, and Liu et al. (1980) analyzed its implications for tropospheric chemistry. Cunnold et al. (1975) presented a model study of the photochemistry of stratospheric ozone using simplified chemistry.

As of this writing, three-dimensional models have only been applied in a limited number of aeronomic studies. Besides the computational demands, another factor influencing the use of three-dimensional models for aeronomic purposes is the difficulty associated with making meaningful comparisons between three-dimensional model studies and observational data, if the data are available only at limited points in space and time. The atmosphere as represented by such a model may not exactly correspond to the real atmosphere at a given place at any particular time. For example, the model calculated temperatures and geopotential height structure at, for example, 10 mb at 60N, 100W, on January 9 will probably not be the same as those observed at the same place and time. Indeed, these observed dynamical parameters will generally not be exactly equal to those obtained in the real atmosphere in a different year. It may be possible to find a nearby date in the model simulation for which the actual and modeled atmospheric states are roughly comparable, but in general it seems unreasonable to expect that even a three-dimensional model will reproduce the specific behavior of the atmosphere at the time and place where a particular observation happened to be made. Where satellite data are available, providing extensive time and spatial coverage, statistical methods can be used to cast the data in a form comparable to the model. However, for comparison to limited observational data, three-dimensional results are often reduced to two dimensions by averaging in longitude, and the resulting zonal averages are compared to the average of the observations. Part of the justification for such an approach is that the atmosphere is observed to behave in a largely predictable and repetitive way from year to year with respect to seasonal zonal mean temperatures and ozone distributions, for example, except perhaps during stratospheric warmings (see §3.7 for definition). Therefore, the model results can be more readily compared to zonally averaged observations than to a limited set of observations at particular points.

Further, a full three-dimensional treatment may not be necessary to resolve many of the important features of the distributions of chemical species. The time constant associated with transport by the zonal wind is relatively rapid, of the order of days. Many chemical species with which we shall be concerned have time constants much longer than this, and the effects of transports by the zonal wind on those constituents is such that they may be considered to be "well mixed" in the zonal direction. This is illustrated, for example, in the three-dimensional study of N_2O by Levy et al. (1979). As shown in Figure 3.10, the model calculated variability in the zonal direction was found to be rather small (less than 10%) for this species in the stratosphere. This is to be expected because its photochemical time constant is very much longer than the time constant for transport by the zonal wind.

In view of the computational difficulties associated with three dimensional models, and the fact that comparison is often made to zonally averaged data, as well as the relatively small impact of zonal transports on longer lived species, the question of the usefulness of two-dimensional (zonally averaged) models may be raised. Clearly such models can more easily accommodate the computational demands of treating numerous chemical species at once, but a difficulty arises as to how to perform the zonal averaging, particularly in the stratosphere, where zonally asymmetric motions are quite important. In the following section, we present a conceptual two-dimensional view of stratospheric transport, and describe the problems and advantages associated with representing the atmosphere in two dimensions instead of three.

3.6 Dynamics of the stratosphere in two dimensions: a conceptual view

3.6.1 Zonal means and eddies

In the field of atmospheric dynamics it is common to distinguish between zonal mean motions and fluctuations about the zonal mean (called the *eddies).* These fluctuations can be of varied spatial scale - from a few meters to thousands of kilometers. Any such fluctuation can be described in terms of waves, as mentioned above in the discussion of three-dimensional spectral models. The eddy terms describe the effects of the ensemble of atmospheric waves. In the stratosphere, observed wave structure indicates that most of the waves are of large scales, of zonal wavenumber three or less, as discussed in §3.2. Note that the use of the word "eddy" in the context of atmospheric dynamics is thus rather different than the idea of a whirlwind, which is probably a more common usage of the same word (Oxford English Dictionary). Any atmospheric quantity χ can be expressed as the sum of the zonal mean:

$$\bar{\chi} = \frac{1}{2\pi} \int_0^{2p} \chi(\lambda) d\lambda \qquad (3.41)$$

and the eddy terms, χ', as follows:

$$\chi(\lambda) = \bar{\chi} + \chi'(\lambda) \qquad (3.42)$$

where λ represents longitude. We shall see below that the understanding of the physical nature of mean and eddy transports represents a formidable obstacle to describing the dynamics of the stratosphere. Using (3.41) and (3.42), the zonal mean momentum, continuity, and thermodynamic equations can be derived from the primitive equations given above:

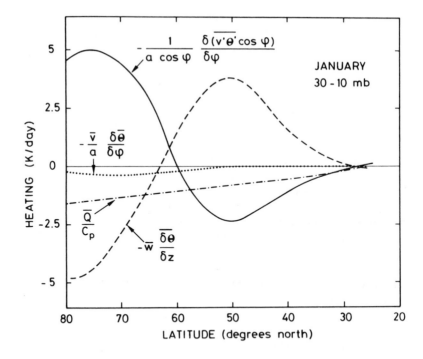

Fig. 3.11. Magnitudes of the terms in the thermodynamic equation as a function of latitude at 10 mb. From O'Neill (1980).

$$\frac{\partial}{\partial t}\overline{u} + \frac{\overline{v}}{a\cos\phi}\frac{\partial\overline{u}\cos\phi}{\partial\phi} + \overline{w}\frac{\partial\overline{u}}{\partial z} - f\overline{v} = \overline{F}_\lambda \qquad (3.43)$$

$$-\frac{1}{a\cos^2\phi}\frac{\partial}{\partial\phi}(\overline{u'v'}\cos^2\phi) - \frac{1}{\rho_s}\frac{\partial}{\partial z}(\rho_s\overline{u'w'})$$

$$\frac{1}{a\cos\phi}\frac{\partial\overline{v}\cos\phi}{\partial\phi} + \frac{1}{\rho_s}\frac{\partial}{\partial z}(\rho_s\overline{w}) = 0 \qquad (3.44)$$

$$\frac{\partial\overline{\Theta}}{\partial t} + \frac{\overline{v}}{a}\frac{\partial\overline{\Theta}}{\partial\phi} + \overline{w}\frac{\partial\overline{\Theta}}{\partial z} = \left(\frac{p_0}{p}\right)^\kappa\frac{\overline{Q}}{c_p} \qquad (3.45)$$

$$-\left(\frac{1}{a\cos\phi}\frac{\partial}{\partial\phi}(\overline{v'\Theta'}\cos\phi) + \frac{1}{\rho_s}\frac{\partial}{\partial z}(\rho_s\overline{w'\Theta'})\right)$$

(See, e.g., Newell, 1963).

Note that equations (3.43) to (3.45) do not constitute a closed system; they contain products of eddy quantities (such as $\overline{v'\Theta'}$) which are not calculated within this system. These eddy terms can be an important part of the heat and momentum budgets. For example, Figure 3.11 shows the relative magnitudes of the most important terms in the thermodynamic equation as a function of latitude at 10 mb. Note that the largest terms are $\overline{w}\partial\overline{\Theta}/\partial z$ and $1/a\cos\phi\ \partial/\partial\phi\overline{(v'\Theta')}\cos\phi$, which are nearly in balance with one another. It is this balance which has proved to be a significant obstacle to the physical understanding of two dimensional representations of the atmosphere, as we shall illustrate in the next section.

3.6.2. Descriptions of the mean meridional stratospheric circulation

Some of the earliest proposed descriptions of stratospheric transport were based on observations of chemical species. The first observations of stratospheric water vapor densities showed that the stratosphere was extremely dry, exhibiting mixing ratios of the order of a few parts per million by volume, in marked contrast to the troposphere, where water vapor abundances are of the order of a few percent. Brewer (1949) suggested that the dryness of the stratosphere was determined primarily by condensation, and that the water vapor content of an air parcel rising from the troposphere to the stratosphere would therefore be determined by the lowest temperature experienced by the parcel, which would normally correspond to the tropopause. He also noted that the tropopause temperatures in the tropics were approximately low enough to yield stratospheric water vapor densities as low as those observed, while the middle and high latitude tropopause was much too warm to explain the observed dryness. Thus he suggested a circulation exhibiting rising motion only in the tropics, and descending motion at extra-tropical latitudes (Such a pattern is sometimes referred to as a "Hadley cell"). Similarly, Dobson (1956)

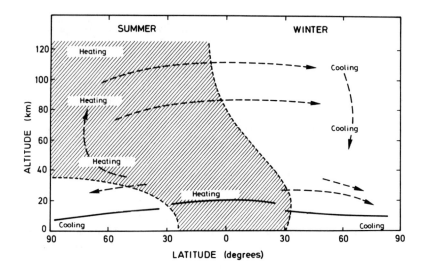

Fig. 3.12. Meridional circulation based on the study by Murgatroyd and Singleton (1961). From Murgatroyd (1971), copyrighted by Reidel Publishing Company.

noted that although the region of maximum ozone production is found in the tropical upper stratosphere, the largest observed total column densities were observed at high latitudes, suggesting that a downward-poleward transport was also required to explain the observed distribution of this species.

Murgatroyd and Singleton (1961) presented the first calculated mean circulation for the middle atmosphere by using an analysis of radiative heating rates. They derived a circulation much like the one suggested by the tracer studies discussed above. However, they noted that discrepancies existed in the angular momentum budget of the derived circulation, which were presumably due to eddy transport processes. The meridional circulation derived from that study is shown in Figure 3.12.

A few years later, observational studies using the angular momentum and/or heat budgets led to mean circulations of the lower stratosphere based on meteorological data, (specifically, observations of the dominant eddy terms in eqns. 3.43 and 3.45; e. g., Reed et al, 1963; Miyakoda, 1963; Julian and Labitzke, 1965; Vincent, 1968). These studies produced circulations which were quite different from the one derived by Murgatroyd and Singleton or those deduced from tracer studies (see, e. g. Fig. 3.13, which is based on a model calculation). In particular, *rising motion* (the "Ferrell cell") was suggested at high latitudes in winter. The reason that the Murgatroyd and Singleton circulation was different was indeed due to their neglect of the eddy transports.

The heating due to eddy transport was shown to be much larger than that due to radiation (Q) in the stratosphere, as can be seen in Figure 3.11. This term is represented as $1/a\cos\phi\,\frac{\partial}{\partial\phi}(\overline{v'\Theta'}\cos\phi)$ in the thermodynamic equation, (3.45).

Reed and German (1965) and Demazure and Saissac (1962) suggested that the eddy transport of heat used to derive the circulation discussed above could be described in terms of two-dimensional turbulent diffusion coefficients of the form

$$\overline{v'\Theta'} = -\left(K_{yy}\frac{\partial\overline{\Theta}}{\partial y} + K_{yz}\frac{\partial\overline{\Theta}}{\partial z}\right) - \left(K_{zy}\frac{\partial\overline{\Theta}}{\partial y} + K_{zz}\frac{\partial\overline{\Theta}}{\partial z}\right) \qquad (3.46)$$

where K_{yy}, K_{zz}, K_{zy} and K_{yz} represent the eddy diffusion coefficients in the y, z and off-diagonal directions. This description assumes that eddy transport of heat occurs via a diffusive process (if $K_{yz} = K_{zy}$), and that some particular mixing length can be assumed to represent the distance travelled by a parcel in the eddy before it mixes irreversibly with its surroundings. Reed and German also explained that a similar form could be adopted for the eddy transport of chemical constituents, simply by replacing potential temperature Θ with the mixing ratio of the species in question in the above equations. This description was widely adopted in two-dimensional photochemical models in

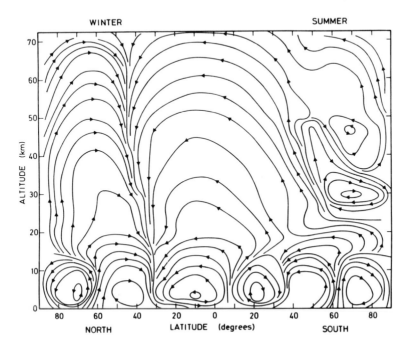

Fig. 3.13. Meridional circulation deduced by the theoretical study of Cunnold et al. (1975). (Copyright by the American Meteorological Society).

the 1970's (see, e. g. Crutzen, 1975; Vupputuri 1979; Hidalgo and Crutzen, 1977). It was found, however, that the numerical values of the K fields derived by Reed and German had to be empirically adjusted (often by multiplication by a factor of two) in order to obtain reasonable agreement between observations of trace constituents and calculations. Which depiction of the mean meridional circulation is correct, Figure 3.12 or Figure 3.13? The answer is that they both are, but that they do not represent the same thing. To a first approximation, Figure 3.12 represents the mean motion of air parcels in the meridional plane, while Figure 3.13 represents only one part of the net transport, with the other component being the eddy portion. As we will show below, the net motion in the latter case is the sum of the mean circulation and the eddy terms, and to a first approximation this derived net motion is equivalent to that shown in Figure 3.12. The differences between the two circulations are a result of the zonal averaging procedure applied in the presence of wave disturbances (the "eddies"). It can be shown mathematically that the eddy heat transport and the mean meridional heat advection produced by steady conservative waves cancel exactly (Andrews and McIntyre, 1976). This property of Eulerian zonal averages is known as the *non-interaction theorem* and is of great importance in understanding the evolution of zonally averaged fields of wind and temperature in the presence of wave motion.

A physical understanding of how the cancellation between the eddy heat transport and the wave induced mean meridional circulation arises can be gained from Figure 3.14. This figure depicts in schematic form the structure

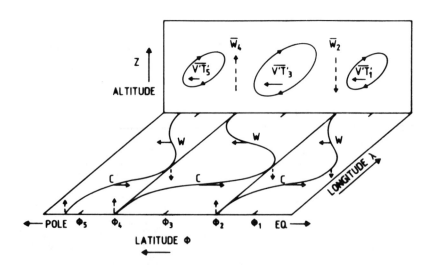

Fig. 3.14. Schematic diagram of the idealized planetary wave structure and the resulting eddy transport of heat. Adapted from Matsuno (1980).

of a vertically propagating planetary wave which is assumed to be steady and conservative. Streamlines of the flow in the latitude-longitude plane are shown as solid lines that meander north and south; wave induced velocities are represented by thin solid arrows (v') in the meridional direction, and by thin dashed arrows (w') in the vertical direction, and the perturbation potential temperature field (Θ') is denoted by the dashed circles labeled "warm" and "cold". The projection onto the height-latitude plane of the parcel trajectories following a given streamline are shown as ellipses, with the arrows indicating the sense of the parcel motion (it may be useful to imagine these projections as those of a three-dimensional elliptical corkscrew). Note, first of all, that for the wave depicted here the horizontal velocity and the temperature perturbations are in phase (i.e., positive velocities coincide with positive temperature perturbations, and vice versa). Thus, the average, $\overline{v'\Theta'}$ taken along any of the latitude circles $\phi_{1,3,5}$ is positive. It is largest at ϕ_3 because the wave has maximum amplitude in mid-latitudes. As a result, there is a *divergence* of zonally averaged eddy heat flux, $\overline{v'\Theta'}$ at ϕ_2 and a *convergence* at ϕ_4. These must be balanced by compression heating (sinking motion) and expansion cooling (rising motion), respectively, and so one obtains negative $\overline{w}$ at ϕ_2 and positive $\overline{w}$ at ϕ_4 (thick dashed arrows). This is the "Ferrell cell" pattern characteristic of the Eulerian meridional circulation in winter. However, since the projections of the parcel trajectories are closed ellipses, there is in fact no net parcel motion in the latitude-height plane as a result of these processes, and the Ferrell cell arises purely as a result of the zonal averaging procedure at the fixed points $\phi_{1,3,5}$. This is essentially the state of affairs described by the non-interaction theorem. A detailed explanation of the relationship between Eulerian eddy heat fluxes, meridional circulations, and parcel trajectories is given by Matsuno (1980).

The foregoing arguments demonstrate qualitatively that as long as wave motion is steady and conservative, a *net* meridional mass circulation can only be produced by the zonally averaged diabatic heating (Q in eqn. 3.45). The large scale planetary waves which dominate the wave structure of the stratosphere probably conform to this description at least in part. Therefore no net transport is produced in the meridional plane through the effects of these eddies. Thus the apparent discrepancy between the mean circulations derived by Brewer, Dobson, and Murgatroyd and Singleton, versus those computed by the later studies is reconciled when it is understood that the former studies represent *net* transport, while the latter correspond to a mean circulation only, which is canceled to a large degree by eddy transports, when these terms are evaluated in the standard fashion using a reference frame which is fixed in latitude and height.

The eddy-mean flow cancellation represents a difficulty for two-dimensional descriptions of stratospheric transport. Perhaps most importantly, it suggests that the mean and eddy transports are intimately coupled, and that a proper representation of atmospheric transport must employ a consistent set (see, for example, Harwood and Pyle, 1975; Rood and Schoeberl,

1983). This has often not been the case in photochemical studies, in which the K fields proposed by Reed and German are sometimes adopted with differing mean meridional circulations, and empirical adjustments are commonly made. Further, it suggests that the use of eddy diffusion coefficients to characterize all types and scales of eddy processes must be questioned, since the stratospheric eddies appear to be at least partly steady and conservative (i. e. non-diffusive in character, see e. g., Matsuno, 1980). We again emphasize that the word "eddy" means any deviation from the zonal mean, regardless of its size. However, it is unfortunate and confusing that the term "eddy diffusion" has been conventionally applied to describe the transport properties of all eddies, including those which are steady and conservative, and therefore do not have the physical nature of diffusion over large spatial scales. It must be emphasized that diffusion certainly occurs through the effects of those eddies which are transient and dissipating. This will be discussed in more detail in the next section.

The difficulties associated with the eddy-mean flow cancellation have motivated the search for a mathematical formalism to describe the zonally averaged structure of the atmosphere in such a way as to provide a more meaningful separation between the eddies and the mean meridional circulation. Andrews and McIntyre (1978) introduced the so-called *Lagrangian mean averaging*, wherein horizontal averages are computed not along latitude circles, but along a path that follows the wave trajectory in latitude and longitude. Referring again to Figure 3.14, it should be clear that if v' is evaluated along any of the wave streamlines shown, then no eddy fluxes appear. Although the Lagrangian mean formalism is mathematically elegant and physically meaningful, it suffers from technical problems that limit is applicability to situations where wave amplitudes are very large (McIntyre, 1980).

A more practical approach is simply to derive the net mean circulation directly from the diabatic heating rate, $\overline{Q}$, as suggested by Dunkerton (1978). The thermodynamic and continuity equations can be written as follows:

$$\frac{\partial \overline{T}}{\partial t} + \frac{\overline{v}}{a}\frac{\partial \overline{T}}{\partial \phi} + (\Gamma_d - \Gamma)\overline{w} = \overline{Q} - \frac{1}{a\cos\phi}\frac{\partial}{\partial \phi}(\overline{v'T'}\cos\phi) - \frac{1}{\rho_s}\frac{\partial}{\partial z}(\rho_s\overline{w'T'}) \quad (3.47)$$

and

$$\frac{1}{a\cos\phi}\frac{\partial}{\partial \phi}(\overline{v}\cos\phi) + \frac{1}{\rho_s}\frac{\partial}{\partial z}(\rho_s\overline{w}) = 0 \quad (3.48)$$

The first of these equations can be simplified by assuming $\partial \overline{T}/\partial t$ and $(\overline{v}/a)(\partial \overline{T}/\partial \phi)$ are small, and that the vertical eddy heat fluxes are negligibly small. See Dunkerton (1978) for justification. Equation (3.47) can now be written:

$$(\Gamma_d - \Gamma)\overline{w} = \overline{Q} - \frac{1}{a\cos\phi}\frac{\partial}{\partial \phi}(\overline{v'T'}\cos\phi) \quad (3.49)$$

A transformation can be introduced to represent the net mean vertical and meridional velocities $\overline{w}^*$ and $\overline{v}^*$, based on the assumption of cancellation of the

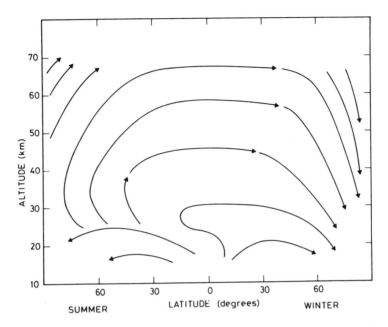

Fig. 3.15. Streamlines of the diabatic circulation obtained by Dunkerton (1978). (Copyright by the American Meteorological Society).

$$\overline{w}^* = \overline{w} + \frac{1}{a\,\cos\phi}\,\frac{\partial}{\partial\phi}\left(\frac{\overline{v'T'}\,\cos\phi}{\Gamma_d - \Gamma}\right) \qquad (3.50)$$

$$\overline{v}^* = \overline{v} - \frac{1}{\rho_s}\,\frac{\partial}{\partial z}\left(\frac{\rho_s\,\overline{v'T'}\,\cos\phi}{\Gamma_d - \Gamma}\right) \qquad (3.51)$$

with these definitions the thermodynamic and continuity equations become

$$(\Gamma_d - \Gamma)\overline{w}^* = \overline{Q} \qquad (3.52)$$

$$\frac{1}{a\,\cos\phi}\,\frac{\partial}{\partial\phi}(\overline{v}^*\cos\phi) + \frac{1}{\rho_s}\,\frac{\partial}{\partial z}(\rho_s\overline{w}^*) = 0 \qquad (3.53)$$

Note that the eddy heat flux does not appear in the transformed thermo-dynamic equation. From these equations, $\overline{v}^*$ and $\overline{w}^*$ may be derived directly from a knowledge of the diabatic heating rate. We emphasize that this is, in fact, just what Murgatroyd and Singleton (1961) did, because those authors neglected the eddy heat flux. As was much later shown by Dunkerton (1978), this neglect is justified provided that the eddies are steady and dissipationless. Figure 3.15 shows the mean streamlines (representing the advection path) derived in this manner by Dunkerton (1978). It is important to note that Mur-gatroyd and Singleton and Dunkerton computed the meridional circulation from the thermodynamic and continuity equations only. This can be done only if the temperature distribution is known a priori. However, a fully self-consistent mean meridional circulation can only be obtained when the momentum budget is considered together with the thermodynamic and continuity equations. Therefore, it is perhaps best to refer to the circulation obtained by Dunkerton as the *diabatic* circulation, since the adjective emphasizes the fact that only the thermodynamic budget is used to infer the circulation. Similar circulations which also avoid the eddy-mean flow cancellation but which involve solution of both the thermodynamic and momentum equations are usually called transformed or residual Eulerian. The important role played by the momentum budget will be discussed in more detail in the next section. Figure 3.16 shows the transformed Eulerian streamlines from the model of Garcia and Solomon (1983), overlaid on the calculated net diabatic heating distribution obtained in that study. Upward transport occurs where the net diabatic heating (Q) is positive, while downward transport is found in regions where Q is negative.

Another coordinate system which provides some conceptual advantages for understanding stratospheric transport is the isentropic coordinate system, as pointed out by Tung (1982) and Mahlman et al. (1983). Instead of using altitude or pressure as the vertical coordinate, isentropic surfaces (corresponding to potential temperature surfaces) are employed. The net circulation slopes

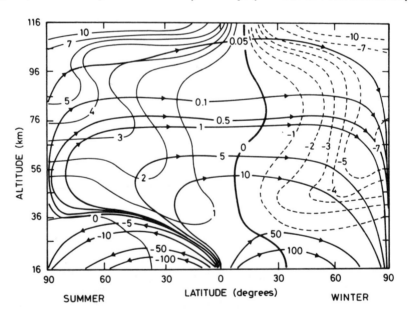

Fig. 3.16. Transformed Eulerian streamlines from the model of Garcia and Solomon (1983). Light lines represent net diabatic heating and cooling. (Copyright by the American Geophysical Union).

more steeply than the isentropic surfaces, reflecting the importance of diabatic heating. Since large scale eddies are quasi-adiabatic, there are no vertical eddy displacements associated with wave disturbances in the isentropic system (although horizontal eddy displacements do occur). The diabatic heating rate is directly related to the vertical velocity, as is the transformed or residual Eulerian velocity when altitude or pressure is used as the vertical coordinate.

Thus far we have emphasized how the formalism used to compute averages affects the mathematical expression of the equations that govern atmospheric dynamics and thermodynamics, and how it changes the interpretation of the mean meridional circulation obtained from these equations. However,

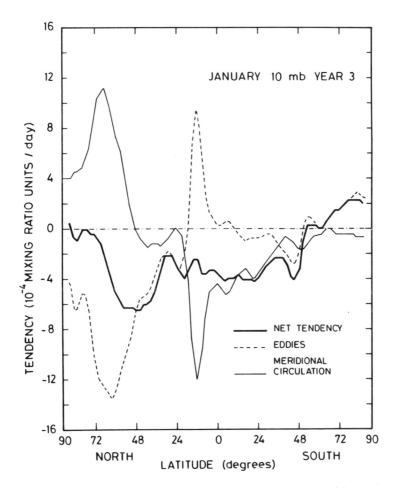

Fig. 3.17. Latitude dependence of the net, Eulerian mean and eddy transports for an inert tracer at 10 mb. From Mahlman and Moxim (1978). (Copyright by the American Meteorological Society).

these ideas are also of great importance in developing a mathematical and conceptual framework for the transport of chemical species in two dimensions. The thermodynamic equation is a conservation law for potential temperature, and, as we have seen, it states that potential temperature does not change following the motion of an air parcel unless there is diabatic heating or cooling. The Eulerian zonally averaged thermodynamic equation contains eddy terms which imply that potential temperature can be affected by eddy transport of heat, but these transports are cancelled to a large degree by the mean meridional circulation induced by the eddies themselves (see also, Fig. 3.11). With the introduction of the transformed Eulerian velocities, the thermodynamic equation takes the simpler form of equation (3.52). Chemical species in the atmosphere should behave much as potential temperature does, i. e., they should be conserved in the absence of chemical sources or sinks. It follows that the zonally averaged equation for a chemical species takes a form analogous to (3.45) if conventional Eulerian averages are used, and a form analogous to (3.52) in the transformed Eulerian framework. Figure 3.17 shows the zonally averaged budget of an inert tracer computed using conventional Eulerian averages. Compare the cancellation between the eddy mean meridional transports with the very similar behavior of their counterparts in the thermodynamic budget shown in Figure 3.11. On the other hand, if we work with the transformed Eulerian circulation (and neglecting the contribution from small scale diffusion for the moment), the equation for chemical species i is simply

$$\frac{\partial n_i}{\partial t} + \frac{\bar{v}^*}{a}\frac{\partial n_i}{\partial \phi} + \bar{w}^*\frac{\partial n_i}{\partial z} = \sum_i P_i - \sum_i L_i n_i \qquad (3.54)$$

provided that chemical production/loss along wave trajectories can be neglected. If the chemical lifetime of the species changes rapidly over a distance comparable to the physical scale of the wave, and if that lifetime is comparable to the time it takes an air parcel to move through the wave, then the effects of production and loss along wave trajectories cannot be ignored. Under these special circumstances, the wave can move air parcels into regions of different chemical lifetime, introducing "chemical eddy transport". This effect can be parameterized in terms of wave structure and the photochemical lifetime of the species considered (see Plumb, 1979). The importance of this effect has been shown, for example, by Garcia and Hartmann (1980), Hartmann (1981), Strobel (1981) and Kawahira (1982). One of the species most strongly influenced by chemical eddy transports is stratospheric ozone in high latitude winter. Figure 3.18 shows the influence of chemical eddy transports on ozone as calculated with the model of Garcia and Solomon (1983).

The conceptual advantage of the diabatic or transformed Eulerian circulations is that the eddy-mean flow cancellation is avoided, and a first approximation to the net transport needed for aeronomic studies is therefore readily obtained. The most important features of this transport description are that air generally enters the lower stratosphere in the tropics and exits at middle and high latitudes in both hemispheres. In the upper stratosphere and

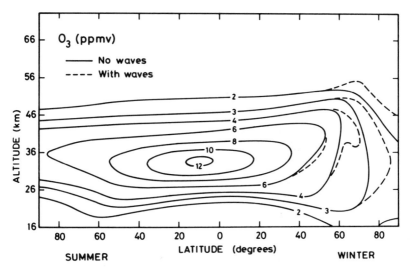

Fig. 3.18. Calculated ozone distribution in ppmv for northern hemisphere winter solstice, with and without chemical eddy transports.   From Garcia and Solomon (1983). (Copyright by the American Geophysical Union).

mesosphere, air flows from the summer hemisphere to the winter hemisphere, and downward near the winter pole. This transport description, which we now recognize as the diabatic or transformed Eulerian circulation, is qualitatively very similar to those proposed by Brewer and Dobson based on early observations of photochemical species. Indeed, this very simple picture provides a reasonable qualitative description of transport of chemical species, at least in the monthly or seasonal mean, and can successfully explain many of the observed features of the distributions of chemical constituents. Pyle and Rodgers (1980a) presented the first photochemical model using this description of transport; Miller et al. (1981) and Garcia and Solomon (1983) later presented similar models.

3.7 The importance of wave transience and dissipation

The preceding section shows that the meridional circulation derived from the transformed thermodynamic equation by neglecting wave transience and dissipation is quite useful, since it corresponds approximately to the net atmospheric transport. The validity of the assumptions of no transience or dissipation must, however, be examined before this simple picture can be assumed to be a complete description. In particular, additional insight to the great importance of transience and dissipation can be obtained from an analysis of the atmospheric momentum budget. (Recall that the diabatic

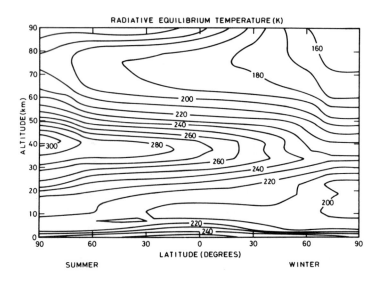

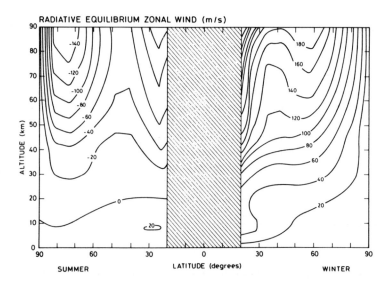

Fig. 3.19. Calculated radiative equilibrium temperatures and zonal winds. From Geller (1983). (Copyright by Reidel Publishing Company).

circulation is based only on the thermodynamic budget and uses observed, rather than calculated, temperature fields. Therefore it does not represent a fully self-consistent solution to the dynamical structure.)

When the effects of wave transience and dissipation are neglected in a model calculation including both the thermodynamic and momentum equations, the atmosphere eventually reaches a radiative equilibrium something like the calculations displayed in Figure 3.19. The temperatures are quite different from those observed, particularly at the tropical tropopause and in the mesosphere, and the zonal winds are very much faster than are observed (see Figs. 3.2 and 3.4). Similar radiative equilibrium calculations including more detailed treatments of radiative transfer yield even more dramatic departures from observations (for example, the calculated 1 mb radiative equilibrium temperature near the winter pole as obtained in the study by Mahlman and Umscheid, 1983, is only 150K, while observed values are about 250K).

What is lacking in this picture is a mechanism for deceleration of the zonal winds. Differential heating alone does not produce a realistic thermodynamic structure or circulation. Leovy (1964), and later Schoeberl and Strobel (1978) and Holton and Wehrbein (1980) showed that some mechanism of deceleration of the zonal winds was required to explain observed atmospheric structure, and they suggested that this process must involve wave transience and dissipation.

Analysis of the relevant equations clearly shows why this must be important. Assuming that wave transience and dissipation can be neglected in the thermodynamic (but not the zonal momentum) equation, and making use of the transformed Eulerian velocities as defined in equations (3.50) and (3.51), we obtain the following set of equations:

$$\frac{\partial \bar{u}}{\partial t} + \frac{\bar{v}^*}{a \cos\phi}\frac{\partial \bar{u}\cos\phi}{\partial \phi} + \bar{w}^*\frac{\partial \bar{u}}{\partial z} - f\bar{v}^* = \frac{\nabla \cdot \vec{F}}{\rho_s a \cos\phi} + \bar{F}_\lambda \tag{3.55}$$

$$\frac{1}{a \cos\phi}\frac{\partial \bar{v}^*\cos\phi}{\partial \phi} + \frac{1}{\rho_s}\frac{\partial}{\partial z}(\rho_s \bar{w}^*) = 0 \tag{3.56}$$

$$\frac{\partial \bar{\Theta}}{\partial t} + \bar{w}^*\frac{\partial \bar{\Theta}}{\partial z} = \left(\frac{p_o}{p}\right)^\kappa \frac{\bar{Q}}{c_p} \tag{3.57}$$

where $\vec{F}$ is the *Eliassen-Palm flux*, whose components are

$$F_\phi = -\rho_s a \cos\phi\,\overline{u'v'} + \rho_s a \cos\phi\,\bar{u}_z\frac{\overline{v'\Theta'}}{\Theta_z} \tag{3.58}$$

$$F_z = \rho_s a \cos\phi\left(f - \frac{1}{a\cos\phi}\frac{\partial \bar{u}\cos\phi}{\partial \phi}\right)\frac{\overline{v'\Theta'}}{\Theta_z} - \rho_s a \cos\phi\,\overline{u'w'} \tag{3.59}$$

For planetary waves in the stratosphere, quasi-geostrophic scaling can be applied, in which case expressions (3.58) and (3.59) reduce to

$$F_\phi = -\rho_s a \cos\phi\,\overline{u'v'} \tag{3.60}$$

and

$$F_z = \rho_s a \cos\phi\, f\frac{\overline{v'\Theta'}}{\Theta_z} \tag{3.61}$$

Scale analysis of equation (3.55) reveals that the dominant terms must be

$$-f\bar{v}^* = \frac{\nabla \cdot \vec{F}}{\rho_s a \cos\phi} + \vec{F}_\lambda \tag{3.62}$$

In the thermosphere (above 100 km), ion drag becomes very large because the molecular viscosity increases rapidly with altitude. This represents a true frictional process ($\vec{F}_\lambda$ in eqn. 3.62). In this case, the lowest order steady state momentum balance is given by

$$-f\bar{v}^* = -K_{ion}\bar{u} \tag{3.63}$$

where K_{ion} is the ion drag coefficient (see, e. g., Kasting and Roble, 1981). However, throughout most of the range of altitudes of interest to us, true frictional processes are nearly absent, and the term $\nabla \cdot \vec{F}$ must provide the necessary momentum balance.

The F defined by equations (3.60) and (3.61) is identical with the quasi-geostrophic potential vorticity flux due to planetary waves, a quantity which can be calculated from the perturbation version of equation (3.34). $\nabla \cdot \vec{F}$ represents the momentum forcing by the eddies, and has the important property that it vanishes for steady, conservative waves. However, if the eddies are completely steady and dissipationless, then $\nabla \cdot \vec{F} = 0$, and $\bar{v}^*$ must also be zero. Therefore, the strength of the mean meridional circulation in a rotating frame of reference such as the Earth depends crucially on the ability of the transient, dissipating eddies to act as a momentum sink (or source) to balance the Coriolis torque ($f\bar{v}^*$) produced when parcels move in the meridional direction.

Thus, we have seen that in the absence of true frictional processes (e.g., ion drag), the zonally averaged momentum budget must be balanced by the convergence or divergence of momentum flux due to wave perturbations. We may think of this wave momentum deposition as a *wave drag,* but it should be emphasized that it is a very different process from the frictional deceleration caused by ion drag. True frictional drag involves a transfer of momentum at the molecular level, whereas wave drag is a macroscopic process that can be brought about by transience (growth or decay of wave amplitude) or by dissipation (thermal damping, turbulent diffusion). Indeed, the distinction is similar to that drawn earlier between molecular and turbulent diffusion.

An understanding of the mechanisms that produce wave drag is essential to the study of the mean meridional circulation of the middle atmosphere. It is also important for chemical modeling because chemical species can be advected by the mean meridional circulation and also transported by the diffusive mixing which probably accompanies wave drag. Before proceeding to discuss in detail how wave drag is produced, it is important to emphasize two points: First, wave drag can be produced in the absence of turbulent diffusion by wave transience and thermal damping; second, the diffusion that is referred to in connection with wave drag is a crude parameterization of the very complex non-linear interaction between scales of motion whereby the distribution of a fluid property acquires ever finer spatial and time scales.

Modeling studies (e.g., Schoeberl and Strobel, 1978; Holton and Wehrbein, 1980) generally use a "Rayleigh" or linear friction coefficient to crudely parameterize the effects of wave drag in the zonal momentum equation. Thus, a deceleration coefficient K_r is assumed to act linearly on $\bar{u}$, and is substituted for the wave drag term $\nabla \cdot \vec{F}$ as follows:

$$\frac{\partial \bar{u}}{\partial t} - f\bar{v}^* = -K_r\bar{u} \qquad (3.64)$$

where K_r is the *Rayleigh friction coefficient*. Although the formulation in equation (3.64) parallels exactly the ion drag that is important in the thermosphere, we emphasize that the $K_r\bar{u}$ term in equation (3.64) must be regarded as no more than a crude parameterization. In spite of this, the dynamic and thermal structure obtained by models that include Rayleigh friction is in reasonably good agreement with observations as long as K_r^{-1} is of the order of 100 days in the stratosphere and a few days in the mesosphere. Figure 3.20 shows the zonal winds and temperatures obtained in the model study by Geller (1983) including such a parameterization (compare, Fig. 3.19). The faster values of K_r required for the mesosphere as compared to the stratosphere indicate that wave drag is much more important for the zonally averaged distribution of winds and temperature there. In fact, except at the tropical tropopause and in the polar night, the stratosphere may be considered to be approximately in radiative equilibrium, while the mesosphere is very far from radiative equilibrium. The most dramatic manifestation of the importance of meridional motions for the temperature structure of the middle atmosphere occurs at the summer polar mesopause, where the temperature (about 140K) is the lowest observed anywhere on Earth, although it receives much more solar radiation than does the winter mesopause.

It has long been suspected that the dissipation of gravity waves might provide the required wave drag in the mesosphere (e. g., Lindzen, 1967). The gravity waves of importance to the mesospheric momentum balance are rather small scale, with horizontal wavelengths of the order of 100 km, in contrast to the very large planetary waves (wavelength $\approx$ 10,000 km) that were discussed earlier. Gravity waves are believed to be produced in the troposphere by a variety of mechanisms, including air flow over surface topographic features. As long as they are not being dissipated, the gravity wave amplitudes will grow as the inverse square of the density. This can be understood by considering that, for example, the kinetic energy density, $1/2\,\rho_s |V|^2$, must remain constant in the absence of dissipation. It follows that $|V(z)| = c\rho_s^{-1/2}$, where c is a constant. Since ρ_s varies as $\rho_0\exp(-z/H)$, then $|V(z)| = c/\rho_0\exp(z/2H)$, i. e., wave amplitude grows exponentially with height. At some altitude, the gravity wave will have grown so large that its temperature perturbation will produce a superadiabatic lapse rate. At this point, the wave becomes unstable and is said to "break". Recently, Lindzen (1981) and Weinstock (1982) have

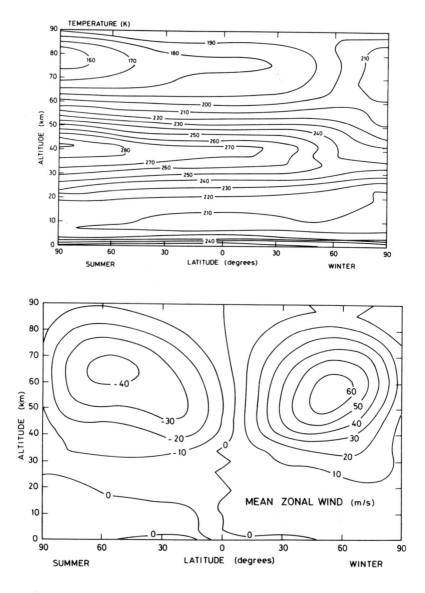

Fig. 3.20. Calculated temperatures and mean zonal winds including a Rayleigh friction parameter. From Geller (1983). (Copyright by Reidel Publishing Company).

parameterized the effects of the breaking process by assuming that, above the breaking level, the wave must produce enough turbulent diffusion to prevent further amplitude growth with height. This assumption leads to expressions for the turbulent diffusion, and for the wave drag, that are strong functions of

$\bar{u} - c$, the difference between the zonally averaged wind and the phase speed (c) of the wave:

$$K_{turb} = \frac{k(\bar{u} - c)^4}{2HN^3} \tag{3.65}$$

$$\bigtriangledown \cdot \vec{F} = -\frac{k(\bar{u} - c)^3}{2HN} \tag{3.66}$$

where k is the horizontal wavenumber of the gravity wave, H is the scale height, and N is the Brunt-Vaisala frequency (see, e.g., Holton, 1979). Equation (3.66) is a much more satisfactory description of the effect of wave drag in the mesosphere than is the simple Rayleigh friction law used in equation (3.64), since it is based on a simple but physically motivated description of the wave breaking process. Equation (3.66) can produce positive or negative acceleration of the zonal wind depending on the sign of $\bar{u} - c$. From the viewpoint of chemical transport, the gravity wave parameterization is useful because, in addition to specifying the wave drag, it yields a consistent value for the accompanying turbulent diffusion.

It should also be noted that equations (3.65)-(3.66) introduce the possibility of seasonal variability of wave drag and diffusion. This occurs because the propagation of gravity waves through the atmosphere depends on the

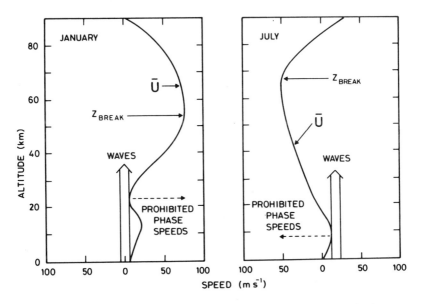

Fig. 3.21. Approximate altitude profiles of the mean zonal winds in summer and in winter. The permitted phase speeds of mesospheric gravity waves, and their breaking levels, are also shown. Adapted from Lindzen (1981). (Copyright by the American Geophysical Union).

zonal wind distribution, which varies markedly with season. Specifically, it can be shown that when the wave phase speed is equal to the zonal wind speed, the wave will be absorbed. The point where this occurs is referred to as a *critical level*. Figure 3.21 schematically shows how this absorption effect influences gravity wave propagation for mid-latitudes, in summer and in winter. Holton (1982; 1983) has shown that the filtering produced by stratospheric winds does indeed result in seasonal differences in mesospheric wave drag, and that these differences in turn produce a mesospheric wind distribution that is in good agreement with that observed. Clearly, if wave drag varies seasonally, so too must the turbulent diffusion produced by wave breaking, a phenomenon that must have consequences on the transport of chemical species in the mesosphere.

The origin and magnitude of wave drag in the stratosphere is also an important question for stratospheric dynamics, but no simple parameterization of wave drag in the stratosphere has yet been proposed. Geller (1983) suggested that gravity wave breaking may provide much of the frictional dissipation in the stratosphere as well as in the mesosphere. However, planetary wave transience and dissipation must also be important, at least in winter.

There are a few special cases in which the effects of stratospheric wave transience and dissipation are particularly large, and are likely to have very significant effects on chemical constituents. For example, large variations of the zonal wind are observed in the tropical lower stratosphere, with periods of 20 to 40 months, but exhibiting a mean period of 26 months. This phenomenon is called the quasi-biennial oscillation (Veryand and Ebdon, 1961; Reed et al., 1961; Reed, 1965; etc.). Theoretical studies indicate that its origin lies in the vertical transport of momentum associated with certain types of tropical waves (Lindzen and Holton, 1968; Holton and Lindzen, 1972). Total ozone exhibits a variation in the tropics which is clearly related to the quasi-biennial oscillation (e. g., Hilsenrath and Schlesinger, 1981). Perhaps a more important phenomenon, however, is that of stratospheric sudden warmings. In late winter and spring, large temperature increases are observed to occur in the lower stratosphere, accompanied by cooling in the mesosphere and upper stratosphere (e. g., Labitzke, 1980). Under these circumstances, the stratopause may be found to be as low as 35 km. The polar night jet decreases in strength and finally reverses. Matsuno (1971) and Holton (1976) have simulated sudden warmings in numerical model studies by considering the interaction of large scale tropospheric waves with the stratosphere and mesosphere. More recently, Matsuno and Nakamura (1979), Dunkerton et al. (1981), and Palmer (1981), have used the Lagrangian or transformed Eulerian equations to show that sudden warmings are probably related to strong planetary wave drag due to wave transience. This induces a mean meridional circulation with strong downward motion at high latitude and upward motion in the tropics, producing the observed high latitude warming as a result of adiabatic compression. Sudden warmings probably have significant effects on the annual cycle of total ozone, and perhaps also on many other constituents whose distributions are not as well known, such as NO_2 (see Noxon et al., 1979).

A simple two-dimensional parameterization of the planetary wave drag that leads to sudden stratospheric warmings is not feasible because warmings are sporadic phenomena that take place during periods of strong wave transience. The latter can be simulated only by explicitly computing the evolution of the wave field and its interaction with the zonally averaged wind field. A model which includes equations for both the zonal mean state and the perturbations to it (i.e., the waves) is not a two-dimensional, but rather a three-dimensional model. Although highly truncated models that consider only the zonal mean and the one or two longest planetary waves can provide fairly realistic simulations of sudden warmings (e. g., Holton and Wehrbein, 1980), these models cannot describe the very large dispersion of air parcels that accompanies the warming (Hsu, 1980). This dispersion must have a diffusive effect on atmospheric constituents, as demonstrated by McIntyre and Palmer (1983), who studied the evolution of the potential vorticity field during a warming episode, and were thus able to trace transport and dispersion of high latitude air parcels to lower latitudes. The complete simulation of such a process is impossible without resort to very high resolution three-dimensional models, but it may prove possible to parameterize its effects on the distribution of trace substances by defining an eddy diffusion parameter that depends on the state of the truncated planetary wave field (i.e., wavenumbers one and two).

The question of what processes are most important in producing wave drag in the stratosphere must be studied carefully in order to determine the effects of transports on the distribution of chemical species. The use of a Rayleigh friction coefficient in dynamical models has not been the subject of much scrutiny, probably because the real stratosphere is close to radiative equilibrium, so that adiabatic effects on the temperature and zonal wind fields are small (compare, Figs. 3.2 and 3.4, and Fig. 3.19). Thus, reasonable representations of these fields are obtained even if the wave drag (and hence the meridional circulation) is not modeled accurately. However, for chemical species, transport by the meridional circulation and diffusion must be realistically represented, and this cannot be done properly unless the processes that contribute to wave drag are understood. An an example, consider the situation in which only planetary wave drag is of importance to the momentum budget of the stratosphere. In this case, we would expect a much weaker meridional circulation (and much less diffusion) in summer, when planetary wave activity is much diminished compared to winter. If, on the other hand, gravity wave breaking is also of importance in generating wave drag in the stratosphere, then the summer and winter meridional circulations would not be expected to be radically different.

Let us briefly summarize the connection between our previous discussion of two-dimensional representations of the atmosphere and the importance of wave transience and dissipation. Dunkerton (1978) showed that the diabatic circulation was equivalent to the true net motion of air parcels (or Lagrangian mean motion) *provided the eddies were steady and dissipationless.* But we have seen that in the momentum equation, the assumption of no wave transience or dissipation cannot be made, even to a first approximation, and that it plays a crucial role in determining the strength of the mean meridional circulation. Further, this transience and dissipation implies that some amount of turbulence and diffusion is present in the stratosphere, although probably much reduced from the estimates of Reed and German, who assumed that all eddy transport is diffusive in nature (see also, Kida, 1983).

Thus while the transformed Eulerian circulation provides a convenient framework for a qualitative understanding of transport in the middle atmosphere, the question of the role of transient, dissipating eddies in determining the distributions of chemical species must also be carefully considered. To understand the effects of these processes on chemical constituents, we must know what kind of waves are involved, and the physical mechanisms responsible for their transience and dissipation. As we have discussed, these are presently not well known in the stratosphere.

One of the advantages of three- dimensional models is that such effects can be explicitly evaluated, at least for the physical scales of eddies resolved by the model. In two-dimensional diabatic or transformed Eulerian models, these terms can in principle be parameterized as diffusion coefficients in the horizontal (K_{yy}) and vertical (K_{zz}) directions, but these are clearly difficult to estimate without a physical understanding of how they arise.

3.8 One-dimensional representations of the atmosphere

Much of the pioneering work in aeronomy has been performed with one-dimensional models (see e. g., Nicolet, 1965; Strobel et al., 1970; Crutzen, 1974; McElroy et al., 1971; Liu and Donahue, 1974; Stolarski and Cicerone, 1976, Chang et al., 1977, to name but a few). It therefore seems appropriate to ask how realistic this representation is from the dynamical point of view.

In one-dimensional models averaging is performed over longitude, latitude, and to varying degrees over time, so that such a model ideally represents a global average. Tuck (1979) has pointed out that careful averaging must be performed in the evaluation of the photochemical production and loss terms in such models; a great deal of effort has been devoted to careful solution of that aspect of the problem (e. g., Miller et al., 1979, for example). A more difficult problem is the question of the validity of such a representation of the dynamical processes of the atmosphere. In one dimension, all transport must be modeled as vertical diffusion. This globally averaged vertical diffusion is usually represented as an eddy diffusion coefficient, K_z, which is often derived from measurements of long lived species such as N_2O and CH_4. Figure 3.22 presents vertical profiles of the K_z coefficients used in several one-dimensional models. The globally averaged vertical flux of species i in such models is:

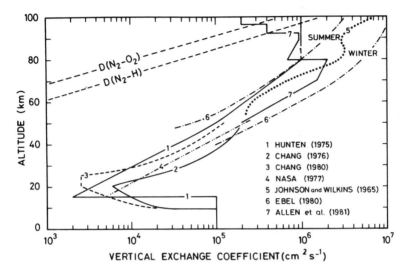

Fig. 3.22. One-dimensional eddy diffusion coefficients used in various aeronomic models.

$$F_z = - K_z(M) \frac{\partial f_i}{\partial z} \qquad (3.67)$$

where f_i is the mixing ratio of species i.

The effect of the vertical diffusion coefficient is to reduce vertical mixing ratio gradients. This is clearly a poor description for transport by the winds, which can only transport material in the direction of flow, regardless of the gradient. As we have discussed above, while transport by diffusion probably occurs to some degree (particularly in the mesosphere), transport by advection (winds) is also quite important.

Perhaps this difficulty can be avoided if it is admitted that transport is probably not truly diffusive in character, but it is assumed that the one-dimensional eddy diffusion coefficient is only an empirical parameter to describe the effects of transport in an approximate way, deriving its value from observations of long lived tracers as mentioned above. Unfortunately this is also not very satisfactory - if the diffusion coefficient is to represent the "averaged" effects of advective processes which transport material both vertically and horizontally, then it should reflect the nature of the circulation and the photochemical lifetime of the species in question as it travels along such a path (see, e. g., Pyle and Rogers, 1980b). Except for tracers whose lifetime does not vary with altitude or latitude, different effective eddy coefficients will result. Indeed, it has been found that a single species-independent eddy coefficient cannot satisfactorily reproduce the observed global average N_2O and CH_4 distributions in the lower stratosphere (WMO, 1982).

One-dimensional models continue to provide a useful and important tool, particularly for the evaluation of the effects of perturbations to the photochemistry of the atmosphere, such as the addition of chlorofluorocarbons (see Chapter 7). However, it is not clear whether they can reproduce the observed profiles of atmospheric constituents whose distributions are dependent on transport.

In this chapter we have examined different ways of understanding these dynamical processes so that we may examine their impact on aeronomy. Some aeronomic problems may require the use of a three-dimensional picture (see, for example, the discussion of NO_2 in chapter 5), others may be best understood in two dimensions, but attention must be paid to the method used to perform the zonal average. One- dimensional frameworks may also be useful for certain other problems. It is likely that the interaction of chemistry and dynamics will continue to be studied and explored with a hierarchy of models.

References

Allen, M., Y. L. Yung, and J. W. Waters, Vertical transport and photochemistry in the terrestrial mesosphere and lower thermosphere (50-120 km), J. Geophys. Res., 86, 3617, 1981.

Andrews, D. G., and M. E. McIntyre, Planetary waves in horizontal and vertical shear: the generalized Eliassen-Palm relation and the zonal mean acceleration, J. Atmos. Sci., 33, 2031, 1976.

Berggren, R., and K. Labitzke, The distribution of ozone on pressure surfaces, Tellus, 20, 88, 1968.

Boyd, J., The noninteraction of waves with the zonally averaged flow on a spherical earth and the interrelationship of eddy fluxes of energy, heat, and momentum, J. Atmos. Sci., 33, 2285, 1976.

Brewer, A. W., Evidence for a world circulation provided by measurements of helium and water vapor distribution in the stratosphere, Q. J. Roy. Met. Soc., 75, 351, 1949.

Campbell, I. M., Energy and the atmosphere, John Wiley and Sons, (Chichester, G. B.), 1977.

Chang, J. S., A. C. Hindmarsh, and N. K. Madsen, Simulation of chemical kinetics transport in the stratosphere, in Stiff differential systems, R. A. Willoughby, ed., Plenum, (New York), 1974.

Chang, J. S., in Halocarbons: Effects on stratospheric ozone, National Academy of Sciences Report, Washington, D. C., 1976.

Crutzen, P. J., Estimates of possible variations in total ozone due to natural causes and human activity, Ambio, 3, 201, 1974.

Crutzen, P. J., A two-dimensional photochemical model of the atmosphere below 55 km: estimates of natural and man-caused perturbations due to NO_x, in Proc. Fourth Conf. on CIAP, DOT-TSC-OST-38, 1975.

Cunnold, D. M., F. Alyea, N. Phillips, and R. G. Prinn, A three-dimensional dynamical-chemical model of atmospheric ozone, J. Atmos. Sci., 32, 170, 1975.

Cunnold, D. M., F. N. Alyea, and R. G. Prinn, Preliminary calculations concerning the maintenance of the zonal mean ozone distribution in the northern hemisphere, Pure Appl. Geophys., 118, 329, 1980.

Danielsen, E. F., Trajectories: Isobaric, isentropic, and actual, J. Met., 18, 479, 1961.

DeMazure, M., and J. Saissac, Generalisation de l'equation classique de diffusion, Note de l'etablissement d'etudes et de recherches meteorologiques, no. 115, Paris, 1962.

Dobson, G. M. G., Origin and distribution of polyatomic molecules in the atmosphere, Proc. Roy. Soc. Lond. A, 236, 187, 1956.

Dunkerton, T., On the mean meridional mass motions of the stratosphere and mesosphere, J. Atmos. Sci., 35, 2325, 1978.

Dunkerton, T., C. P. F. Hsu, and M. E. McIntyre, Some Eulerian and Lagrangian diagnostics for a model stratospheric warming, J. Atmos. Sci., 38, 819, 1981.

Ebel, A., Eddy diffusion models for the mesosphere and lower thermosphere, J. Atmos. Terr. Phys., 42, 617, 1980.

Fels, S. B., J. D. Mahlman, M. D. Schwarzkopf, and R. W. Sinclair, Stratospheric sensitivity to perturbations in ozone and carbon dioxide: radiative and dynamical response, J. Atmos. Sci., 37, 2265, 1980.

Garcia, R. R., and D. L. Hartmann, The role of planetary waves in the maintenance of the zonally averaged ozone distribution of the upper stratosphere, J. Atmos. Sci., 37, 2248, 1980.

Garcia, R. R., and S. Solomon, A numerical model of the zonally averaged dynamical and chemical structure of the middle atmosphere, J. Geophys. Res., 88, 1379, 1983.

Geller, M. A., Dynamics of the middle atmosphere, Space Science Rev., 34, 359, 1983.

Gill, A. E., *Atmosphere-Ocean Dynamics,* Academic Press, (New York), 1982.

Grose, W. L., and K. V. Haggard, Numerical simulation of a sudden stratospheric warming with a three-dimensional, spectral, quasi-geostrophic model, J. Atmos. Sci., 38, 1480, 1981.

Hare, F. K., and B. W. Boville, The polar circulation, Technical note 70, World meteorological organization, Geneva, Switzerland, 1965.

Harwood, R. S., and J. A. Pyle, A two-dimensional mean circulation model for the atmosphere below 80 km, Q. J. Roy. Met. Soc., 101, 723, 1975.

Hartmann, D. L., Some aspects of the coupling between radiation, chemistry, and dynamics in the stratosphere, J. Geophys. Res., 86, 9631, 1981.

Hidalgo, H., and P. J. Crutzen, The tropospheric and stratospheric composition perturbed by NO_x emissions of high altitude aircraft, J. Geophys. Res., 83, 5833, 1978.

Hilsenrath, E., and B. M. Schlesinger, Total ozone seasonal and interannual variations derived from the 7 year Nimbus 4 BUV data set, J. Geophys. Res., 86, 12087, 1981.

Holton, J. R., and R. S. Lindzen, An updated theory for the quasi-biennial cycle of the tropical stratosphere, J. Atmos. Sci., 29, 1076, 1972.

Holton, J. R., *The dynamic meteorology of the stratosphere and mesosphere*, Met. Mono. 15, American Met. Soc., 1975.

Holton, J. R., A semi-spectral numerical model for wave, mean flow interactions in the stratosphere: application to sudden stratospheric warmings, J. Atmos. Sci., 33, 1639, 1976.

Holton, J. R., *An introduction to dynamic meteorology*, Academic Press, (New York), 1979.

Holton, J. R., and W. M. Wehrbein, The role of forced planetary waves in the annual cycle of the zonal mean circulation of the middle atmosphere, J. Atmos. Sci., 37, 1968, 1980.

Holton, J. R., The role of gravity wave induced drag and diffusion in the momentum budget of the mesosphere, J. Atmos. Sci., 39, 791, 1982.

Holton, J. R., The influence of gravity wave breaking on the general circulation of the middle atmosphere, J. Atmos. Sci., 40, 2497, 1983.

Hsu, C. P., Air parcel motions during a numerically simulated sudden stratospheric warming, J. Atmos. Sci., 37, 2768, 1980.

Hunt, B. G., and S. Manabe, Experiments with a stratospheric general circulation model II. Large scale diffusion of tracers in the stratosphere, Mon. Weath. Rev., 96, 503, 1968.

Hunt, B. G., A generalized aeronomic model of the mesosphere and lower thermosphere including ionospheric processes, J. Atm. Terr. Phys., 35, 1755, 1973.

Hunten, D. M., Vertical transport in atmospheres, in *Atmospheres of Earth and the Planets,* W. McCormac, ed., Reidel Pub., (Dordrecht), 1975.

Johnson, F. S., and E. M. Wilkins, Thermal upper limit on eddy diffusion in the mesosphere and lower thermosphere, J. Geophys. Res., 70, 1281, 1965.

Julian, P. R., and K. B. Labitzke, A study of atmospheric energetics during the January-February 1963 stratospheric warming, J. Atmos. Sci., 22, 597, 1965.

Kasahara, A., T. Sasamori, and W. M. Washington, Simulation experiments with a 12 layer stratospheric global circulation model I. Dynamical effects of earth's orography and thermal influence of continentality, J. Atmos. Sci., 30, 1229, 1973.

Kasting, J. F., and R. G. Roble, A zonally averaged chemical-dynamical model of the lower thermosphere, J. Geophys. Res., 86, 9641, 1981.

Kawahira, K., A two-dimensional model for ozone changes by planetary waves in the stratosphere I. Formulation and the effect of temperature waves on the zonal mean ozone concentration, J. Met. Soc. Jap., 60, 1058, 1982.

Kida, H., General circulation of air parcels and transport characteristics derived from a hemispheric GCM Part 1. A determination of advective mass flow in the lower stratosphere, J. Met. Soc. Jap., 61, 171, 1983.

Labitzke, K., Climatology of the stratosphere and mesosphere, Phil. Trans. Roy. Soc. Lond. A, 296, 7, 1980.

Leovy, C., Simple models of thermally driven mesospheric circulation, J. Atm. Sci., 21, 327, 1964.

Levy, H., J. D. Mahlman, and W. J. Moxim, A preliminary report on the numerical simulation of the three-dimensional structure and variability of atmospheric N_2O, Geophys. Res. Lett., 6, 155, 1979.

Levy, H., J. D. Mahlman, and W. J. Moxim, A stratospheric source of reactive nitrogen in the unpolluted troposphere, Geophys. Res. Lett., 7, 441, 1980.

Lindzen, R. S., Thermally driven diurnal tide in the atmosphere, Q. J. Roy. Met. Soc., 93, 18, 1967.

Lindzen, R. S., and J. R. Holton, A theory of the quasi-biennial oscillation, J. Atmos. Sci., 22, 341, 1968.

Lindzen, R. S., Turbulence and stress owing to gravity wave and tidal breakdown, J. Geophys. Res., 86, 9707, 1981.

Liu, S. C., and T. M. Donahue, The aeronomy of hydrogen in the atmosphere of the earth, J. Atmos. Sci., 31, 1118, 1974.

Liu, S. C., D. Kley, M. McFarland, J. D. Mahlman, and H. Levy II, On the origin of tropospheric ozone, J. Geophys. Res., 85, 7546, 1980.

London, J. L., Radiative energy sources and sinks in the stratosphere and mesosphere, Proc. NATO Advanced Study Institute on Atmospheric Ozone, FAA-EE-80-20, NTIS, Springfield, Va., 1980.

Mahlman, J. D., and W. J. Moxim, Tracer simulation using a global general circulation model: results from a mid-latitude instantaneous source experiment, J. Atmos. Sci., 35, 1340, 1978.

Mahlman, J. D., H. Levy II, and W. J. Moxim, Three-dimensional tracer structure and behavior as simulated in two ozone precursor experiments, J. Atmos. Sci., 37, 655, 1980.

Mahlman, J. D., and L. J. Umscheid, Dynamics of the middle atmosphere: successes and problems of the GFDL "SKIHI" general circulation model, in *Proceedings of the U. S.-Japan seminar on middle atmosphere dynamics,* Terra Scientific Pub., (Tokyo), 1983.

Mahlman, J. D., D. G. Andrews, D. L. Hartmann, T. Matsuno, and R. J. Murgatroyd, Transport of trace constituents in the stratosphere, in *Proceedings of the U. S.-Japan seminar on middle atmosphere dynamics,* Terra Scientific Pub., (Tokyo), 1983.

Manabe, S., and J. D. Mahlman, Simulation of seasonal and interhemispheric variations in the stratospheric circulation, J. Atmos. Sci., 33, 2185, 1976.

Matsuno, T., A dynamical model of the stratospheric sudden warming, J. Atmos. Sci., 28, 1479, 1971.

Matsuno, T., and K. Nakamura, The Eulerian and Lagrangian mean meridional circulations in the stratosphere at the time of a sudden warming, J. Atmos. Sci., 36, 640, 1979.

Matsuno, T., Lagrangian motion of air parcels in the stratosphere in the presence of planetary waves, Pure Appl. Geophys., 118, 189, 1980.

McElroy, M. B., and J. C. McConnell, Nitrous oxide: a natural source of stratospheric NO, J. Atmos. Sci., 28, 1095, 1971.

McIntyre, M. E., Towards a Lagrangian mean description of stratospheric circulations and chemical transports, Phil. Trans. Roy. Soc. Lond. A, 296, 129, 1980.

McIntyre, M. E., and T. N. Palmer, Breaking planetary waves in the stratosphere, Nature, 305, 593, 1983.

Miller, C., D. L. Filkin, and J. P. Jesson, The fluorocarbon-ozone theory VI. Atmospheric modeling: calculation of the diurnal steady state, Atmos. Env., 13, 381, 1979.

Miller, C., D. L. Filkin, A. J. Owens, J. M. Steed, and J. P. Jesson, A two-dimensional model of stratospheric chemistry and transport, J. Geophys. Res., 76, 202, 1981.

Murgatroyd, R. J., and F. Singleton, Possible meridional circulations in the stratosphere and mesosphere, Q. J. Roy. Met. Soc., 87, 125, 1961.

Murgatroyd, R. J., in *The global circulation of the atmosphere,* G. A. Corby, Ed., Roy. Met. Soc., (London), 1969.

Murgatroyd, R. J., Dynamical modelling of the stratosphere and mesosphere, in *Mesospheric models and related experiments,* G. Fiocco, ed., Reidel Publishing Co., (Dordrecht), 1971.

Murgatroyd, R. J., An introduction to studies of the general characteristics of the stratosphere and mesosphere, Proc. NATO Advanced Study Institute on Atmospheric Ozone, FAA-EE-80-20, NTIS, Springfield, Va., 1980.

NASA, National Aeronautics and Space Administration, *Chlorofluoromethanes and the stratosphere,* NASA Reference Publication 1010, 1977.

Nastrom, G. D., B. B. Balsley, and D. A. Carter, Mean meridional winds in the mid and high latitude mesosphere, Geophys. Res. Lett., 9, 139, 1982.

Newell, R. E., The general circulation of the atmosphere and its effects on the movement of trace substances, J. Geophys. Res., 68, 3949, 1963.

Newson, R. L., An experiment with a tropospheric and stratospheric three-dimensional general circulation model, Proc. Third Conf. on CIAP, DOT-TSC-OST-74-15, U. S. Dept. of Transportation, NTIS, Springfield, Va., 1974.

Nicolet, M., Nitrogen oxides in the chemosphere, J. Geophys. Res., 70, 679, 1965.

Noxon, J. F., E. Marovich, and R. B. Norton, Effect of a major warming upon stratospheric NO_2, J. Geophys. Res., 84, 7883, 1979.

O'Neill, A., Dynamical processes in the stratosphere: wave motion, Proc. NATO Advanced Study Institute on Atmospheric Ozone, FAA-EE-80-20, NTIS, Springfield, Va., 1980.

Palmer, T. N., Diagnostic study of a wavenumber 2 stratospheric sudden warming in a transformed Eulerian mean formalism, J. Atmos. Sci., 38, 544, 1981.

Pedlosky, J., *Geophysical fluid dynamics*, Springer-Verlag, (New York), 1979.

Phillips, N. A., Geostrophic motion, Rev. Geophys., 1, 123, 1963.

Plumb, R. A., Eddy fluxes of conserved quantities by small amplitude waves, J. Atm. Sci., 36, 1699, 1979.

Pyle, J. A., and C. F. Rogers, A modified diabatic circulation model for stratospheric tracer transport, Nature, 287, 711, 1980a.

Pyle, J. A., and C. F. Rogers, Stratospheric transport by stationary planetary waves - The importance of chemical processes, Q. J. Roy. Met. Soc., 106, 421, 1980b.

Reed, R. J., W. J. Campbell, L. A. Rasmusson, and D. G. Rogers, Evidence of a downward propagating annual wind reversal in the equatorial stratosphere, J. Geophys. Res., 66, 813, 1961.

Reed, R. J., J. L. Wolfe, and H. Nishimoto, A spectral analysis of the energetics of the stratospheric sudden warming of early 1957, J. Atmos. Sci., 20, 256, 1963.

Reed, R. J., and K. E. German, A contribution to the problem of stratospheric diffusion by large scale mixing, Mon. Weath. Rev., 93, 313, 1965.

Rind, D., R. Suozzo, A. Lacis, G. Russell, and J. Hansen, 21 layer troposphere-stratosphere climate model, submitted to Mon. Weath. Rev., 1984.

Rood, R. B., and M. R. Schoeberl, A mechanistic model of Eulerian, Lagrangian mean, and Lagrangian ozone transport by steady planetary waves, J. Geophys. Res., 88, 5208, 1983.

Schmidt, M., The influence of large scale advection on the vertical distribution of stratospheric source gases in 44° and 41° North, J. Geophys. Res., 87, 11239, 1982.

Schoeberl, M. R., and D. F. Strobel, The zonally averaged circulation of the middle atmosphere, J. Atm. Sci., 35, 577, 1978.

Schoeberl, M. R., Strobel, D. F., and J. P. Apruzese, A numerical model of gravity wave breaking and stress in the mesosphere, J. Geophys. Res., xx, xx, 1983.

Smagorinsky, J., General circulation experiments with the primitive equations I. The basic experiment, Mon. Weath. Rev., 91, 99, 1963.

Solomon, S., and R. R. Garcia, Simulation of NO_x partitioning along isobaric parcel trajectories, J. Geophys. Res., 88, 5497, 1983.

Stolarski, R. S., and R. J. Cicerone, Stratospheric chlorine: a possible sink for ozone, Can. J. Chem., 52, 1610, 1974.

Strobel, D. F., D. M. Hunten, and M. B. McElroy, Production and diffusion of nitric oxide, J. Geophys. Res., 75, 4307, 1970.

Strobel, D. F., Parameterization of linear wave chemical transport in planetary atmospheres by eddy diffusion, J. Geophys. Res., 86, 9806, 1981.

Trenberth, K. E., Global model of the general circulation of the atmosphere below 75 km with an annual heating cycle, Mon. Weath. Rev., 101, 287, 1973.

Tuck, A. F., A comparison of one, two, and three-dimensional model representations of stratospheric gases, Phil. Trans. Roy. Soc. Lond. A, 290, 9, 1979.

Tung, K. K., On the two-dimensional transport of stratospheric trace gases in isentropic coordinates, J. Atmos. Sci., 39, 2330, 1982.

Veryand, R. G., and R. A. Ebdon, Fluctuations in tropical stratospheric winds, Met. Mag., 90, 125, 1961.

Vincent, D. G., Meridional circulation in the northern hemisphere lower stratosphere during 1964 and 1965, Q. J. Roy. Met. Soc., 94, 333, 1968.

Vupputuri, R. K., The structure of the natural stratosphere and the impact of chlorofluoromethanes on the ozone layer investigated in a 2-D time dependent model, Pure Appl. Geophys., 117, 448, 1979.

Weinstock, J., Vertical turbulent diffusion in a stably stratified fluid, J. Atmos. Sci., 35, 1022, 1978.

Weinstock, J., Nonlinear theory of gravity waves: momentum deposition, generalized Rayleigh friction, and diffusion, J. Atmos. Sci., 39, 1698, 1982.

World Meteorological Organization (WMO), The stratosphere 1981: Theory and measurements, Report no. 11, WMO global ozone research and monitoring project, Geneva, Switzerland, 1982.

Zimmerman, S. P., and E. A. Murphy, Stratospheric and mesospheric turbulence, in *Dynamical and chemical coupling*, D. Reidel, (Dordrecht, Holland), 1977.

Chapter 4

Radiation

4.1 Introduction

From the viewpoint of aeronomy, the atmosphere can be considered to be a mixture of gases exposed to the electromagnetic spectrum of the sun. An understanding of the dynamical and photochemical processes which occur in this environment requires consideration of atmospheric radiative transfer. For example. the rate of reaction between two constituents generally depends on the local temperature (see Chapter 2). which is related to the effects of absorption, scattering, and emission of solar and terrestrial radiation. Further, solar radiation of particular energies can dissociate and ionize atmospheric molecules to produce reactive ions and radicals which participate in many of the important atmospheric chemical processes.

As we have discussed in Chapter 2, atmospheric molecules are characterized by discrete rotational and vibrational energy states, and by their electronic configurations (see Fig. 4.1). The absorption of photons can induce transitions between ground and excited states. The differences between rotational energy levels are much smaller than those of the vibrational states, which are nearly as great as the electronic levels. For those processes which do not result in continuum ionization or photodissociation of the molecule, only the photons of the frequencies corresponding to transitions between energy states of the molecule can be absorbed; the corresponding spectrum therefore appears as a series of distinct lines which are often part of absorption bands.

Due to the relatively high energies needed to dissociate or ionize atmospheric gases, photochemistry is mainly initiated by less than one per cent of the solar photons; more specifically, those whose characteristic wavelength is in the x-ray, ultraviolet, or, for certain molecules, in the visible region.

Solar radiation also produces thermal effects. Solar energy is absorbed primarily by ozone in the stratosphere and mesosphere, and by molecular oxygen in the upper mesosphere and lower thermosphere, and is rapidly converted to thermal energy through chemical reactions (recombination of atomic oxygen in the presence of a third body) which follow the photodissociation of O_2

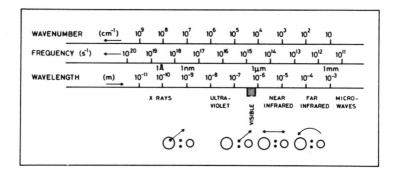

Fig. 4.1. Spectral regions and their effect on molecules: from left to right: ionization, dissociation, vibration, and rotation.

and O_3. This heating by absorption of solar ultraviolet radiation is balanced by cooling related to infrared emission by CO_2 (15 μm band), and, to a limited extent, by ozone (9.6 μm band) and H_2O (mostly 8.0 μm band). Thus, the thermal structure depends on an equilibrium between these processes as well as other phenomena such as the production of latent heat in the troposphere and the dissipation of atmospheric waves in the mesosphere and thermosphere.

In the study of radiation affecting the middle atmosphere it is often convenient to distinguish between two distinct spectral regions (Fig. 4.2): (1) wavelengths less than 4μm, which represent radiation of solar origin; and (2) wavelengths longer than 4μm, which is made up of radiation of both terrestrial and atmospheric origin. Above 4 μm, the solar flux can be neglected relative to the radiation emitted by the earth's surface. This regime is therefore quite different from the shorter wavelengths because of the sharp limit imposed by the spectral function of the terrestrial emission temperature. In purely aeronomic applications, analysis is limited almost exclusively to energetic solar radiation, while for meteorological applications the infrared radiation plays an essential role.

The penetration of solar radiation into the Earth's atmosphere depends on the absorption by each constituent in the atmosphere. Since the absorption coefficients of minor constituents are functions of wavelength, the penetration depth is dependent on the shape of the spectrum (Fig. 4.3).

For wavelengths below 100 nm, radiation is almost completely absorbed above 100 km by molecular and atomic oxygen and by molecular nitrogen. Only x-rays of wavelength less than 1 nm penetrate to the middle atmosphere. These x-rays can photoionize, or indirectly dissociate the major species, N_2 and O_2, and are responsible for certain sporadic ionospheric perturbations. These aspects of the photochemistry will be discussed in Chapter 6.

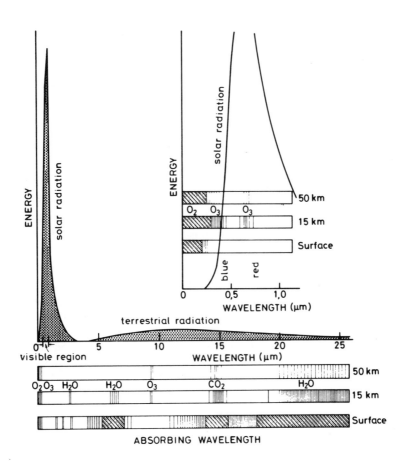

Fig. 4.2. Spectral distribution of solar and terrestrial radiation. Atmospheric absorption at different wavelengths. After Iribarne and Cho (1980). (Copyright by Reidel Publishing Company).

At wavelengths greater than 100 nm, solar ultraviolet radiation can photodissociate atmospheric molecules. The solar Lyman α line at 121.6 nm is very intense and is situated in an optical window; that is, a spectral region where absorption by the atmosphere is quite weak. Therefore, it can penetrate into the upper part of the middle atmosphere, where it effectively dissociates water vapor, carbon dioxide, and methane. Further, it photoionizes nitric oxide, providing the principal source of ionization in the D-region (see, Chapter 6). At longer wavelengths, the solar spectrum is subdivided into regions of absorption by the principal absorbing species, O_2 and O_3. The first of these absorbs only the radiation at wavelengths less than about 240 nm. The latter, abundant in the stratosphere, absorbs primarily between 200 and 300 nm, but also to some extent in the visible and even in the infrared. Even so, in the visible (wavelengths greater than 310 nm), the greater part of the solar photons reach the troposphere and the surface. Thus, in this region the effects of molecular scattering and cloud and surface albedo must be considered. Table 4.1 indicates the different spectral regions which must be considered in the photochemical effects of solar radiation on the neutral atmosphere.

4.2 Solar radiation at the top of the atmosphere

• *The solar atmosphere*

The production of radiation by the sun depends strongly on the physical and chemical structure of the solar atmosphere. Therefore, in order to understand the nature and variations of the radiation incident on the Earth, we briefly present some of the important aspects of the sun's atmosphere.

The sun is primarily composed of hydrogen and helium, along with smaller amounts of heavier elements such as calcium, iron, magnesium, aluminum, nickel, etc. The temperature in its interior is believed to be as high as 2×10^7 K, due to a chain of nuclear reactions which convert H into He. This energy is radiated to the upper convective levels, undergoing a series of

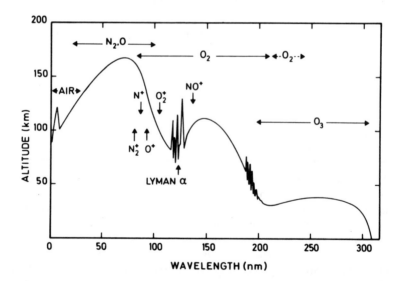

Fig. 4.3. Depth of penetration of solar radiation as a function of wavelength. Altitudes correspond to an attenuation of 1/e. The principle absorbers and ionization limits are indicated.

Table 4.1 Spectral regions of photochemical importance in the atmosphere

Wavelength	Atmospheric absorbers
121.6 nm	Solar Lyman α line, absorbed by O_2 in the mesosphere; no absorption by O_3
100 to 175 nm	O_2 Schumann Runge continuum. Absorption by O_2 in the thermosphere. Can be neglected in the mesosphere and stratosphere.
175 to 200 nm	O_2 Schumann Runge bands. Absorption by O_2 in the mesosphere and upper stratosphere. Effect of O_3 can be neglected in the mesosphere, but is important in the stratosphere.
200 to 242 nm	O_2 Herzberg continuum. Absorption by O_2 in the stratosphere and weak absorption in the mesosphere. Absorption by the O_3 Hartley band is also important; both must be considered.
242 to 310 nm	O_3 Hartley band. Absorption by O_3 in the stratosphere leading to the formation of $O(^1D)$.
310 to 400 nm	O_3 Huggins bands. Absorption by O_3 in the stratosphere and troposphere leads to the formation of $O(^3P)$.
400 to 850 nm	O_3 Chappuis bands. Absorption by O_3 in the troposphere induces photodissociation even at the surface.

absorption and emission processes.

Most of the energy reaching the Earth's atmosphere originates from a relatively thin layer (about 1000 km thick) called the *photosphere*. This layer defines the visible volume of the sun, and although the entire star is in the gaseous state, the photosphere is generally considered to be the "surface" of the sun. Its effective temperature is about 6000 K. Observations indicate that the brightness of the photosphere is not continuous, but is characterized by granules which are uniformly distributed over the solar disk. These granules are probably associated with strong convective processes in and below the photosphere.

Transient phenomena such as sunspots and faculae appear in the photosphere, and influence the variability in the solar emission at short wavelengths. Sunspots are relatively dark regions with a temperature of about 3000 K and a typical diameter of under 50000 km. These spots are usually grouped in certain places, called *active regions* on the solar disk. The lifetime of a sunspot is variable (from a few days to a few months). Faculae are bright events generally associated with active regions, at least at middle and

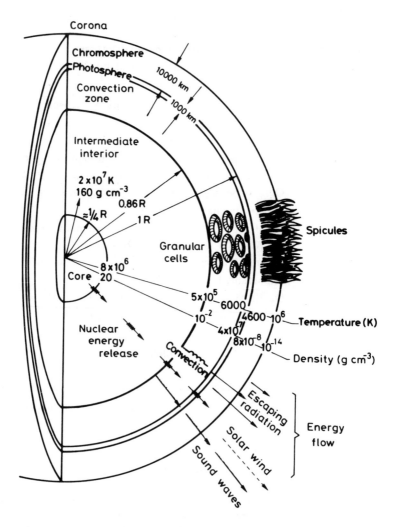

Fig. 4.4a. Schematic diagram of the solar atmosphere.

low solar latitudes. They also appear independently of active regions at high solar latitudes. Observations indicate a fairly regular variation in the occurrence of active regions, with a periodicity of about 11 years, which is called the *solar cycle*.

The emission of the photosphere is a continuum, superimposed with relatively dark lines called the *Fraunhofer lines*. These are produced by selective absorption and reemission of radiation in the upper photosphere where the temperature is as low as 4000 to 5000 K. Some of these lines also originate from higher levels, but their effective emission temperature is lower than the kinetic temperature.

The layer above the photosphere extends to 5000 to 10000 km, and is called the *chromosphere*. This layer can sometimes be seen during total solar eclipses. Its temperature is $10^5 - 10^6$K at the upper levels. Radiation originating from the chromosphere is composed predominantly of emission lines (H, He, Ca) and the visible emission is weak.

The region above the chromosphere is called the *corona*, which extends outward for several solar diameters. Its temperature is about 10^6 K. Several emission lines come from this region. Its free electrons scatter photospheric light.

Active regions in these upper layers are characterized by variations in chromospheric plages, spicules, prominences, solar flares, and enhanced coronal emission. Plages are bright areas in the chromosphere, and are usually observed by their intense calcium K line emission. They usually precede sunspots and last after the sunspots disappear. Spicules are small protuberances which appear continuously at the top of the chromosphere, even in quiet regions. Their lifetime is only a few minutes. Prominences are large, rather stable clouds of bright gas in the upper part and sometimes above the chromosphere. Solar flares are intense eruptive phenomena occurring in the active regions of the chromosphere. They are accompanied by a rapid increase in brightness and an intense enhancement in the emission of ionizing radiation (EUV and x-rays). Flares last from a few minutes to more than an hour, and are divided into different classes in order of increasing importance to the Earth's atmosphere (class 1 to 3). The most intense flares are accompanied by ejection of large quantities of high energy particles.

The sun is also a source of radio waves. These emissions vary with the solar cycle, and are enhanced (radio burst) during chromospheric or coronal events. Since these emissions can easily be recorded (e. g. at 10.7 cm or 2.8 GHz), they are often used as an indicator of solar activity.

Figure 4.4a presents a schematic diagram illustrating the features of the solar atmosphere described above.

• *The solar constant*

The solar constant is defined as the total radiative energy flux outside the earth's atmosphere. This parameter is used to characterize the total solar radiation input. The exact value and the magnitude of the fluctuations of this constant are not yet known with certainty despite many observations performed over the last 25 years. The work by Brusa and Frohlich (1982) suggests a value for the solar constant S of 1367 W m^{-2}, or 1.96 cal cm^{-2} min^{-1} at 1 AU (astronomical unit), which corresponds to an effective solar temperature of 5780 K.

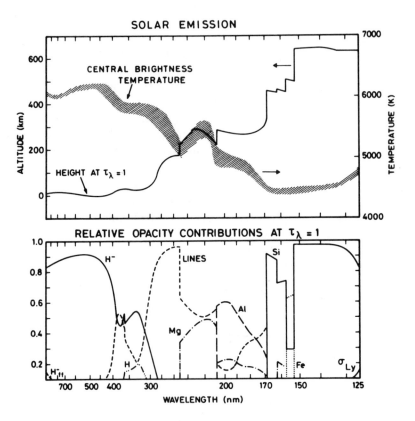

Fig. 4.4b. Observed continuum brightness temperature of the sun as a function of wavelength and corresponding relative height above $\tau_{500nm} = 1$ of the solar emission layers (upper panel). Relative contributions to the opacity at this height (lower panel). After Vernazza et al. (1976).

4.2.1 The sun as a black body

Let us assume for the purpose of illustration that the sun behaves as a black body. In this case, the emitted energy flux M emerging from the surface (also called the exitance) is proportional to the fourth power of the temperature, as indicated by the Stefan-Boltzmann law:

$$M_E = \sigma T^4 \tag{4.1}$$

where $\sigma = 5.67 \times 10^{-8}\,\mathrm{W\ m^{-2}\ K^{-4}} = 8.13 \times 10^{-11}\,\mathrm{cal\ cm^{-2}\ min^{-1}\ K^{-4}}$. The theoretical value of the solar constant is obtained by multiplying the solar emission by a dilution factor β_s which accounts for the earth-sun distance:

$$S_C = \beta_s M_E \qquad (4.2)$$

or

$$\beta_s = \frac{R_s^2}{r_s^2} \qquad (4.3)$$

where R_s represents the radius of the sun and r_s the sun-earth distance. Due to the annual variation of r_s, the mean dilution factor β_s is 2.164×10^{-5}, while the maximum value attained in January $\beta_{s_{max}}$ is 2.350×10^{-5} and the minimum value occurring in July is $\beta_{s_{min}}$ is 2.093×10^{-5}. The energy input from the sun thus varies annually by 6.6 percent.

The spectral distribution of the exitance emitted by a black body, $M_\nu = dM_E/d\nu$ is given by Planck's law:

$$M_\nu = \frac{2\pi h \nu^3}{c^2} \frac{1}{e^{h\nu/kT} - 1} \quad (W \; m^{-2} \; Hz^{-1}) \qquad (4.4a)$$

where $k = 1.38 \times 10^{-23} J \; K^{-1}$ is Boltzmann's constant and $h = 6.63 \times 10^{-34} J \; s$ is Planck's constant. It is also often expressed in terms of wavelength as

$$M_\lambda = \frac{dM}{d\lambda} = \frac{c_1}{\lambda^5} \frac{1}{e^{c_2/\lambda T} - 1} \qquad (4.4b)$$

where $c_1 = 2\pi h c^2 = 3.74 \times 10^8 W m^{-2} \mu m^4$ and $c_2 = hc/k = 1.44 \times 10^4 \mu m K$ are the first and second radiation constants, respectively. Thus, if one approximates the sun as a black body, the number of photons of frequency between ν and $\nu + d\nu$ impinging on the top of the atmosphere per unit surface and per unit time is

$$q_\nu d\nu = \frac{\beta_s M_\nu d\nu}{h\nu} = \frac{2\pi \beta_s \nu^2}{c^2} \frac{d\nu}{e^{h\nu/kT} - 1} \qquad (4.5a)$$

The corresponding flux of photons between wavelength λ and $\lambda + d\lambda$ (expressed for example in photons $cm^{-2} \; s^{-1} \; \mu m^{-1}$) is

$$q_\lambda d\lambda = \frac{c_1' \beta_s}{\lambda^4} \frac{d\lambda}{e^{c_2/\lambda T} - 1} \qquad (4.5b)$$

where $c_1' = 2\pi c = 1.88 \times 10^{23} cm^{-2} s^{-1} \mu m^3$. The total number of photons of frequencies between ν_1 and ν_2 is therefore given by:

$$q(\nu_1 - \nu_2) = \frac{2\pi \beta_s}{c^2} \int_{\nu_1}^{\nu_2} \frac{\nu^2}{e^{h\nu/kT} - 1} d\nu \qquad (4.6)$$

In the ultraviolet where the factor $(h\nu/kT)$ is much larger than one:

$$q(\nu_1 - \nu_2) = \frac{2\pi \beta_s}{c^2} \left(\frac{kT}{h} \right)^3 \left[e^{-h\nu/kT} \left(\left(\frac{h\nu}{kT} + 1 \right)^2 + 1 \right) \right]_{\nu_1}^{\nu_2} \qquad (4.7)$$

In practice, the sun cannot be very well approximated by a black body with a single emission temperature T. A better description is obtained by adopting an equivalent temperature which varies with frequency, although it is also necessary to account for the solar emission lines in the visible, near ultraviolet and far ultraviolet (see below).

4.2.2. The observed solar spectrum

Since this volume focuses on the terrestrial atmosphere below 100 km, we shall concern ourselves only with the radiation penetrating to or below that level (i. e., the ultraviolet, visible and infra-red, as well as x-rays of wavelengths less than 1 nm).

The visible and infrared portion of the solar spectrum is essentially a continuum. Absorption lines superimposed on the continuum become increasingly more pronounced in the ultraviolet at wavelengths less than 300 nm. At wavelengths shorter than 208 nm, a sharp increase in the solar flux occurs. This feature is associated with the Al I ionization edge. Other less pronounced edges with their related continua are due to other elements in the solar atmosphere such as H, Mg, Si, Fe, and C. Below the aluminum edge, the importance of the solar emission lines increases rapidly, while the absorption lines disappear from the spectrum below 150 nm. At wavelengths shorter than 140 nm, the emission by chromospheric and coronal lines begins to dominate the emission in the continuum. The source of the solar emission thus changes from the photosphere to the chromosphere as the wavelength decreases from 300 to 120 nm. As shown in Figure 4.4b, the effective emitting height of the radiation moves up with shorter wavelength as the absorption in the solar atmosphere increases. The minimum in the brightness temperature between 180 and 150 nm characterizes the transition region between the photosphere and the chromosphere. The variation in the brightness temperature (6000K above 300 nm, 5000K at 200 nm, 4500K between 170 and 130 nm) is reflected in the irradiance values at different wavelengths.

The solar spectrum has been the subject of numerous rocket experiments undertaken since the 1940's. The first spectrum measured above the ozone layer dates from 1946 (Baum et al., 1946), and not until the 1950's was a solar spectrum observed from an altitude of 100 km (Johnson et al., 1952). Today the solar irradiance is routinely observed by spectrometers onboard rockets, balloons, aircraft or satellites.

X-rays of wavelength less than 1 nm constitute an important source of ionization in the D region, becoming the dominant source during high levels of solar activity. The value of the corresponding irradiance varies considerably: between 0.1 and 0.8 nm, it has been noted that variations as large as a factor of 1000 may occur over the solar cycle. Similarly, important fluctuations are observed over a solar rotation (27 days) and substantial differences occur from one solar rotation to another. During a solar flare, the x-ray irradiance may become 10000 times greater than that observed during quiet periods. The order of magnitude expected for the solar irradiance in various spectral regions is shown in Figure 4.5.

In the spectral region at wavelengths longer than 100 nm, the solar hydrogen emission line at Lyman α represents an important source of ionization and dissociation and which, during quiet periods, contains more energy than the rest of the spectrum at shorter wavelengths. Measurements of the irradiance in this line have been reviewed by Vidal-Madjar (1977) and discussed by Simon (1978; 1981). The total flux of this line as well as its shape vary with solar activity. The irradiance varies between a minimum of 2.5 to 3.0 x 10^{11} photons $cm^{-2}s^{-1}$ for quiet solar activity and a maximum of about 4.0 to 6.0 x 10^{11} during high solar activity. Correlation studies between Lyman α and the decametric solar flux at 10.7 cm have been performed by Bossy and Nicolet (1981) and Bossy (1983). Using data obtained by the Atmospheric Explorer E satellite (Hinterregger, 1981), the OSO 5 satellite (Vidal-Madjar, 1975), and the SME satellite (Rottmann et al., 1982) Bossy (1983) gives the following expression for the variation of Lyman α over the 11 year solar cycle:

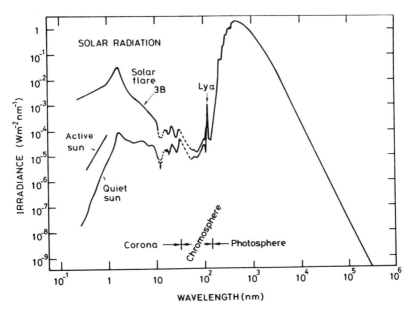

Fig. 4.5. Spectral distribution of the solar irradiance, and its variation with solar activity. The logarithmic representation emphasizes the contribution of x-rays and extreme ultraviolet radiation. After Smith and Gottlieb (1974).

$$q_{L\alpha} = 2.91 \times 10^{11} \left[1 + 0.20 \left(\frac{F_{10.7} - 65}{100} \right) \right] \text{ photons cm}^{-2}\text{s}^{-1} \qquad (4.8)$$

where $F_{10.7}$ is the 10.7 cm flux in 10^{-22} W m^{-2} Hz^{-1}. Since $F_{10.7}$ varies from about 65 to 360 units between solar minimum and solar maximum, the suggested variation in the Lyman α line is less than a factor of 2, probably about 60%. It must be noted that these values are lower than those suggested by Hinteregger (1981). Rottman et al. (1982; 1983) detect variations of about 10 to 40% over the 27 day rotation period of the sun.

The analysis of the spectral region at wavelengths greater than 120 nm is complex, particularly near the transition region at 140 nm. Observations of the solar irradiance in this part of the spectrum have been reported for solar cycle 20 by Heroux and Swirbalus (1976) and by Samain and Simon (1976), and for solar cycle 21 by Mount et al. (1980), Mount and Rottman (1981; 1983), and Rottman (1981). Differences of a factor of two in the data at 140 nm can probably be attributed to long term variability as indicated by the AE-E satellite observations (Hinteregger, 1981).

In the spectral range from 175 to 200 nm, observations have been reported by Brueckner et al. (1981; 1983), Samain and Simon (1976), Heath

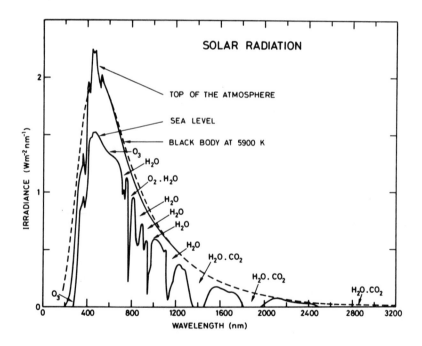

Fig. 4.6. Spectrum of solar radiation (UV, visible, IR) outside the earth's atmosphere and at sea level. (Adapted from Coulson, 1975).

(1980), and Mount and Rottman (1981; 1983). Differences between these experiments beyond 185 nm are probably due to uncertainties in the measurements since the variations between them are larger than the expected solar variability.

The solar irradiance beyond 200 nm has been measured by Broadfoot (1972), Heath (1980), Mentall et al. (1981), Mount and Rottman (1981; 1983), and Simon et al. (1982a,b). The differences between experiments are less than 20% between 205 and 240 nm, 15% from 240 to 270 nm and 10% between 270 and 295 nm, but these variations are again significantly greater than those expected to result from solar variability. These variations must therefore be considered as measurement uncertainties, and can have important effects on the calculated photodissociation rates and densities of atmospheric species. Beyond 330 nm, the values reported by Neckel and Labs (1981) are in good agreement with the observations by Heath (1980). Figure 4.6 shows a representation of the entire solar spectrum while Figure 4.7 presents a reference spectrum between 120 and 310 nm as suggested by Brasseur and Simon (1981).

The variability in the solar irradiance in the range from 120 to 300 nm is not yet well known. A few observations of the variability over the 27 day solar rotation period show that the amplitude of the variation decreases rapidly with wavelength (about 30% at 140 nm; less than 10% at 175 nm). The variability with the 11 year solar cycle is difficult to estimate because of the differences associated with different instruments used over such a long period of time and because of possible instrument drift in satellite studies.

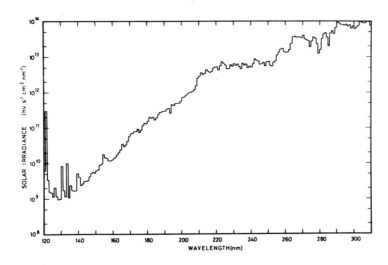

Fig. 4.7. Spectral irradiance in the ultraviolet between 120 and 310 nm. After Brasseur and Simon (1981). (Copyright by the American Geophysical Union).

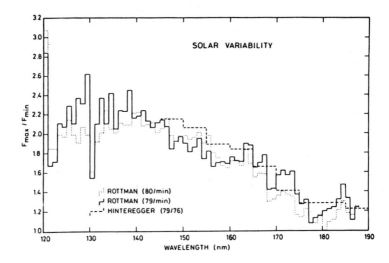

Fig. 4.8. Variation of the solar ultraviolet flux over the 11 year solar cycle from 120 to 190 nm.

Analyses of observed data suggest, however, that the variability over the 11 year solar cycle is of the order of 100% between 140 and 155 nm, 70% at 160 nm, 50% at 170 nm, and 20% at 180 nm. These values are in general agreement with the empirical model of solar variability developed by Lean et al. (1982). This model expresses the total irradiance as the sum of three components: a quiet emission constant over the entire solar disc, a moderately bright emission from active network areas and a strongly enhanced emission from plages. The total flux varies according to the observed size, location and intensity of the plages as well as ground based observations of the Ca II K emission line at 393 nm. This line is assumed to be a good indicator of the solar irradiance in the continuum from 145 to 200 nm.

The solar variability above the Al I edge at 208 nm is extremely difficult to measure since its magnitude is smaller than instrumental error. It is at most a few percent between 210 and 300 nm and becomes less than one percent beyond 300 nm. Figure 4.8 shows the variability between 120 and 190 nm as suggested by the observations of Rottman (1981) and Hinteregger (1981).

4.3 The attenuation of solar radiation in the atmosphere

4.3.1. Absorption

As solar photons penetrate into the earth's atmosphere, they undergo collisions with atmospheric molecules and are progressively absorbed and scattered. The probability of absorption by a molecule depends on the nature of the molecule and the wavelength of the incoming photon. An effective absorption cross section, $\sigma_a(\lambda)$ can be defined for each photochemical species. This quantity is expressed in cm^2 and is independent of the concentration of the species under consideration.

The Beer-Lambert law describes the absorption of a ray of incident intensity $I_{0,\lambda}$ and wavelength λ passing through an infinitesimally thin layer ds (see Fig. 4.9). The variation of intensity is given by

$$dI(\lambda) = - k_a(\lambda)I(\lambda)ds \qquad (4.9)$$

where k_a is the absorption coefficient (expressed for example in cm^{-1}). This coefficient is proportional to the concentration n (cm^{-3}) of the absorbing particles and is related to the effective cross section $\sigma_a(cm^2)$ by the expression

$$k_a(\lambda) = \sigma_a(\lambda)n \qquad (4.10)$$

At standard temperature (273.15K) and pressure (1013.25 mb), the ambient concentration is given by Loschmidt's number $n_0 = 2.687 \times 10^{19} cm^{-3}$. Integrating (4.9) and substituting (4.10) we find

$$I(\lambda) = I_0(\lambda)\exp\left(-\int_s \sigma_a(\lambda)n(s)ds \right) \qquad (4.11)$$

The optical thickness over the length s is defined as

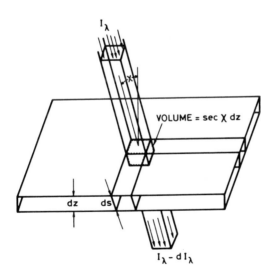

Fig. 4.9. Absorption of solar radiation in an atmospheric layer of unit area.

$$\tau_a(\lambda,s) = \int_s \sigma_a(\lambda)n(s)ds = \int_s k_a(\lambda,s)ds \qquad (4.12a)$$

and the corresponding transmission is

$$T(\lambda,s) = \exp\left(-\tau_a(\lambda,s)\right) \qquad (4.12b)$$

Solar radiation penetrates the atmosphere at an angle of incidence which depends on the local time, season and latitude. The cosine of the local solar zenith angle (χ) is given by

$$\cos(\chi) = \cos(\phi)\cos(\delta)\cos(H) + \sin(\phi)\sin(\delta) \qquad (4.13)$$

where ϕ is the latitude, δ is the solar declination angle, which depends on season ($\pm\,23.5\,°$ for solstice conditions, $0\,°$ for equinox, for example), and H is the hour angle, which is $0\,°$ for local noon, and increments by $15\,°$ for each hour from noon. Neglecting the curvature of the earth,

$$ds = dz \sec(\chi) \qquad (4.14)$$

where dz is a unit of altitude variation. Further, the concentration $n(z)$ is often assumed to vary exponentially with altitude according to a scale height H (see Chapter 3):

$$n(z) = n_0\exp(-z/H) \qquad (4.15)$$

such that, for a medium containing only one absorbing gas, the variation of monochromatic solar radiation with altitude can be written (omitting the index λ):

$$I(z) = I(\infty)\exp\left(-\sec\chi \int_z^\infty \sigma_a n_0 e^{-z/H}\, dz'\right) \qquad (4.16)$$

$$= I(\infty)\exp\left(-\sec\chi\, \sigma_a n_0\, H\, e^{-z/H}\right) \qquad (4.17)$$

where $I(\infty)$ represents the solar intensity outside the earth's atmosphere. The rate of formation of ions, the rates of photodissociation and production of heat, are all directly linked to the rate of energy deposition in the atmosphere by absorption. This last quantity can be written as

$$r = -\frac{dI}{ds} = \frac{dI}{dz}\cos\chi \qquad (4.18)$$

$$= \sigma_a n_0 I(\infty)\exp\left(-(\frac{z}{H} + \tau_0\exp -z/H)\right) \qquad (4.19)$$

where

$$\tau_0 = \sigma_a n_0\, H \sec\chi \qquad (4.20)$$

represents the optical depth of the entire atmosphere for zenith angle χ. The rate of energy deposited in the atmosphere exhibits a maximum at the altitude

$$z_m = H \ln \tau_0 = H \ln(\sigma_a n_0 H \sec \chi) \qquad (4.21)$$

For an overhead sun, this maximum is located at the altitude:

$$z_0 = H \ln(\sigma_a n_0 H) \qquad (4.22)$$

The rate of energy deposited at this altitude,

$$z_m = z_0 + H \ln \sec \chi \qquad (4.23)$$

is thus equal to

$$r_m = \sigma_a n_0 I(\infty) \cos \chi \, \exp\left(-1 - \frac{z_0}{H}\right) \qquad (4.24)$$

For an overhead sun, one obtains

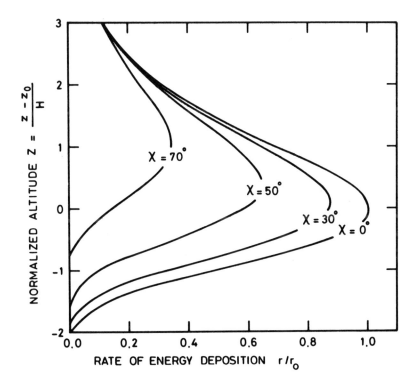

Fig. 4.10. Appearance of a characteristic layer due to absorption of solar energy. Distribution obtained by Chapman theory for different values of the zenith angle.

$$r_0 = r_m(\chi = 0) = \sigma_a \, n_0 \, I(\infty) \exp\left(-1 - \frac{z_0}{H}\right) \qquad (4.25)$$

Making use of the dimensionless variable

$$Z = \frac{z - z_0}{H} \qquad (4.26)$$

the rate of energy deposition becomes

$$\frac{r}{r_0} = \exp\left(1 - Z - \sec \chi \exp(-Z)\right) \qquad (4.27)$$

which is represented in Figure 4.10. It should be noted that the penetration of solar radiation and its interaction with the atmosphere leads to the formation of layers whose characteristic altitude z_m is independent of the intensity of the radiation, but strongly dependent on the physical characteristics of the atmospheric medium (nature and concentration of the absorbing gas), the solar zenith angle, and the wavelength of the radiation (because of the spectral distribution of the absorption coefficient). Thus for polychromatic radiation for which regions of strong absorption by one or more absorbing gases occur at different wavelengths, several absorbing layers are found at different altitudes.

The theoretical discussion just presented was first suggested by Chapman (1931). This theory provides an explanation for the behavior of the layers of ionization in the thermosphere or of photodissociation in the middle atmosphere. The production of ozone through photodissociation of molecular oxygen exhibits a maximum near 45 km dependent on the insolation. The rate of heating through absorption of ultraviolet radiation by ozone similarly leads to a maximum near the stratopause. Numerous examples of such layers can be found in the neutral and ionized atmosphere.

In practice, however, the analytic calculations by Chapman should be replaced by an expression which accounts for the combined effects of several absorbing gases. The Beer-Lambert law can also be written:

$$q_\lambda(z,\chi) = q_\lambda(\infty) \exp\left(-\tau_a(\lambda,z,\chi)\right) = q_\lambda(\infty) T(\lambda,z,\chi) \qquad (4.28)$$

where $T = \exp -\tau_a$ is the transmission function, but in this case the value of the optical depth results from the sum of several terms which vary with the wavelength of incident radiation and the column amount of absorbing gases. For example, in the middle atmosphere, ultraviolet radiation is absorbed by molecular oxygen, or ozone, or both (see Table 4.1), such that one generally writes:

$$\tau_a(\lambda,z,\chi) = \sec\chi \left[\int_z^\infty \sigma_a(O_2,\lambda)(O_2)dz + \int_z^\infty \sigma_a(O_3,\lambda)(O_3)dz\right] \qquad (4.29)$$

When the solar zenith angle exceeds 75 degrees (sunrise or sunset), the effect of the earth's curvature can no longer be neglected and the secant must be replaced by a more complicated function (Chapman function), which depends not only on χ but also on the altitude where the absorption occurs (see

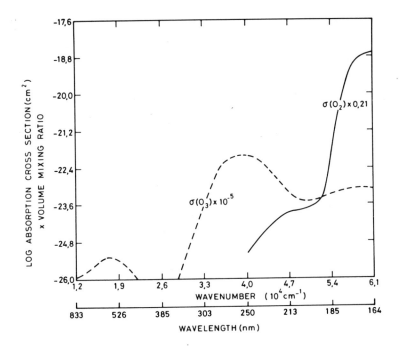

Fig. 4.11. Relative contributions of ozone and molecular oxygen to absorption by the atmosphere for wavelengths between 164 and 833 nm. (After Brewer and Wilson, 1965, copyrighted by the Royal Meteorological Society).

insert). Figure 4.11 provides an estimate of the relative contributions of absorption by O_2 and O_3 for different wavelengths. The cross sections of each of these gases must be multiplied by their respective column abundances to compare their absorbing capacities. The effect of molecular oxygen dominates for wavelengths less than 190 nm but ozone absorption dominates near 250 nm and slightly in the visible.

In the visible region, the absorption by nitrogen dioxide can also contribute to the optical depth, especially in the lower stratosphere and troposphere. Moreover, molecular oxygen has two weak bands in the red region of the solar spectrum near 0.7 μm.

In the near infrared, the absorption of solar radiation is due chiefly to vibrational and rotational transitions of several atmospheric molecules. The most important absorber is water vapor with several absorbing bands between 0.8 and 3.2 nm (see Fig. 4.6). Carbon dioxide also exhibits absorption bands such as the 2.7 μm band which overlaps with the 2.7 μm band of water vapor and the 4.3 μm band, located in a spectral region where both solar and terrestrial radiation is weak. Since these bands consist of many narrow lines, the global transmission function corresponding to a given spectral interval cannot be described in terms of the Beer-Lambert law and empirical formulas must be used in practical applications.

The attenuation of solar radiation in the atmosphere depends on the solar zenith angle because of the appearance of sec χ in expression (4.14, 4.29, etc.). When the sun is near the horizon, more precisely, for solar zenith angles $\chi > 75$, the solar zenith angle must be replaced by the Chapman function, $\text{Ch}(\chi)$, in order to account for the earth's curvature. This function represents the ratio of the total amount of absorbant along the oblique angle (χ with respect to the vertical) versus the amount of total absorbant in the vertical. For an isothermal atmosphere, and assuming that the absorbing gas has a constant scale height H, the value of the Chapman function can be estimated with the following expressions (Swider and Gardner, 1967):

$$\text{Ch}(x, \chi \leqslant \frac{\pi}{2}) = (\frac{\pi x}{2})^{1/2} \left[1 - \text{erf}(x^{1/2}\cos\frac{\chi}{2}) \right] \exp\left[x \cos^2(\frac{\chi}{2}) \right]$$

$$\text{Ch}(x, \chi \geqslant \frac{\pi}{2}) = (\frac{\pi x}{2}\sin\chi)^{1/2} \left[1 + \text{erf}\left(-\cot g\chi (\frac{x \sin\chi}{2})^{1/2} \right) \right] (1 + \frac{3}{8x \sin\chi})$$

where $x = (a + z)/H$, a being the earth's radius, z the altitude and H is the atmospheric scale height. For $\chi = \frac{\pi}{2}$, (sunrise or sunset), $\text{Ch}(x, \frac{\pi}{2}) = (\pi x/2)^{1/2}$. This function is equal to about 40 in the middle atmosphere.

4.3.2. Scattering by molecules and aerosol particles

In the denser layers of the atmosphere, solar radiation undergoes multiple scattering due to the presence of large concentrations of air molecules and aerosol particles. Scattering is a physical process by which a particle in the path of an electromagnetic wave abstracts energy from this incident wave and reradiates that energy in all directions. The optical depth describing the attenuation of radiation is thus the result of two contributions (absorption and scattering):

$$\tau(\lambda, z, \chi) = \tau_a(\lambda, z, \chi) + \tau_s(\lambda, z, \chi) \tag{4.30}$$

where

$$\tau_s(\lambda,z,\chi) = \sec\chi \int_z^\infty k_s(\lambda,z')dz' = \sec\chi\,\sigma_s(\lambda)n(M;z)H(z) \qquad (4.31)$$

k_s being the scattering coefficient, $\sigma_s(\lambda)$ being the corresponding effective cross section, $n(M;z)$ is the total concentration and H is the scale height. The albedo for single scattering can be defined as

$$\Omega = \frac{k_s}{k_a + k_s} = \frac{k_s}{k} \qquad (4.32)$$

which describes the relative importance of scattering versus the total attenuation. The parameter k is also called the extinction coefficient.

The atmospheric particles responsible for scattering cover a range of sizes from gas molecules $(\approx 10^{-8}$ cm$)$ to large raindrops and hail particles $(\approx$ cm$)$. The relative intensity of the scattered light depends strongly on the ratio of the particle size to the wavelength of the incident radiation. When this ratio is small, the scattered light is distributed equally into the forward and backward directions (Rayleigh scattering). When the particles are large, an increasing portion of the light is concentrated in the forward direction. In this case (Mie scattering), the distribution of scattered light intensity with scattering angle becomes very complex.

The scattering of unpolarized light by air molecules is generally described by the Rayleigh theory. In this case, the scattering coefficient for anisotropic molecules is given by (see e.g. Penndorf, 1957)

$$k_s = \frac{8\pi^3}{3}\frac{(\mu^2-1)^2}{\lambda^4 n(M)}D \qquad (4.33)$$

where μ represents the index of refraction of air and D is the depolarization factor which expresses the influence of molecular anisotropy. For air, a value of D of 1.06 is recommended (Rayleigh, 1919; Penndorf, 1957; Elterman, 1968). Adopting an analytic expression for the wavelength dependence of the refractive index, the Rayleigh scattering cross section can be written as (adapted from Nicolet et al., 1982):

$$\sigma_s = \frac{4.0\times10^{-28}}{\lambda^{3.916 + 0.074\lambda + .05/\lambda}} \quad cm^2 \qquad (4.34)$$

where λ is in μm.

The aerosols (solid or aqueous particles in the atmosphere) also interact with the radiation field. However, Rayleigh theory can no longer be applied to these particles, because the size of the aerosols are of the same order of magnitude or larger than the wavelength of the incident light.

The basic theory for the study of the scattering of light by aerosols was presented by Mie (1908). This theory generally assumes an ensemble of identical spherical particles and requires considerable computing resources for its solution. Modifications to allow for cylindrical and ellipsoidal particles have also been developed. In practice, simplified formulas are often used (see, for example, Van de Hulst, 1957). The total extinction coefficient at a particular wavelength can be determined by integrating the absorption and scattering by aerosols of distribution f(r). Thus, if N represents the total number of particles per unit volume,

$$k = N \int_0^\infty \pi r^2 Q_E(r,\mu) f(r) dr \qquad (4.35)$$

where $Q_E(r,\mu)$ is the efficiency factor for attenuation by a particle of radius r and index of refraction μ.

4.4 Radiative transfer

4.4.1 General equations

The study of the interaction of radiation with the atmosphere and the resulting aeronomic and thermal effects, requires that the radiation field be defined at each point in space and for each wavelength which plays a role in these processes. Thus, before examining in detail the specific effects of solar and terrestrial radiation on the middle atmosphere, we present the general laws and equations which govern radiative transfer in the atmosphere.

The radiation field is defined by the radiance L which represents the amount of energy dE traversing a unit of surface dS per unit time dt in a pencil of solid angle dω inclined at an angle θ relative to the normal to the surface (see Fig. 4.12a). We can then write

$$L = \frac{dE}{dt \, dS \, d\omega \, \cos\theta} \qquad (4.36)$$

This quantity is often expressed in $Wm^{-2}sr^{-1}$, and is defined at each point $\bar{r}$ in space and for each direction $\bar{\omega}$ of the pencil of light.

In order to evaluate the rate of radiative heating due to absorption of energy by a unit volume, the net flux density traversing the surface must be determined. The flux density across the surface at angle $\bar{\omega}$ due to all the pencils of direction $\bar{\omega}'$ is given by the integral:

$$F(\bar{r},\bar{\omega}) = \int_{4\pi} L(\bar{r},\bar{\omega}') \cos(\bar{\omega},\bar{\omega}') \, d\bar{\omega}' \qquad (4.37)$$

In atmospheric problems, we are most often concerned with the energy transfer between horizontal layers, and can therefore write:

$$F(z) = \int_0^{2\pi} d\phi \int_{-1}^{1} \mu L(\mu,\phi) d\mu \qquad (4.38)$$

where $\mu = \cos(\bar{\omega},\bar{\omega}')$, and ϕ are the parameters describing the zenith and azimuthal directions, respectively, of the propagation (see Fig. 4.12b).

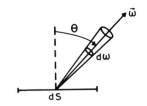

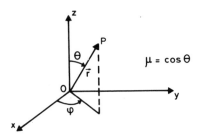

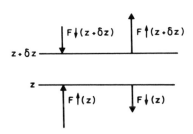

Fig. 4.12a (top). Geometry of a pencil of light in solid angle $d\omega$ traversing a unit of surface dS'. θ is the inclination of the beam relative to the surface. Fig. 4.12b (center). Definition of the azimuthal and zenith angles.

Fig. 4.12c (bottom). Balance of energy fluxes for a thin layer between altitude z and z + δz.

In the case of a plane parallel atmosphere at altitude z, the net flux can be separated into its upward propagating and downward propagating components:

$$F\uparrow(z) = \int_0^{2\pi} d\phi \int_0^1 \mu L(z;\mu,\phi)d\mu \qquad (\mu > 0) \qquad (4.39)$$

$$F\downarrow(z) = \int_0^{2\pi} d\phi \int_0^{-1} \mu L(z;\mu,\phi)d\mu \qquad (\mu < 0) \qquad (4.40)$$

such that the net flux is given by

$$F(z) = F\downarrow(z) - F\uparrow(z) \qquad (4.41)$$

To determine the heating rate at the layer between altitude z and $z+\delta z$, the energy balance at each boundary of the layer must be solved (see Fig. 4.12c). As the thickness δz approaches zero, the energy absorbed per unit volume is given by the net flux divergence dF/dz. Therefore, the variation of the temperature in the layer per unit time is given by

$$\frac{dT}{dt} = \frac{-1}{\rho c_p}\frac{dF}{dz} = \frac{g}{c_p}\frac{dF}{dp} \qquad (4.42)$$

where c_p is the specific heat at constant pressure, ρ is the total air density, g is the acceleration of gravity and p is the pressure.

In order to study the photochemical effects of radiation, the amount of energy penetrating a unit volume of space at point $\bar{r}$ must be evaluated; this is called the *irradiance* defined by

$$\Phi(\bar{r}) = \int_{4\pi} L(\bar{r};\bar{\omega})d\bar{\omega} \qquad (4.43)$$

Thus, each photon arriving at point $\bar{r}$ is given equal weight regardless of the direction of propagation. The dissociation or ionization of atmospheric molecules occurs regardless of the direction of the incident photon. In the case of a plane stratified atmosphere, one obtains:

$$\Phi(z) = \int_0^{2\pi} d\phi \int_{-1}^1 L d\mu \qquad (4.44)$$

In practice, it is necessary to consider the propagation of radiation at each frequency ν, wavenumber $\bar{\nu}$, or wavelength λ. For each quantity X, a spectral density $dX/d\nu$, $dX/d\bar{\nu}$, or $dX/d\lambda$ must be evaluated.

The analysis of the radiation field involves the evaluation of the spectral density of the radiance, $L_\nu = dL/d\nu$. This can, in principle, be determined from the equation of radiative transfer (Chandrasekhar, 1950; Kourganoff, 1952; Sobolev, 1963; Lenoble, 1977) which is an expression of the energy balance in each unit volume of the atmosphere, including absorption, scattering, and emission. In the general case of a horizontally stratified medium, the following expression can be used to describe the radiative transfer in a layer bounded by two infinite parallel planes (Lenoble, 1977):

$$\frac{\mu dL_\nu(z;\mu,\phi)}{dz} = -k_\nu(z)\left[L_\nu(z;\mu,\phi) - J_\nu(z;\mu,\phi)\right] \qquad (4.45)$$

where $k_\nu(z)$ is the extinction coefficient and $J_\nu(z;\mu,\phi)$ is called the source function at altitude z and for the direction μ,ϕ. J_ν expresses the incoming radiation due either to scattering from all other directions (μ',ϕ'), to solar radiation, or to thermal emission by atmospheric molecules. The importance of these contributions varies with the spectral domain under consideration. When the source function can be neglected, a simple exponential extinction expression is obtained. It should also be noted that in many studies, the altitude variable z is replaced by the equivalent variable τ_ν, which represents the optical depth for vertically incident radiation $(d\tau_\nu = -k_\nu dz)$. In this case, the radiative transfer equation can be written as:

$$\frac{\mu\,dL_\nu(\tau_\nu;\mu,\phi)}{d\tau_\nu} = L_\nu(\tau_\nu;\mu,\phi) - J_\nu(\tau_\nu;\mu,\phi) \qquad (4.46)$$

When the atmosphere is assumed to be stratified and plane-parallel, it is common to introduce the radiance $L_\nu\uparrow$ and $L_\nu\downarrow$ describing upward and downward propagating photons, as was presented previously for the flux (see 4.39 and 4.40). In this case, integrating (4.46), we obtain:

$$L_\nu\uparrow(\tau_\nu;\mu,\phi) = L_\nu\uparrow(\tau_{\nu,1};\mu,\phi)e^{(\tau_{\nu,1}-\tau_\nu)/\mu} - \frac{1}{\mu}\int_{\tau_\nu}^{\tau_{\nu,1}} J_\nu(t;\mu,\phi)e^{(t-\tau_\nu)/\mu}dt \quad (4.47)$$

for $\mu > 0$

$$L_\nu\downarrow(\tau_\nu;\mu,\phi) = L_\nu\downarrow(0;\mu,\phi)e^{-\tau_\nu/\mu} + \frac{1}{\mu}\int_0^{\tau_\nu} J_\nu(t;\mu,\phi)e^{(t-\tau_\nu)/\mu}dt \qquad (4.48)$$

for $\mu < 0$

$$L_\nu(\tau_\nu;\mu,\phi) = J_\nu(\tau_\nu;\mu,\phi) \qquad (4.49)$$

for $\mu = 0$

where the source function J_ν depends on the radiance L_ν, at least when scattering cannot be neglected (see §4.4.2). To solve these equations, certain boundary conditions must be specified. For example, at the outside of the atmosphere one can assume that the only downward propagating radiation is that emitted by the sun. At lower altitudes (surface or clouds) $(\tau = \tau_{\nu,1})$ one can assume that short wavelength incident radiation is isotropically reflected by the surface of albedo A $(0 < A < 1)$. A more accurate treatment allows for an angular dependence of A. For long wave radiation, it is usually assumed that the Earth's surface emits as a black body at temperature T_s.

4.4.2 Solution of the equation of radiative transfer for wavelengths less than 4μm: Multiple scattering

As indicated previously, short wave radiation below 4 μm originates from the sun. When the direct flux penetrates into the atmosphere, it is progressively attenuated by absorption (τ_a) and scattering (τ_s). In this case, the source function must account for the diffuse radiation scattered into the beam from other directions (μ',ϕ') as well as the direct sunlight scattered into the direction of the pencil of radiation. If $\Phi_\nu(\infty)$ is the solar irradiance at the top of the atmosphere and (μ_o,ϕ_o) the direction of the sun, the source function is given by (Lenoble, 1977):

$$J_\nu(\tau_\nu;\mu,\phi) = \frac{\Omega_\nu(\tau_\nu)}{4\pi} \int_0^{2\pi} d\phi' \int_{-1}^{1} p_\nu(\tau_\nu;\mu,\phi;\mu',\phi') L_\nu(\tau_\nu;\mu',\phi') d\mu'$$

$$+ \frac{\Omega_\nu(\tau_\nu)}{4\pi} p_\nu(\tau_\nu;\mu,\phi;\mu_o,\phi_o) \Phi_\nu(\infty) e^{-\tau_\nu/\mu_o} \qquad (4.50)$$

where $\Omega(\tau_\nu)$ is the albedo for single scattering (see expression 4.32) and $p_\nu(\tau_\nu;\mu,\phi;\mu',\phi')$ is the phase function defining the probability that a photon propagating in the direction (μ',ϕ') is scattered in the direction (μ,ϕ).

If the atmosphere is perfectly clear (neglecting the presence of solid and liquid particulates), the scattering is due only to air molecules and Rayleigh theory can be applied. The corresponding attenuation coefficient is given by the expression (4.33), and the phase function is given by:

$$p(\Theta) = \frac{3}{4}(1 + \cos^2\Theta) \qquad (4.51)$$

where

$$\cos\Theta = \mu\mu' + (1 - \mu^2)^{1/2}(1 - \mu'^2)^{1/2}\cos(\phi - \phi') \qquad (4.52)$$

Θ being the angle between incident and scattered light.

The scattering by solid and liquid particulates constitutes a more difficult problem (see, e. g., McCartney, 1976). If the particles may be assumed spherical, Mie theory can be invoked. Thus for aerosols of radius r, the phase function is given by

$$p(r,\Theta) = \frac{\lambda^2}{2\pi k_s(r)}(S_1 S_1^* + S_2 S_2^*) \qquad (4.52)$$

where λ is the wavelength of incident light and the other coefficients are given by the following expressions:

$$S_1(\Theta) = \sum_{n=1}^{\infty} \frac{2n+1}{n(n+1)}\left(a_n \frac{dP_n^1}{d\Theta} + b_n \frac{P_n^1}{\sin\Theta}\right) \qquad (4.53)$$

$$S_2(\Theta) = \sum_{n=1}^{\infty} \frac{2n+1}{n(n+1)}\left(a_n \frac{P_n^1}{\sin\Theta} + b_n \frac{dP_n^1}{d\Theta}\right) \qquad (4.54)$$

where P_n^1 is the associated Legendre polynomial and a_n and b_n are coefficients depending on the index of refraction and on the Mie parameter $\alpha_a = 2\pi r/\lambda$. S^* is the complex conjugate of S and k_s is the scattering coefficient

$$k_s(r) = \frac{2\pi r^2}{\alpha_a^2} \sum_{n=1}^{\infty} (2n+1)(a_n^2 + b_n^2) \tag{4.55}$$

The number of significant terms in this series is of the order of $2\alpha_a + 3$.

In the presence of an ensemble of aerosols characterized by a distribution function $f(r_a)$, the phase function $p(\Theta)$ is deduced from a function related to a specific radius r_a by the relation

$$p(\Theta) = \frac{\int_0^{\infty} p(r_a,\Theta)\sigma(r_a)f(r_a)dr_a}{\int_0^{\infty} \sigma(r_a)f(r_a)dr_a} \tag{4.56}$$

Mie theory can be extended to non-spherical particles (such as ice crystals, for example) but it is generally sufficient to assume that randomly oriented irregular particles scatter light in the same way that spherical particles do.

Several analytic methods have been proposed to solve the equation of radiative transfer in an absorbing and scattering atmosphere, but they can only be applied for the most simple cases. To obtain quantitative solutions, numerical methods are generally used, such as the Monte-Carlo method, DART method, iterative Gauss, discrete ordinate method, etc. A complete summary of these techniques is provided by Lenoble (1977), and a detailed

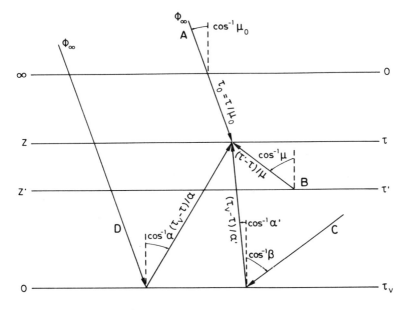

Fig. 4.13. Schematic diagram of the scattering model by Meier et al. (1982). (Copyright by Pergamon Press).

discussion of multiple scattering processes in plane parallel atmospheres is given in the book by Liou (1980).

Since multiple scattering, surface reflection, clouds and aerosols have a significant effect on radiative intensities at photodissociative wavelengths and consequently on the composition of the middle atmosphere (see, e. g., Luther et al., 1978), simplified radiative transfer schemes have been incorporated in photochemical models. Fiocco et al. (1978), for example, have utilized a code based on the work by Grant and Hunt (1969) and Wiscombe (1976), which supplies the direct and diffuse component of the solar irradiance as a function of altitude and wavelength. In this method, the monochromatic radiative transfer equation is solved for the azimuthally averaged radiance, considering the absorption by O_3 and H_2O, the scattering by molecules and the absorption and scattering by aerosols for a given size distribution and complex refractive index. The calculation applied to ozone shows that an increase in the aerosol load always leads to enhanced photodissociation in the upper regions, while the effect of extinction leads to reduced rates in the lower region. The deviation of the O_3 photodissociation coefficient with a heavy aerosol load compared to that of a purely molecular atmosphere is of the order of 15% in the lower stratosphere.

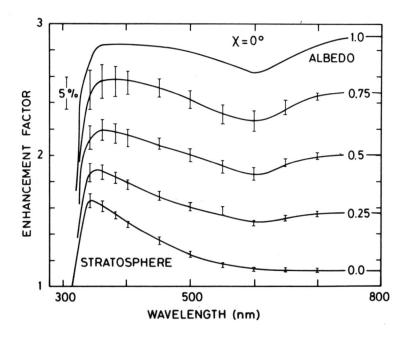

Fig. 4.14. Stratospheric enhancement factor calculated by Meier et al. (1982) for different values of the albedo. (Copyright by Pergamon Press).

More simplified models that can easily be incorporated in photochemical calculations have been developed, for example, by Luther and Gelinas (1976), and Isaksen et al. (1977). These methods are based on the assumption that all scattering takes place in the direction of the solar radiation, with one half scattered toward the sun and the other half away from the sun. The radiance is calculated successively for different orders of scattering up to the fifth or sixth order. The first order radiances are expressed in terms of the direct solar irradiance. The second order radiance is then obtained from expressions containing the first order flux. Similarly, all scattered fluxes of higher orders are given by fluxes of lower orders. Finally the total flux reaching an altitude level i is given by addition of the fluxes of all orders. The calculation of the photodissociation coefficients including molecular scattering and ground albedo effects, which appear later in this chapter, are made with a code

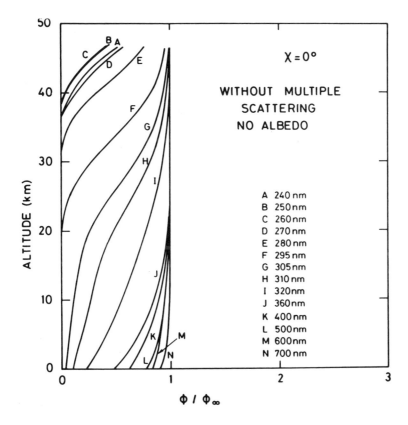

Fig. 4.15 a,b,c. Stratospheric enhancement factor as a function of altitude calculated by Meier et al. (1982). a: absorption only (Copyright by Pergamon Press).

provided by Luther (private communication). All of these models are based on the determination of the radiance L as a function of the azimuthal and zenith angles, although the angular dependence of the scattered light is usually oversimplified. This problem can be overcome by solving a form of the radiative transfer equation which is integrated over all solid angles. Such an approach (which has been applied by Meier et al., 1982 and Nicolet et al., 1982) provides the solar irradiance Φ at wavelength λ or the enhancement factor or source function, which is defined by $S = \Phi/\Phi_\infty$, where Φ_∞ is the solar irradiance at the top of the Earth's atmosphere. A schematic diagram of their model is shown in Figure 4.13. It can be shown that the calculated source function S at altitude z results from four contributions: 1) the direct solar flux, 2) the multiply-scattered flux (both by molecules and by particles), 3) the ground reflection of the multiply scattered flux, and 4) the ground reflection of the direct flux. The mean enhancement factor in the stratosphere

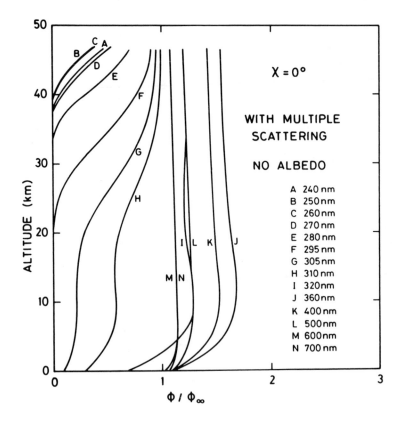

Fig. 4.15 b: Same as Fig. 4.15a but for absorption and multiple scattering. (Copyright by Pergamon Press).

between 300 and 800 nm for different values of the albedo and for an overhead sun is shown in Figure 4.14. The arrows indicate the variability of the S factor between the tropopause and the stratopause, suggesting that a single enhancement factor can be applied to the entire atmosphere with an inaccuracy of less than 5 percent at all wavelengths. The S factor as a function of altitude for wavelength intervals from 240 to 700 nm is shown in Figure 4.15a, b, c for different cases. Figure 4.15a shows the results ignoring the effect of multiple scattering and reflection at ground level. In this case, the enhancement factor is simply the transmission factor which characterizes the attenuation by absorption as well as by Rayleigh and Mie scattering. In the visible part of the spectrum, the attenuation occurs near ground level, while in the UV region, the irradiance is reduced by the presence of ozone at higher altitudes. The effect of multiple scattering, neglecting ground albedo, appears in Figure 4.15b. The contribution of the scattered light to the radiation field appears to

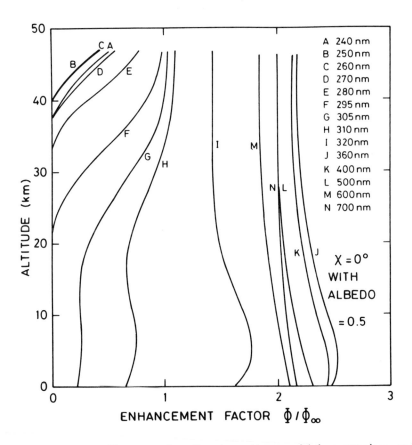

Fig. 4.15 c: Same as Fig. 4.15a but for absorption, multiple scattering, and surface albedo (A = 0.5). (Copyright by Pergamon Press).

be large except at short wavelengths where pure absorption is strong. At 350 to 400 nm, for example, the scattering provides an additional irradiance of about 50 percent. Finally Figure 4.15c shows the S factor when both multiple scattering and ground albedo ($A = 0.5$) are taken into account. The enhancement factor is now larger than two even at higher altitudes for radiation of wavelength larger than 320 nm. The multiple scattering and albedo will therefore considerably affect the photodissociation of molecules whose absorption cross sections are large above 300 nm (O_3, NO_2, N_2O_5, NO_3, $ClONO_2$). Below 300 nm, however, absorption accounts for most of the radiative transfer process although the scattered flux in the Schumann-Runge bands (see Fig. 4.16) could be as large as 10% of the direct solar flux at 40 km (Herman and Mentall, 1982).

The presence of clouds in the troposphere will modify somewhat the radiation field in the stratosphere (Lacis and Hansen, 1974) by altering the albedo and introducing a highly scattering medium. Since the reflectance properties of the clouds vary considerably with each cloud type, numerical models dealing with this problem will have to define statistical properties of the cloud distribution.

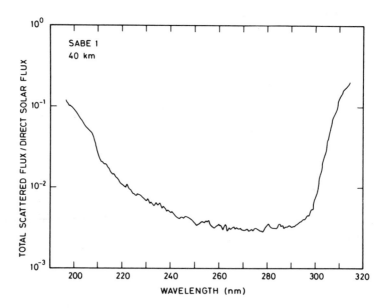

Fig. 4.16. Observed ratio of scattered to direct solar flux as a function of wavelength at 40 km. From Herman and Mentall (1982). (Copyright by the American Geophysical Union).

*4.4.3. Solution of the radiative transfer equation at wavelengths
longer than 4μm: Absorption and emission of infrared radiation*

In the infrared region above 4 μm, the contribution from terrestrial radiation clearly dominates the solar input. Significant thermal emission and absorption by atmospheric gases and aerosols take place in this spectral region. Scattering can be neglected. The detailed study of radiative transfer requires complete knowledge of the important bands and a compilation of a large body of spectroscopic data (see, e. g., McClatchey et al., 1973; Rothman et al., 1980; Chedin et al., 1980).

The radiatively active trace gases which are the most important from the thermal point of view are carbon dioxide, ozone and water vapor. But other gases, whose sources are partly related to anthropogenic activity, such as CH_4, N_2O, and chlorofluorocarbons can in part contribute to the radiation budget.

Carbon dioxide is a linear molecule which has a relatively simple absorption spectrum. One of the strongest bands is the ν_2 fundamental at 15 μm, which contributes significantly to the energy budget of the atmosphere because it is located in the spectral region where the emission of the terrestrial environment is very intense. The ν_3 band at 4.3 μm is also strong but has a marginal influence since it is located at a wavelength where both solar and

Table 4.2 Important infrared atmospheric bands (From WMO, 1982)

Trace gas	Band	Band center cm^{-1}	Band intensity cm^{-2} atm^{-1} (296K)
CO_2	ν_2 (15 μm)	667	$\approx$ 220
CO_2	ν_3 (4.3 μm)	2348	$\approx$ 2440
O_3	ν_3 (9.6 μm)	1041	312
H_2O	rotation	0-1650	1306
H_2O	ν_2(6.3μm)	640-2800	257
CH_4	ν_4 (7.66 μm)	1306	134
N_2O	ν_1 (7.78 μm)	1285	218
N_2O	ν_2 (17 μm)	589	24
$CFCl_3$	ν_4 (11.8 μm)	846	1813
$CFCl_3$	$\nu_1 + 2\nu_2$	1085,2144	679
CF_2Cl_2	ν_8 (10.9 μm)	915	1161
CF_2Cl_2	ν_1 (9.13μm)	1095	1141
CF_2Cl_2	ν_6 (8.68μm)	1152	777

terrestrial emissions are weak. Additional weak bands centered in the 12-18 μm, 10 μm, and 7.6 μm regions must be considered in the CO_2 climate problem, as discussed by Augustsson and Ramanathan (1977).

Ozone is a non-linear molecule, and possesses strong rotational structure as well as 3 fundamental vibration bands, ν_1, ν_2, and ν_3 at 9.066, 14.27, and 9.597 μm, respectively. The ν_2 band is masked by the CO_2 band but the ν_1 and ν_3 both absorb to form the important 9.6 μm band. Another strong ozone band at 4.7 μm is located in a spectral region with weak solar and terrestrial fluxes.

Water vapor is a non-linear molecule with a complex vibration-rotation spectrum. The ν_2 fundamental band centered at 6.25 μm is overlayed with a series of rotational transitions to make a broad intense band centered at 6.3 μm. A wide pure rotation band extends from about 18 to beyond 100 μm with varying intensity. The ν_1 and ν_3 fundamentals at 2.74 and 2.66 μm absorb a significant amount of solar energy, particularly in the troposphere. Table 4.2 lists the principal spectral bands which are important for climate

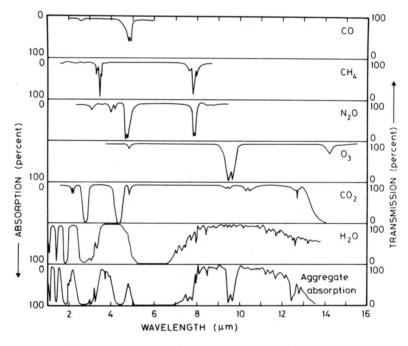

Fig. 4.17. Spectral distribution of the absorption by several radiatively active gases (from the Earth's surface to infinity). The aggregate spectrum due to all atmospheric gases is shown in the bottom panel (adapted from Shaw, 1953).

studies, including the bands from radiatively active constituents which play a secondary role in the radiation budget. Figure 4.17 shows the absorption of radiant energy by the gases which contribute significantly to radiative transfer in the atmosphere. This figure shows that each gas has a particular absorption signature which is wavelength dependent. The observation of absorbed or emitted terrestrial radiation in these wavelength bands can yield important information about the distributions of these gases in the atmosphere.

Throughout most of the middle atmosphere, the primary energy loss occurs via radiative emission in the ν_2 band of CO_2 at 15 μm. The contribution of this band to the total cooling has been studied in detail below 75 km. Dickinson (1973) and Ramanathan (1976), for example, indicated the importance of the CO_2 hot bands in the calculation of the cooling rate. In the region of the atmosphere where the conditions of local thermodynamic equilibrium (LTE) apply, (that is when collisions are sufficiently frequent that the energy levels are populated according to a Boltzmann distribution, see eqn. 2.29) then the source function is:

$$J_\nu(\tau_\nu) = B_\nu(T(\tau_\nu)) \tag{4.58}$$

where $B_\nu(T(\tau_\nu))$ is given by Planck's law:

$$B_\nu(T(\tau_\nu)) = \frac{2h\nu^3}{c^2} \frac{1}{e^{h\nu/kT} - 1} \tag{4.59}$$

Thus, the local emission corresponds to a black body radiance at the temperature of the point under consideration.

The radiance over most of the middle atmosphere (below 75 km) can thus be derived from the transfer equation (called Schwarzchild's equation, see Liou, 1980):

$$\mu \frac{\partial L_\nu(\tau_\nu;\mu,\phi)}{\partial \tau_\nu} = L_\nu(\tau_\nu;\mu,\phi) - B_\nu(\tau_\nu) \tag{4.60}$$

In other words, assuming azimuthal symmetry (no dependence of the radiance on ϕ) and making use of equations (4.47) and (4.48), the upward and downward components of the radiance will be evaluated from equations of the following type:

$$L_\nu\uparrow(z;\mu) = B_\nu(T_s)\,T_\nu(z,0;\mu) + \int_0^z B_\nu(z')\frac{\partial T_\nu(z',z;\mu)}{\partial z'}\,dz' \tag{4.61a}$$

and

$$L_\nu\downarrow(z;\mu) = -\int_z^\infty B_\nu(z')\frac{\partial T_\nu(z',z;\mu)}{\partial z'}\,dz' \tag{4.61b}$$

where T_s is the temperature of the Earth's surface and z is the altitude, while

$$T_{\nu}(z,z_0;\mu) = \exp\left(-\int_{z_0}^{z} k_{\nu}dz'/\mu \right) = \exp(-\sigma_{\nu}N) \qquad (4.62)$$

is the transmission function (see eqn. 4.12b) for an optical path from z_0 to z with inclination μ. The quantity N represents the integrated amount of absorbing gas with effective absorption cross section at frequency ν is σ_{ν}.

The solution of equations (4.61a,b) is complex since the absorption spectrum exhibits structure characterizing the numerous rotation and vibration-rotation lines of atmospheric molecules. The absorption coefficient or the transmission function T_{ν} varies therefore considerably over small regions of frequency (or wavenumber) and in principle a line-by-line integration is required.

The determination of the transmission function versus wavenumber depends strongly on knowledge of the detailed structure of the absorption spectrum, and in particular, the shape of the absorption lines (For reviews on the subject, see Mitchell and Zemansky, 1934; Penner, 1959; Goody, 1964; Kondratyev, 1969; Liou, 1980). In the absence of molecular collisions and molecular motions, the natural width of a spectral line is very narrow and is determined by the radiative lifetime of the excited state. This can be entirely neglected compared to the Doppler and Lorentz broadening due, respectively, to the thermal motion of the molecules and the collisions between them. In the troposphere and lower stratosphere the width of the lines of the absorbing molecules such as carbon dioxide and water vapor is determined essentially by collision. In this case, one can assume a Lorentz line shape:

$$\sigma_{\bar{\nu}} = \frac{S\gamma_L}{\pi\left((\bar{\nu}-\bar{\nu}_0)^2 + \gamma_L^2\right)} \qquad (4.63)$$

where $\sigma_{\bar{\nu}}$ represents the effective absorption cross section and $\bar{\nu}$ is the wavenumber, following the convention of spectroscopists. γ_L indicates the half width at half maximum of the line (expressed in wavenumber), $\bar{\nu}_0$ is the midpoint of the line and

$$S = \int_{0}^{\infty} \sigma_{\bar{\nu}}d\bar{\nu} \qquad (4.64)$$

is the integrated intensity of the line. Since the molecular collision frequency varies with pressure and temperature, one may write, (see, Tiwari, 1978):

$$\gamma_L = \gamma_{L,0}\frac{p}{p_0}\left(\frac{T_0}{T}\right)^{0.5} \qquad (4.65)$$

where $\gamma_{L,0}$ represents the width of the line at standard pressure p_0 and temperature T_0. For most gases, $\gamma_{L,0}$ is of the order of 0.1 cm^{-1}. At high altitude, collisions become much less frequent, and the effect of pressure can be neglected, which results essentially in a Doppler linewidth. In this case, the profile can be represented as a Gaussian:

$$\sigma_{\bar{\nu}} = \frac{S}{\gamma_D \sqrt{\pi}} \exp\left(-\left[\frac{\bar{\nu} - \bar{\nu}_0}{\gamma_D}\right]^2\right) \qquad (4.66)$$

where γ_D is the half width of the line measured at a height equal to $1/e$ of the maximum intensity. A typical value of γ_D (which is independent of pressure) is 10^{-3} cm^{-1}. In the transition region between the Lorentz and Doppler regimes, the lineshape is determined by both broadening processes. If these are assumed to be independent, they may be combined to give the Voigt profile (Mitchell and Zemansky, 1934; Penner, 1959):

$$\sigma_\nu = \frac{a}{\gamma_D} \frac{S}{\pi^{3/2}} \int_{-\infty}^{+\infty} \frac{\exp(-y^2)}{a^2 + (v-y)^2} dy$$

where $a = \gamma_L/\gamma_D$, $v = (\bar{\nu} - \bar{\nu}_0)/\gamma_d$, and $y = (\bar{\nu} - \bar{\nu}_0)/\gamma_D$. The Voigt profile is a good approximation to the observed line shape. At high pressure, when $a \to \infty$, and at low pressure, when $a \to 0$, it approaches the Lorentz and Doppler profiles, respectively.

To determine the upward flux $F\uparrow$ and the downward flux $F\downarrow$ in a one dimensional representation, an integration of (4.61a,b) must be performed over all solid angles according to (4.39) and (4.40). It is convenient to introduce the "diffuse transmission function"

$$T^*(z',z) = 2 \int_0^1 T(z',z;\mu)\mu \, d\mu \qquad (4.67)$$

which provides a weighted average of the contribution of light propagating along various inclination angles. It can be shown (see Rodgers and Walshaw, 1966) that the scattered transmission is equivalent to the propagation of a collimated beam at about a 53 degree angle ($\mu = 3/5$). In other words, the calculation of T^* can be performed with an air mass whose depth is $5/3$ its vertical thickness. Thus

$$T^*(z';z) \approx T(z',z;\mu = 3/5) \qquad (4.68)$$

One then finds

$$F_{\bar{\nu}}\uparrow(z) = \pi B_{\bar{\nu}}(T_s)T_{\bar{\nu}}^*(z_0,z) + \int_0^z \pi B_{\bar{\nu}}(z') \frac{\partial T_{\bar{\nu}}^*(z',z)}{\partial z'} dz' \qquad (4.69a)$$

$$F_{\bar{\nu}}\downarrow(z) = -\int_z^\infty \pi B_{\bar{\nu}}(z') \frac{\partial T_{\bar{\nu}}^*(z,z')}{\partial z'} dz' \qquad (4.69b)$$

Thus, in theory, by knowing the location, strength, and shape of all absorption lines, the flux density at any wavelength (or frequency) can be derived, and the total flux corresponding to an entire band could be obtained by a lengthly integration over the spectrum. In order to simplify the computation, and to derive the flux over a larger spectral range, the previous equations can be integrated over an interval $\delta\bar{\nu}$ which is sufficiently narrow that the Planck function can be assumed constant ($\delta\bar{\nu} < 50\text{cm}^{-1}$), but containing the largest possible number of emission lines ($\delta\bar{\nu} > 5\text{cm}^{-1}$). In this case, the flux for interval i is given by the following expression (after integration by parts of the expression above):

$$F_i\uparrow(z) = \pi B_i(z) - \int_0^z \pi \frac{dB_i(z')}{dz'} T_i^*(z',z)dz' \qquad (4.70a)$$

$$F_i\downarrow(z) = -\pi B_i(z=\infty)T_i^*(z,\infty) + \pi B_i(z) + \int_z^\infty \pi \frac{dB_i(z')}{dz'} T_i^*(z',z)dz' \quad (4.70b)$$

where $B_i(z)$ represents the Planck function in the frequency interval i for the temperature of the air and the altitude z, and where

$$T_{\delta\nu}^*(z',z) = \frac{1}{\delta\bar{\nu}} \int_{\delta\bar{\nu}} T^*(z',z)d\bar{\nu} \qquad (4.71)$$

is an effective transmission function corresponding to interval i. The determination of this effective transmission still requires, in principle, a line by line integration including all the complex structures of the various absorbing lines and their overlap. Such a calculation not only requires a great deal of computer time but is somewhat inaccurate since some of the spectral parameters associated with these lines are poorly known. This direct approach is therefore usually replaced by an approximate method in which the effective transmission function for a band or a portion of a band is expressed in terms of global parameters such as the mean line strength, the mean spacing between the individual lines, and the mean line width. Models have been developed to treat either sections of bands (narrow band models) or entire bands (wide band models). See Tiwari, (1978) for an excellent review of these models. The choice of any particular formulation of the transmission function or of the absorptance, which is defined by

$$A_{\delta\bar{\nu}}(w) = \int_{\delta\bar{\nu}} \left(1-T(w)\right)d\bar{\nu} \quad (\text{cm}^{-1}) \qquad (4.72)$$

(where $w = \int_z^{p_i} \rho dz$ is the optical path, and p_i is the partial pressure of the absorber i) is dictated by the degree to which the model approximates experimental data. The spectral parameters are usually adjusted to optimize the fit to observed absorption. The narrow band model of Elsasser (1942) represents the spectrum by a series of regularly spaced Lorentz lines of the same size and intensity. This model is best applied to linear triatomic molecules such as CO_2 and N_2O. The model by Goody (1964) is based on the idea that the lines are randomly spaced over a particular wavelength interval, with some exponential distribution of line strength. This model can readily be applied to water vapor and to carbon dioxide. If an exponential distribution of the line intensities is assumed (i.e., the probability of finding a line intensity S_j in a given intensity range decreases exponentially with the value of S_j), the effective transmission function in the spectral interval $\delta\bar{\nu}$ is given by

$$T_{\delta\bar{\nu}} = \exp\left(-\left[\frac{S}{d}\right]_{\delta\bar{\nu}} w\left[1 + \left(\frac{S}{\pi\gamma}\right)_{\delta\bar{\nu}} w\right]^{-1/2}\right) \qquad (4.73a)$$

where S $(cm^{-2}atm^{-1})$ is the mean strength of the lines for the interval $\delta\bar{\nu}$, γ the mean half width (cm^{-1}), and d the mean line spacing, w the amount of absorber $(atm\ cm^{-1})$. Malkmus (1967) introduced a distribution of the lines in S_j^{-1}. The statistical model then yields

$$T_{\delta\bar{\nu}} = \exp\left(\frac{-(S/d)_{\delta\bar{\nu}}}{2(S/\pi\gamma)_{\delta\bar{\nu}}}[(1 + 4w\,(S/\pi\gamma)_{\delta\bar{\nu}}^{1/2} - 1]\right) \qquad (4.73b)$$

Since lines in vibration-rotation bands are arranged neither at regular nor at random intervals, other models have been suggested, such as the random Elsasser model, which is constructed by the random superposition of several different Elsasser bands. Each of the superposed bands may have different average line intensities, line widths, and line spacing. Such a model provides a more accurate representation of the absorption for bands whose lines have relatively constant intensities over a narrow band region, but different intensities from one of these regions to another. These models were originally developed for Lorentz lines, but have also been extended for the Doppler and Voigt line shapes (see Tiwari, 1978).

The above formulations of the effective transmission function have been derived for homogeneous (constant pressure and temperature) optical paths. In the atmosphere, the pressure and the temperature vary with altitude, and the value of the spectral parameters (line width, line strength) should be adjusted accordingly. Curtis (1952) and Godson (1953) have presented an approximation in which the transmission of an inhomogeneous atmospheric path is replaced by the transmission of a homogeneous path with an effective line strength

$$\bar{S} = \int_0^w \frac{S(T)dw}{w} \qquad (4.74a)$$

and an effective half width of the line

$$\bar{\gamma} = \int_0^w S(T)\gamma(p,T)dw \Big/ \int_0^w S(T)dw \qquad (4.74b)$$

after which the band models for the homogeneous case can be applied. This approximation provides fairly good results for species whose optical mass decreases with pressure (CO_2, H_2O), but is rather poor for constituents such as ozone, whose concentration increases with altitude (Liou, 1980). In another widely used method, the atmosphere is divided into a number of layers which are thin enough to assume homogeneous and isothermal conditions.

Wide band models generally provide a value of the absorptance for a whole band as a function of the optical path. Thus, for example, Cess and

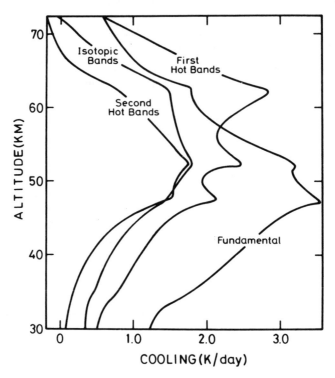

Fig. 4.18. Long wave CO_2 cooling showing the contribution from fundamental, hot, and isotopic bands. From Dickinson (1973). (Copyright by the American Geophysical Union).

Ramanathan (1972) and Ramanathan (1976) suggested that for Lorentzian lines the following could be used:

$$A = 2A_o \ln \left[1 + \frac{u}{[4 + u(1 + \frac{1}{\beta})]^{1/2}} \right] \qquad (4.75)$$

where $u = Sw/A_o$ is a dimensionless optical depth, and $\beta = 4\gamma/d$ is a mean line width parameter, A_o is the effective band width in cm^{-1}, and S is the band strength $(cm^{-2}atm^{-1})$ (see Table 4.3). It should be noted that, in order to account for the angular integration of the radiant energy in all directions, the u factor should be multiplied by 5/3, (1.66), if w is obtained from an integration along the vertical. This model has been adjusted to account for the presence of the hot bands and isotopic bands of CO_2 and to account for the overlap of certain spectral features.

In this case, equation (4.75) is modified (Edwards, 1965) and becomes, assuming that $\mu/\beta \gg 1$ (strong line limit):

Table 4.3. Parameters involved in the calculation of the absorptance.
(Models by Ramanathan, 1976; and Donner and Ramanathan, 1980).

---

1. Carbon dioxide - Fundamental band at 15 μm.

$A_o = 21.3(T/273)^{0.5}$ cm^{-1}
- Lorentz broadening
  $S(300K) = 194.0$ cm^{-2} atm^{-1}
  $\gamma_L(STP) = 0.064$ cm^{-1}
  $d = 1.56$ cm^{-1}
  $\beta = 0.164(273/T)^{0.5}(p/1013)$ (p in mb)
- Doppler broadening
  $\gamma_d = 3.60 \times 10^{-5}T^{0.5}$ cm^{-1}
  $d = 1.56$ cm^{-1}

---

2. Ozone - 9.6 μm band.

$A_o = 39(T/273)^{0.5}$ cm^{-1}
- Lorentz broadening
  $S(STP) = 387$ cm^{-2} atm^{-1}
  $\gamma_L(STP) = 0.079$ cm^{-1}
  $d = 0.1$ cm^{-1}
  $\beta = 0.316(273/T)^{0.5}(p/1013)$ (p in mb)
- Doppler broadening
  $\gamma_D = 5.38 \times 10^{-5}T^{0.5}$ cm^{-1}
  $d = 0.07$ cm^{-1}

---

3. Other minor constituents (Lorentz broadening, p in mb)

3.1 Methane ν_4 band at 1306 cm^{-1}
  $A_o = 52(T/300)^{0.5}$ cm^{-1}
  $\beta = 0.17(300/T)^{0.5}(p/1013)$
3.2 Nitrous oxide ν_1 band at 1285 cm^{-1}
  $A_o = 20.4(T/300)^{0.5}$ cm^{-1}
  $\beta = 1.12(300/T)^{0.5}(p/1013)$
3.3 Nitrous oxide ν_2 band at 589 cm^{-1}
  $A_o = 23(T/300)^{0.5}$ cm^{-1}
  $\beta = 1.08(300/T)^{0.5}(p/1013)$

$$A = 2A_o \ln[1 + \sum_{i+} \sum_j \varsigma_{ij}^{0.5}] \tag{4.76}$$

where i is the index for the CO_2 isotope and j the index for the band type (fundamental and hot bands). Moreover,

$$\varsigma_{ij} = \beta_{ij}\, u_{ij} = 1.66\frac{4\gamma_{L,0}}{A_o d_{ij}} q_i f_{CO_2} \int S_{ij} n_j \left(\frac{p}{p_o}\right)^2 dz \tag{4.77}$$

where q_i is the fractional abundance of the ith isotope (see Table 4.4), f_{CO_2} is the mixing ratio of CO_2 (320 ppmv), p is the pressure, and p_o its value at the Earth's surface, while S_{ij} is the band strength, d_{ij} the mean line spacing (which is equal to 4B for bands with alternate lines missing and equal to 2B for all other bands; $B = 0.39$ cm^{-1} is the rotational constant) and n_j is the population in the lower vibrational state relative to the ground state.

When all these parameters are substituted in equation (4.76), it can be shown that the absorptance of the CO_2 bands near 15 μm, between pressures p and p' is given by (Ramanathan et al., 1983):

$$A(p,p') = 2A_o \ln\left[1 + c_1(1 + 3.41e^{-480/T} + 7.05e^{-960/T})\left[\left(\frac{p}{p_o}\right)^2 - \left(\frac{p'}{p_o}\right)^2\right]^{1/2}\right] \tag{4.78}$$

where $A_o = 22(T/300)^{0.5}$ cm^{-1} and

$$c_1 \approx 26.54\left(\frac{300}{T}\right)^{1/2}\left(\frac{f_{CO_2}}{320}\right)^{1/2}$$

The largest contribution to the total cooling to space is due to the fundamental band (Ramanathan and Coakley, 1978), and to a lesser extent, to the first hot bands of $^{12}C^{16}O_2$. The second hot bands and isotopic bands are considerably weaker as illustrated by Figure 4.18 (Dickinson, 1973). Above a given altitude, the previous description cannot be used since, as discussed by Cess

Table 4.4. Percentage abundance of atmospheric CO_2 isotopes. From Goody (1964).

Isotope	Percent composition
$^{12}C^{16}O^{16}O$	98.420
$^{13}C^{16}O^{16}O$	1.108
$^{12}C^{16}O^{18}O$	0.408
$^{12}C^{16}O^{17}O$	0.064

(1974), Doppler broadening of rotational lines within a vibration-rotation band becomes the dominant process at low pressure. In this case, the absorptance can be replaced (see Cess, 1974, and Ramanathan, 1976) by

$$A = A_o u[1 - 0.18u/\delta] \qquad u/\delta \leqslant 1.5 \qquad (4.79a)$$

$$A = 0.753A_o \delta \left(|\ln(u/\delta)^{3/2} + 1.21\right) \qquad u/\delta \geqslant 1.5 \qquad (4.79b)$$

where $\delta = a\pi^{1/2}\gamma_D/d$, and γ_d is the Doppler linewidth. a is equal to 2 for bands whose alternate rotational lines are missing (most CO_2 bands) and 1 for all other bands. Thus, it is possible to determine the radiative fluxes for each band if the 4 parameters S, γ_o, d and A_o are specified (see Table 4.3). For example, for a pure CO_2 atmosphere, the radiative fluxes are given in terms of absorptance by (Kiehl and Ramanathan, 1983):

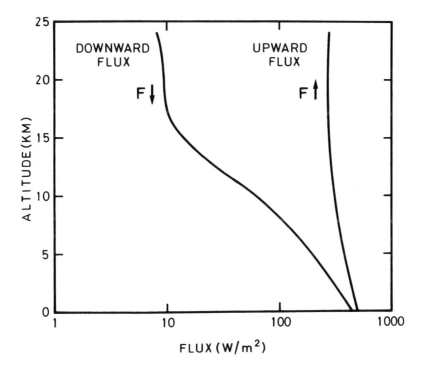

Fig. 4.19a. Upward and downward fluxes of terrestrial radiation in the troposphere and lower stratosphere (From Ramanathan, 1976). (Copyright by the American Meteorological Society).

$$F\uparrow = \sigma\,T_s^4 - B_{15}(T_s)A(\mid u - u_0 \mid) - \int_{u_0}^{u} B_{15}[T(u')]\,dA(\mid u - u_0 \mid) \quad (4.80a)$$

$$F\downarrow = -\int_0^u u\,B_{15}[T(u')]\,dA(\mid u - u' \mid) \qquad (4.80b)$$

where $B_{15}(T)$ is the Planck function evaluated at the center of the 15 μm band (667.4cm^{-1}), σ is the Stefan-Boltzmann constant, u is the reduced optical depth measured from the top of the atmosphere and u_0 is the value of u corresponding to the Earth's surface. This optical depth must be multiplied by a factor of 5/3 to account for the angular integration in all directions, as discussed above. The effect of other bands is introduced in a similar manner.

Radiative calculations must also deal with the overlap of the various spectral bands. This occurs, for example, between the water vapor and the CO_2, CH_4, and N_2O bands, respectively, at 15 μm, 7.66 μm, and 7.78 μm, and between the CH_4 band at 16.98 μm and the CO_2 band at 15 μm. This overlap is generally accounted for by multiplying the calculated absorptance by the mean transmissivity of the overlapping species in the spectral region under consideration.

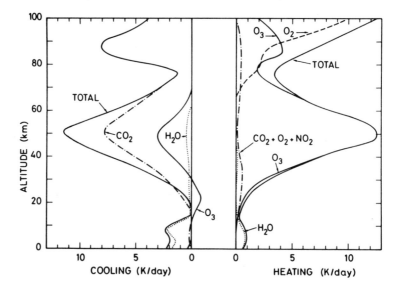

Fig. 4.19b. Vertical distribution of solar short wave heating rates by O_3, O_2, NO_2, H_2O, CO_2, and of terrestrial long wave cooling rates by CO_2, O_3, and H_2O. From London (1980).

Figure 4.19a presents the total upward and downward fluxes calculated from the model by Ramanathan (1976), and Figure 4.19b shows the contribution of different atmospheric constituents to the gross heating and cooling rates.

Band models are useful for practical applications but have some basic limitations; for example, they are only capable of yielding low spectral resolution and they do not always account for the effects of band wings. Kiehl and Ramanathan (1983) have shown, for example, that in the case of the CO_2 band at 15 μm the Goody and Malkmus formulation can only yield accurate results if the spectral interval $\delta\bar{\nu}$ is less than 10 cm^{-1}, and that the wide band models described above are probably more suitable for radiative transfer formulations in the middle atmosphere.

The previous discussion is based on the assumption that collisions are frequent and that the vibrational and rotational energy levels of all radiatively active molecules remain populated according to a Boltzmann distribution (which is determined by a single kinetic temperature). The gas is then considered to be in local thermodynamic equilibrium (LTE) and Kirchoff's law can be applied locally, so that the source function is given by Planck's law (eqn. 4.59). The validity of this assumption rests on the idea that the relaxation time due to collisions is considerably shorter than the radiative lifetime of the excited rotational and vibrational states. The radiative lifetimes of the vibrational states with which we are concerned is of the order of 10^{-1} s $(4\times10^{-1}$ s for the CO_2 15 μm band, 7×10^{-2} s for the ozone 9.6 μm band, and 6×10^{-2} s for the H_2O 6.3 μm band; Curtis and Goody, 1956). The vibrational *relaxation* time at ground level is considerably smaller and, in the case of the CO_2 15 μm band, is of the order of 2.5×10^{-5} s for the temperature range 150-210 K with an uncertainty of about 50% (Simpson et al., 1977). Since this relaxation lifetime varies inversely with the pressure, the radiative and relaxation time constants become equal at about 68 km, assuming a scale height of 7 km. A similar analysis for collisional and radiative relaxation for rotational levels shows that the critical factor in the applicability of the LTE assumption is the relative lifetimes associated with vibrational (not rotational) states.

Near and above 65-70 km, the source function cannot be expressed by Planck's law, but must be derived in a more general form which takes into account radiative absorption between energy levels, spontaneous and induced emission, as well as energy transitions due to collisions with other particles. Assuming that transitions occur between two energy levels, it can be shown (Kuhn and London, 1969; Houghton, 1977) that the source function is given by

$$J_\nu = \frac{EB_\nu + \dfrac{1}{4\pi}\int L_\nu d\omega}{1 + E} \qquad (4.81)$$

where B_ν is the Planck function, c is the velocity of light, and E is the ratio between the probabilities for collisional de-excitation and spontaneous transitions from the upper to the lower level. The contribution to the source function of additional vibrational levels has been considered by Kuhn and London (1969). Below 60 km, when collisions dominate, $E \to \infty$, and

$$J_\nu = B_\nu \qquad (4.82a)$$

and LTE conditions apply. At high altitude, when the population of the energy levels is achieved by radiative processes only, $E \to 0$, and the source function becomes

$$J_\nu = \frac{1}{4\pi} \int L_\nu d\omega \qquad (4.82b)$$

This case is referred to as monochromatic radiative equilibrium (MRE) in which, for each frequency, the absorbed radiant energy is balanced by the emitted radiant energy.

The importance of non-LTE processes in the atmosphere was first pointed out by Milne (1930), and the first method for treating the problem in the mesosphere was presented by Curtis and Goody (1956). Later on, these questions were discussed by Houghton (1969; 1977), Williams (1971), and others. Accuracy in the calculations of heating and cooling rates in the mesosphere and thermosphere remains limited by uncertainties in the relaxation times and in their temperature dependencies. Williams (1971) has shown, for example, that at 90 km the cooling rate due to transfer in the 15 μm band of CO_2 is of the order of 2 K/day for a relaxation time of 3×10^{-5} s, but as much as 10 K/day for a value of 2×10^{-6} s.

Other factors could influence the heating and cooling in the mesosphere and lower thermosphere. Crutzen (1970) has pointed out that oxygen atoms can assist transfer between translational and vibrational energy on collision with CO_2, and lead to an additional cooling for the 15 μm band. Further, it should be noted that oxygen atoms release energy upon recombination through the processes:

$$O + O + M \to O_2 + M$$

and

$$O + O_2 + M \to O_3 + M$$

The released energy is 119 kcal/mole and 25 kcal/mole, respectively (see Chapter 2). The consequences of these processes on the thermal budget near the mesopause will be discussed in the next section.

4.5 The thermal effects of radiation

The thermal effects of solar and atmospheric radiation have been studied using mathematical models, generally by solving the radiative transfer equations which we have just described (Ramanathan, 1976; Donner and Ramanathan, 1980; Ramanathan and Coakley, 1978; Manabe and Wetherald, 1967; Manabe and Strickler, 1964; Owens et al., 1983; Groves and Tuck, 1980).

The most elaborate models, particularly those with fine spectral resolution, require rather large computing resources, and cannot be coupled with chemical or dynamical models using present computers unless they are greatly simplified.

In most cases, only the mean vertical temperature distribution is desired. The distribution obtained by assuming local radiative equilibrium between the energy absorbed and the energy emitted at a given altitude yields a very large vertical gradient at low altitude (Manabe and Moeller, 1961): it becomes necessary to consider convection and vertical transport of non-radiative energy. To resolve this problem, one-dimensional *radiative convective models,* as first presented by Manabe and Strickler (1964) are widely used. In such models the

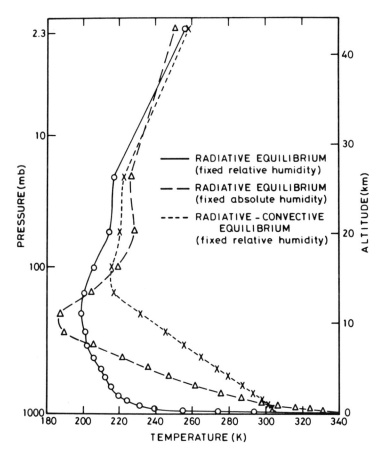

Fig. 4.20. Vertical temperature profile calculated for pure radiative and radiative-convective equilibrium conditions. From Manabe and Wetherald (1967). (Copyright by the American Meteorological Society).

lapse rate is adjusted to a representative value whenever the computed value exceeds this level. This adjustment is performed until an equilibrium is reached in each atmospheric layer. The net incoming solar energy flux is thus balanced by the net outgoing long wave energy flux at the top of the atmosphere. At the Earth's surface, a balance must exist between the net energy gain by radiation and the loss of heat by convective transfer into the atmosphere. Whenever the lapse rate is subcritical, the model assumes that pure radiative equilibrium is achieved. Such model techniques require an iteration procedure in which the layers of the atmosphere are scanned repeatedly until these conditions are satisfied. Figure 4.20 presents the temperature profile obtained by Manabe and Wetherald (1967) assuming pure radiative equilibrium and radiative convective equilibrium. The differences are particularly striking in the troposphere, but are also important up to the mid-stratosphere.

Radiative convective models may include a few important feedback mechanisms between the components of the climate system (Wang et al., 1981; Hansen et al., 1983; Hummel and Kuhn, 1981). These mechanisms occur mostly in the troposphere, but can have significant impact on the middle atmosphere because they determine the radiation budget near the tropopause and consequently the temperature profile in the lower stratosphere. Further, these complex tropospheric mechanisms regulate the amplitude of climatic responses to perturbations occurring at higher levels, through, for example, the effect of a change in stratospheric composition.

The surface temperature T_s is particularly sensitive to the cloud distribution, the ice and snow cover, and the water vapor profile. Clouds have a significant influence on both solar and terrestrial radiative fluxes. They absorb, scatter, and transmit direct solar radiation but they also absorb and emit in the infra-red. Clouds will thus have two opposite effects on the surface temperature T_s (cooling by reflecting a portion of the incident solar radiation and heating by reducing the escape to space of long wave terrestrial radiation), and the net result is probably a cooling as shown by Ellis (1977). Models generally assume a cloud reflectivity near 0.5 for short wave radiation and a global cloud cover of 50%. In the long wave range clouds are considered to be either black or semi-black. Feedback mechanisms regarding clouds are difficult to express quantitatively, but it is known that climate sensitivity to perturbations will be different if a fixed cloud top temperature is assumed (Cess, 1974), rather than a fixed cloud top altitude (Manabe and Wetherald, 1967; Rasool and Schneider, 1971; Ramanathan, 1976). A few radiative-convective model studies include clouds calculated according to the temperature and humidity profiles (Wang et al., 1981; Hummel and Kuhn, 1981).

Ice albedo feedback can be included by making the surface reflectivity change with the mean surface temperature (e. g., Wang and Stone, 1980). The neglect of this mechanism in most radiative-convective models leads to an underestimate in the response of T_s to external perturbations (Ramanathan and Coakley, 1978).

Although controlled exclusively by dynamics and chemistry in the stratosphere, the distribution of water vapor is highly dependent on temperature in the troposphere. Indeed, from an analysis of summer and winter observations, Manabe and Wetherald (1967) suggested that the troposphere tends to conserve relative humidity rather than the absolute humidity (or mixing ratio). Cess (1976) noted, however, that the hemispherically averaged relative humidity (RH) is usually somewhat larger in summer than in winter and therefore suggested a vertical distribution slightly dependent on the surface temperature T_s:

$$RH = RH_o \left[\frac{\left((p/p_o) - 0.02 \right)}{\left(1 - 0.02 \right)} \right]^X \tag{4.83}$$

with

$$X = 1.0 - 0.03(T_s - 288)$$

p being the pressure, p_o its value at sea level, and $RH_o \approx 0.77$ the relative humidity at the surface. The mixing ratio is derived from

$$f_{H_2O} = \frac{0.622 RHe_s(T)}{p - RHe_s(T)} \quad (g\ H_2O/g\ air) \tag{4.84}$$

where $e_s(T)$ is the saturation vapor pressure of H_2O (see Appendix A). This expression can be used up to the level where stratospheric values of the H_2O mixing ratio are found (e. g. 3 ppmv), and introduces a strong positive water vapor - temperature feedback. An increase in the tropospheric temperature (or in the surface temperature) is accompanied by an increase in the water vapor content. The resulting enhancement in the long wave opacity amplifies the increase in T_s. The changes in the surface temperature due to stratospheric chemical perturbations are therefore amplified by this mechanism.

Radiative convective models have been used extensively to assess climatic changes resulting from perturbations in the concentrations of optically active constituents such as CO_2, O_3, H_2O, N_2O, CH_4, chlorofluorocarbons, and aerosols (see Ramanathan and Coakley, 1978). These possibly anthropogenic perturbations will be discussed in Chapter 7. These models are particularly suitable for such studies since the calculated surface temperature varies in response to the changes in the radiative energy received by the troposphere. The change in this tropospheric energy, however, is constrained by the fact that radiative equilibrium conditions must exist in the stratosphere. Consequently, perturbations to stratospheric composition that affect the radiative input to the troposphere will affect the surface temperature in order to maintain the global balance between the net solar energy input and the net terrestrial energy output. Limitations in these models are due primarily to the fact that they describe only global energy exchange, with parameters (such as the critical lapse rate, the cloud coverage, surface albedo) which are generally prescribed in terms of present day conditions and may be variable with the amplitude of external perturbations.

Besides these empirical radiative convective models which do not treat the convective processes explicitly, semi-empirical techniques have been used to parameterize the non-radiative heat transport. Liou and Ou (1983) have, for example, developed a model in which the vertical transport of heat is parameterized in the thermodynamic equation by a thermal eddy diffusion coefficient.

Two- and three-dimensional models can calculate the spatial and temporal distribution of the net heating rate Q and the temperature not only as a function of altitude but also as a function of latitude, and even of longitude (local time). Such studies must consider the multidimensional transport of heat and the solution of a thermodynamic equation like the one shown in equation (3.10). The effect of waves should also be considered. As pointed out by Apruzese et al. (1983), gravity wave dissipation may play an important role in the mesospheric heat budget. In the multidimensional models, the radiative scheme is generally very simplified and parameterized; the most simple approach is to assume the cool to space approximation, in which it is assumed that exchange of heat between layers can be neglected in comparison to propagation out to space. It is also common to employ the Newtonian cooling approximation (Murgatroyd and Goody, 1958; Dickinson, 1973), in which the cooling rate is linearized in terms of the temperature deviation from a reference profile.

4.5.1 Heating due to absorption of radiation

The absorption of ultraviolet radiation by ozone in the Huggins and Hartley bands constitutes the principal source of heat in the stratosphere and mesosphere. The heating rate reaches as much as 12K/day near the stratopause, with a maximum of about 18K/day near the summer pole. The effect of the Huggins bands in the visible region becomes important in the lower stratosphere, where the resulting heating rate is not quite 1K/day. These numbers are obviously related to the amount of ozone present, and an increase in ozone density would lead to an increase in the stratospheric and mesospheric temperatures, as well as possible changes in the locations of the stratopause and mesopause.

Above about 75 or 80 km, absorption in the Schumann-Runge continuum of molecular oxygen contributes to the heating of the atmosphere and begins to play the dominant role in the lower thermosphere. At 100 km, the mean heating rate is about 10K/day but large variations occur with latitude and season, and the maximum heating may be a factor of ten higher. Ultraviolet radiation leads to the dissociation of molecular oxygen, forming atomic oxygen. At altitudes above 80 km, the lifetime of atomic oxygen exceeds a day, and the energy absorbed during O_2 photolysis may be stored as chemical energy. It is released as thermal energy when the oxygen atom recombines, which may occur at a point removed from the original absorption of a photon. Due to horizontal and vertical transport, much of the chemical energy stored

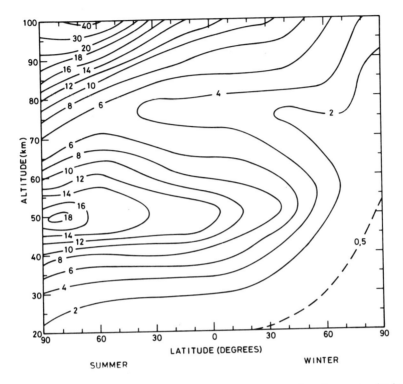

Fig. 4.21. Meridional distribution of the atmospheric heating rate (K/day). Effects of absorption of radiation by ozone and oxygen. After London (1980).

in this manner is released in high latitude winter. This process, along with adiabatic heating as a result of downward vertical velocities, leads to the warm mesopause temperatures observed in winter.

The absorption by molecular oxygen in the Herzberg continuum only contributes slightly to the heating rate in the middle atmosphere, and can be neglected relative to the effects of ozone in the stratosphere. There are, however, some O_2 bands at wavelengths as long as 760 nm in the visible.

Figure 4.21 shows the meridional distribution of the heating rate due to O_2 and O_3, and shows the existence of quite different regimes. The corresponding values are determined by an expression of the form (4.42), i.e.,

$$\frac{dT}{dt} = \frac{\cos\chi}{\rho C_p} \int_\nu \frac{dF_{s,\nu}}{dz} d\nu \tag{4.85}$$

where the incident solar flux is given by

$$F_{s,\nu} = F_{s,\nu}(\infty)\exp-\left[\sigma(O_3)\int_z^\infty n(O_3)dz' + \sigma(O_2)\int_z^\infty n(O_2)dz'\right]\sec\chi \qquad (4.86)$$

considering only the effects of absorption by ozone and oxygen.

One may then write:

Table 4.5 Parameterization of atmospheric heating rates
(from Schoeberl and Strobel, 1978).

Spectral region	Parameterization
Chappuis bands	$\dfrac{Q_c}{(O_3)} = 1.05\times10^{-15}\exp(-2.85\times10^{-21}N_3)$
Huggins bands	$\dfrac{Q_{Hu}}{(O_3)} = \dfrac{1}{N_3}[4.66\times10^3 - 7.8\times10^2\exp(-1.77\times10^{-19}N_3)$
	$\qquad -3.88\times10^3\exp(-4.22\times10^{-18}N_3)]$
Hartley region	$\dfrac{Q_{Ha}}{(O_3)} = 4.8\times10^{-14}\exp(-8.8\times10^{-18}N_3)$
Schumann Runge bands	$\dfrac{Q_{SRB}}{(O_2)} = \dfrac{1}{0.67N_2 + 3.44\times10^9(N_2)^{1/2}}$
	If $N_2 < 10^{18}cm^2$, $\dfrac{Q_{SRB}}{(O_2)} = 2.43\times10^{-19}$
Schumann Runge continuum	$\dfrac{Q_{SRC}}{(O_2)} = \dfrac{1}{N_2}[0.98\exp(-2.9\times10^{-19}N_2)$
	$\qquad -0.55\exp(-1.7\times10^{-18}N_2) - 0.43\exp(-1.15\times10^{-17}N_2)]$

(Note that N_2 and N_3 represent the total slant column abundances (molec cm^{-2}), and (O_2) and (O_3) the number densities (molec cm^{-3}), of O_2 and O_3, respectively. Heating rates in 0.1 W m^{-3}.)

$$\frac{dT}{dt} = \frac{1}{\rho C_p}\left[n(O_3)\int_\nu \sigma(O_3)F_{s,\nu}d\nu + n(O_2)\int_\nu \sigma(O_2)F_{s,\nu}d\nu\right] \qquad (4.87)$$

A parameterization of the atmospheric heating rate by O_3 and O_2 in different spectral regions has been suggested by Schoeberl and Strobel (1978), and is presented in Table 4.5.

The heating due to absorption of the 2.7 and 4.3 μm bands of carbon dioxide, and the absorption of ultraviolet by water vapor, play minor roles in the heating budget. However, the H_2O bands in the visible and near infrared can not be neglected in the troposphere, where the water vapor content is much higher than in the middle atmosphere. The absorption of solar radiation by NO_2 in the region between 300 and 600 nm should also be considered. The mean heating rate depends, however, on the distribution of NO_2 concentration and must also include the effects of multiple scattering and reflection by the surface and by clouds. Its contribution can be considerable in the lower stratosphere in summer, where NO_2 is probably present in large amounts.

4.5.2. Cooling by radiative emission

The cooling produced by radiative emission in the stratosphere and lower mesosphere is principally due to 15 μm band of carbon dioxide. A precise calculation must include the contribution of different isotopes of CO_2, even those which constitute only 1.5% of the atmospheric abundance of this species. The effect of the fundamental and hot bands of these isotopes is large in the mesosphere and upper stratosphere, where a corresponding cooling rate of about 2K/day is calculated (Williams and Rodgers, 1972).

An analysis based on the work of Dopplick (1972), Williams and Rodgers (1972) and Kuhn and London (1969) (Fig. 4.22) shows that the maximum cooling rate by CO_2 is found at the winter mesopause, where the temperature is relatively warm. Flux convergence leads to a slight heating near 80 km in summer, that is, at relatively cold temperatures. The results in the lower thermosphere are subject to some doubt as a result of the uncertainties in the relaxation times of vibrational levels.

The second contribution to infrared cooling in the middle atmosphere results from infrared emission by ozone at 9.6 μm. This contribution is important at a layer near the stratopause, and extends only over about 10 km of altitude.

In the lower stratosphere, the radiative transfer at 9.6 μm leads to a net heating because ozone absorbs some of the terrestrial emission at this frequency at altitudes below the ozone layer. Figure 4.23 shows the meridional distribution of the cooling by O_3 from a number of studies. (Kuhn and London, 1969 and Dopplick, 1972).

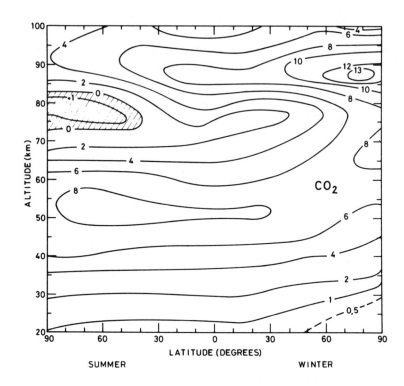

Fig. 4.22. Meridional distribution of the atmospheric cooling rate due to infrared emission by CO_2 (15 μm) After London (1980). (K/day)

The effect of water vapor is manifested in the middle atmosphere through an infrared emission at 80 μm. This contribution is relatively weak as shown in Figure 4.24, based on the work of Kuhn and London (1969) and Dopplick (1972). In the troposphere, the temperatures are warmer than in the stratosphere, and the 6.3μm band of H_2O determines the net cooling rate.

To evaluate the thermal budget of the middle atmosphere, it can be useful to apply the "cool to space" approximation. The cooling rate due to the infrared emission is then proportional to $dF\uparrow/dz$, and using equation (4.69a), in which the contribution of the surface emission is neglected, we can write

$$\frac{dT}{dt} = -C_1 B_\nu(T) \tag{4.88}$$

The coefficient C_1 for the 15 μm band of CO_2 has been determined by Houghton (1977), assuming that the lineshapes are Lorentzian, that their overlap is negligible, and that the absorption is strong (strong line approximation). According to this approximation, the cooling dT/dt will be obtained in $^\circ$K/day if the constant C_1 is:

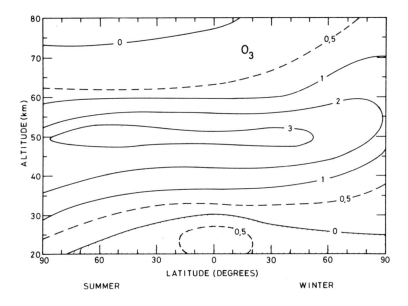

Fig. 4.23. Meridional distribution of the cooling rate. Effect of infrared emission by ozone at 9.6 μm. After London (1980) (K/day).

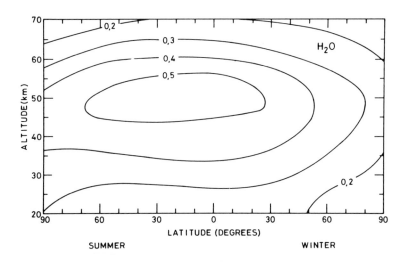

Fig. 4.24. Meridional distribution of infrared cooling by water vapor. After London (1980) (K/day)

$$C_1 = 5.43 \times 10^5 \left(\frac{f_{CO_2}}{3.3 \times 10^{-4}}\right)^{1/2}$$

where f_{CO_2} is the mole fraction of carbon dioxide, and where B_ν is given by

$$B_\nu \approx 3.53 \times 10^{-4} \exp(-960/T) \ (W \ cm^{-2} \ s^{-1})$$

This expression can be applied between 30 and 70 km, and shows that the cooling by CO_2 varies approximately as the square root of its mixing ratio. If the temperature varies little about its reference value T_0, equation (4.88) can be linearized, such that the cooling rate can be written:

$$\frac{dT}{dt} = \left(\frac{dT}{dt}\right)_{T_0} + \alpha(T - T_0) \tag{4.89}$$

where $(dT/dt)_{T_0}$ is determined by detailed calculations. The coefficient α defines the time constant for thermal radiation, $\tau_{rad} = 1/\alpha$. This time is of the order of 15 to 20 days in the troposphere (Manabe and Strickler, 1964; Hering et al., 1967), but is only about of 3 to 5 days in the upper stratosphere (Dickinson, 1973; Strobel, 1979). Transport of heat can only occur if the dynamic time constant, τ_{dyn} is less than or comparable to the radiative time constant, τ_{rad}. Thus, in the troposphere, where $\tau_{dyn} < \tau_{rad}$, the temperature field is dependent on transport, while in the middle atmosphere, where $\tau_{rad} << \tau_{dyn}$,

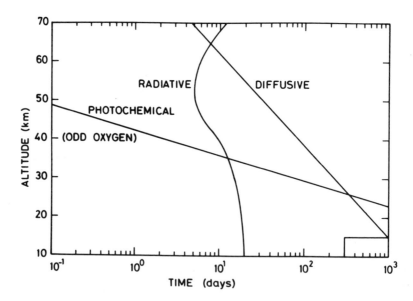

Fig. 4.25. Radiative lifetime in the stratosphere. Approximate lifetimes for vertical transport and odd oxygen photochemistry are shown for comparison.

the thermal structure results from an equilibrium between absorption of ultra-violet radiation and emission in the infrared. Since the radiative lifetime is never less than a few days, the diurnal variation in solar energy input produces only small diurnal temperature variations. Figure 4.25 shows the approximate shape of the radiative lifetime (London, 1980) in comparison to those for vertical transport and odd oxygen photochemistry. Variations in the temperature structure induce modifications in the concentration of chemical species, particularly ozone, and as a result, influence the transfer of radiation. Ghazi et al. (1979) showed that this effect leads to a reduction in the radiative lifetime in the upper stratosphere, and that the total relaxation coefficient can be written:

$$\alpha = \alpha_{NC} + \alpha_{PH} \qquad (4.90)$$

i. e., as the sum of the Newtonian cooling and photochemical relaxation coefficients. The analysis of simultaneous observations of temperature and ozone from satellite shows that at 2 mb (about 43 km), for example, α_{PH} cannot be neglected relative to α_{NC}. This will be discussed in more detail in §7.2.

4.6 Photochemical effects of radiation

4.6.1. General

Absorption of ultraviolet and visible photons by atmospheric molecules can induce transitions into electronically excited states which may then photodissociate (see Chapter 2). The dissociation products play a crucial role in the photochemistry of the middle atmosphere.

In atmospheric problems, it is necessary to estimate the quantitative value of the photodissociation rate of a molecule A. This is given by:

$$\frac{dn(A)}{dt} = -J(A)n(A) \qquad (4.91)$$

where $n(A)$ represents the concentration (cm^{-3}) of the molecule and $J(A)$ the photodissociation probability or coefficient. This coefficient is also called the photodissociation frequency and is expressed in s^{-1}. The inverse of J represents the lifetime of the molecule against photolysis.

For a wavelength interval $d\lambda$, the photodissociation coefficient of molecule A is proportional to the photon flux $q_\lambda d\lambda$, the absorption cross section $\sigma(A;\lambda)$, and the quantum efficiency $\epsilon(A;\lambda)$. When the entire portion of the solar spectrum over which the molecule can dissociate is considered $(\lambda_x - \lambda_y)$, the coefficient is given by:

$$J(A;z;\chi) = \int_{\lambda_x}^{\lambda_y} \epsilon(A;\lambda)\sigma(A;\lambda)q_\lambda(\lambda;z;\chi)d\lambda \qquad \textbf{(4.92)}$$

The spectral distribution of the absorption cross section and the quantum efficiency of atmospheric molecules is determined in the laboratory and can vary with temperature. In most cases, the quantum efficiency ϵ is nearly unity but may be less, especially near the dissociation limit. This limit (maximum wavelength or minimum frequency) corresponds to the minimum energy required to dissociate the molecule (see, e. g. eqn. 2.14).

In most cases, the spectral variation of the cross section is sufficiently gradual that one can integrate equation (4.92) by simple methods. In a few cases (Schumann Runge bands of O_2 and δ bands of NO) the spectrum contains a multitude of narrow lines and more sophisticated methods are needed. This will be discussed in more detail in §4.6.3.

4.6.2 Absorption cross sections of the principal atmospheric molecules

• Molecular nitrogen

The photodissociation of the most abundant gas in the atmosphere plays very little role in atmospheric chemistry below 100 km because the absorption of dissociating radiation by N_2 is very weak. However, production of atomic nitrogen from N_2 photolysis via predissociation does occur following absorption into the Lyman- Birge-Hopfield bands ($a^1\pi_g - X^1\Sigma_g$ transition). The coefficient of photodissociation is about 5×10^{-12} s^{-1} at the top of the atmosphere and decreases rapidly, becoming negligible near the mesopause. The solar CIII line at 97.7 nm plays an important role in this process. (see Richards et al., 1981).

• Molecular oxygen

The photodissociation of molecular oxygen produces oxygen atoms, which play an important role in middle atmosphere chemistry. A potential diagram for the principal electronic states of the O_2 molecule was presented in Chapter 2, and some of the important transitions between them were mentioned. From the aeronomic viewpoint, the most important transitions are the $X^3\Sigma_g^- - A^3\Sigma_u^+$ which constitutes the Herzberg system (forbidden transition resulting in weak absorption from 185 to 242 nm) and the $X^3\Sigma_g^- - B^3\Sigma_u^-$ which constitutes the Schumann Runge system. The latter is characterized by banded structure from 175 to 200 nm and a continuum from 137 to 175 nm. The first of these two transitions is a predissociation, that is, a non-radiative transition to the repulsive $^3\Pi_u^-$ state; it thus leads to the formation of two oxygen atoms in the ground ^3P state. In the second case ($\lambda < 175$nm) one of the atoms is formed in the excited ^1D state. At shorter wavelengths

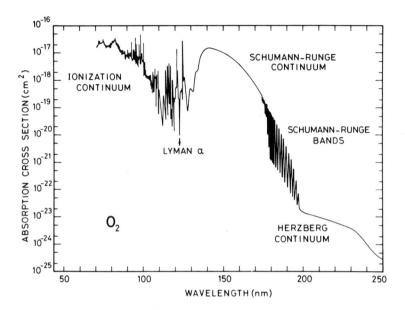

Fig. 4.26. Spectral distribution of the absorption cross section of molecular oxygen.

$(\lambda < 137\text{nm})$ some diffuse bands occur, but these are interspersed with a series of windows, which allow some wavelengths to penetrate relatively deep into the atmosphere. One of these windows coincides by chance with the solar Lyman α line, which plays an important role in atmospheric chemistry. Figure 4.26 shows the general shape of the spectral distribution of the absorption cross section. At wavelengths less than 102.8 nm, photoionization can occur.

The absorption coefficient of molecular oxygen has been the subject of numerous laboratory investigations. In the Herzberg continuum, the measurements of Ditchburn and Young (1962), Blake et al. (1966), Ogawa (1971), Hasson and Nicholls (1971) and Shardanand and Prasad-Rao (1977) indicate an absorption cross section less than 1.5×10^{-23} cm^{-2}, but exhibit some discrepancies in the absolute value, especially at long wavelengths (see Fig. 4.27). Part of the difficulty associated with measuring this cross section in the laboratory arises from the formation of the Van Der Waal's molecule, O_4, for which observational evidence has even been found in the atmosphere (Perner and Platt, 1980). Balloon observations of the solar irradiance in the stratosphere have led to inferences of the O_2 absorption cross section in-situ (Frederick and Mentall, 1982; Herman and Mentall, 1982), and these have resulted in values somewhat lower than the laboratory investigations. The inferred cross section depends sensitively on the accuracy of a simultaneous measurement of the ozone amount and on the adopted value for the ozone cross sections used in the analysis. Rayleigh scattering, which contributes at least 10% to the total

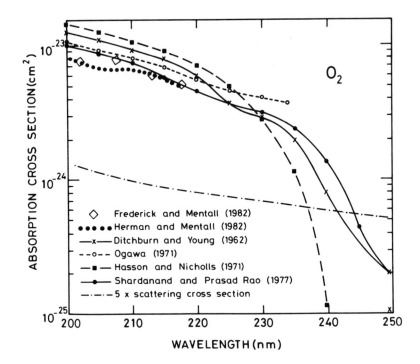

Fig. 4.27. Absorption cross section of molecular oxygen in the Herzberg continuum. Spectral distribution from several laboratory studies and from in-situ observations of the solar irradiance (Herman and Mentall).

attenuation of the direct solar radiation in this wavelength region, must also be considered. The Herzberg continuum extends to approximately 150 nm and therefore lies beneath the Schumann-Runge bands, where it makes a minor contribution to the total cross section. The absorption of solar radiation in this spectral region has important implications in middle atmosphere chemistry, particularly for the distributions of N_2O and chlorofluorocarbons in the stratosphere (Froidevaux and Yung, 1982; Brasseur et al., 1983).

The photodissociation frequency corresponding to this spectral region is about $10^{-9} s^{-1}$ for zero optical depth. In spite of the weakness of this transition, it becomes dominant in the stratosphere (Fig. 4.31). The rate of ozone production in the stratosphere depends critically on the value of the absorption cross sections in this region.

In the region of the Schumann-Runge bands, molecular oxygen is predissociated and provides an important source of $O(^3P)$. The absorption cross section varies by about 5 orders of magnitude between 175 and 205 nm. The calculation of the photodissociation rate requires a detailed study of the absorption cross section. Experimental results obtained in this region, notably by Hudson and Carter (1968,1969), Hudson et al., (1969), Ackerman et al. (1969),

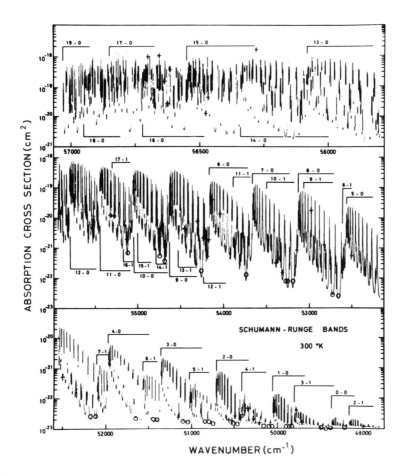

Fig. 4.28. Spectral distribution of the absorption cross section of molecular oxygen in the Schumann Runge bands. From Kockarts (1971). (Copyright by Reidel Publishing Company).

Biaume (1972a,b) have been applied to the atmospheric problem in studies by Kockarts (1971), Hudson and Mahle (1972), Fang et al. (1974), Park (1974), Kockarts (1976), Blake (1979), Nicolet and Peetermans (1980), and Allen and Frederick (1982). Figure 4.28 shows the variation of the absorption cross section of O_2 from 57000 cm^{-1} (175.4 nm) to 49000 cm^{-1} (204.1 nm) as calculated by Kockarts (1971), with a resolution of 0.5 cm^{-1}, (that is, with a resolution corresponding to the width of the rotational lines for O_2 in most of the bands). Comparisons made by Frederick et al. (1981) of the solar irradiance observed at 40 km and calculated using a detailed absorption spectrum indicate errors of the order of 10% above 194 nm, 20% between 189 and 194 nm,

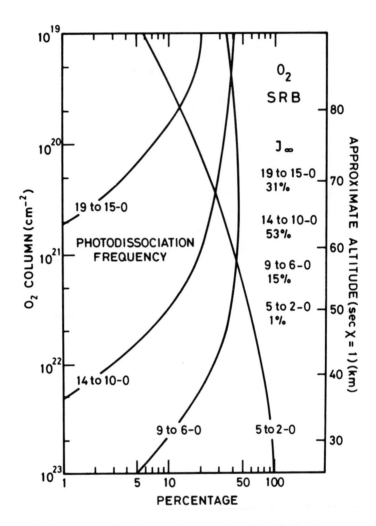

Fig. 4.29. Relative contribution to the O_2 photodissociation frequency in the Schumann-Runge region by 4 groups of bands: the 19 to 15-0, 14 to 10-0, 9 to 6-0, and 5 to 2-0, with different sensitivities to solar activity effects. Ozone absorption has not been considered in the calculation (From Nicolet, 1984).

and 40% below 189 nm. These differences may be due to observational errors, or to uncertainties in the spectral data. Nicolet and Peetermans (1980) have, for example, emphasized the difficulties in obtaining accurate experimental values of the oscillator strength and of the predissociation widths of the rotational lines. The magnitude of these uncertainties is a function of altitude, since, as shown in Figure 4.29, the contribution of the lower v' bands increases with decreasing altitude. These problems lead to an uncertainty in the absolute value of the optical depth in the mesosphere and in the photodissociation coefficient, which is about 10^{-7} s^{-1} at zero optical depth.

The determination of the O_2 photodissociation frequency in the Schumann-Runge bands is obtained from a computation including the detailed rotational structure of the band system (see § 4.6.3). In order to avoid lengthy calculations in photochemical models, expressions have been derived which give nearly the same result as the more complete calculations. Nicolet (1984) has derived, for example, the following expressions: In the thermosphere, when the total oxygen column is less than 1×10^{19} cm^{-2}, and for low solar activity:

$$J(O_2; SRB) = 1.1 \times 10^{-7} \exp[-1.97 \times 10^{-10} N^{0.522}] s^{-1} \qquad (4.93a)$$

with an accuracy better than ± 10 percent, and in the mesosphere and upper stratosphere, where $10^{19} cm^{-2} \leqslant N \leqslant 10^{22} cm^{-2}$,

$$J(O_2; SRB) = 1.45 \times 10^8 N^{-0.83} s^{-1} \qquad (4.93b)$$

with an accuracy of about ± 15 %. Such a formula is inaccurate in the stratosphere, where the transmission of incident radiation is dependent on the ozone density.

The amplitude of the changes in the incident irradiance from quiet to active sun conditions varies with wavelength in the spectral range of the

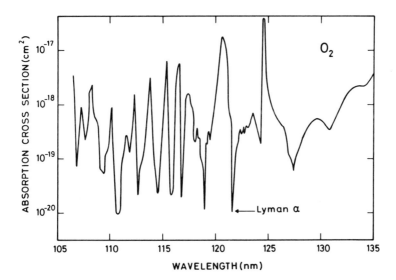

Fig. 4.30. Spectral distribution of the absorption cross section of oxygen from 105 to 135 nm. The cross section exhibits a minimum for the wavelength corresponding to the solar Lyman α line. After Watanabe et al., (1953).

Schumann-Runge bands, being for example, about 20% in the region of the (19-0) band and only 5% in the (2-0) band. Since the contribution of each band to the total value varies with altitude (Fig. 4.29), the overall variation in the photodissociation with solar activity decreases with decreasing altitude. An average variation suggested by Nicolet (1984) is given by:

$$J(O_{2;} \text{ SRB active sun}) = (1.11\pm 0.04)J(O_2; \text{ SRB quiet sun}) \qquad (4.93c)$$

Because of its strong absorption ($\sigma = 1.5 \times 10^{-17} \text{cm}^2$ at 140 nm), the Schumann Runge continuum is only important in the thermosphere, and from the point of view of photolysis, only plays a dominant role above 90 km. Each photolysis event leads to the production of two oxygen atoms, one in the ground ^{3}P state and the other in the excited ^{1}D state. Laboratory measurements of the oxygen cross section in this spectral range have been presented by Watanabe et al. (1953), Watanabe and Marmo (1956), Huffman et al. (1964), Metzger and Cook (1964), Goldstein and Mastrup (1966), Hudson et al. (1966), Ogawa and Ogawa (1975), and Lean and Blake (1981). As shown by Hudson et al. (1966) as well as by Lean and Blake (1981) and Gibson et al. (1982), a temperature dependence of the cross section becomes noticeable for wavelengths longer than 160 nm and is mainly due to the redistribution of the population of rotational states. The Schumann-Runge continuum adjoins the Schumann-Runge band system at 175 nm and underlies the bands up to 183 nm. This continuum therefore contributes to the absorption cross section

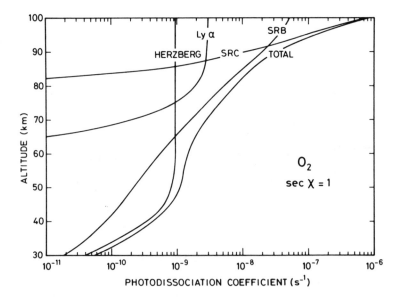

Fig. 4.31. Contribution of each spectral region to the photodissociation of molecular oxygen as a function of altitude.

minima in the band system. Adopting the values of Ackerman (1971), the photodissociation frequency in the Schumann-Runge continuum at zero optical depth is about 2×10^{-6} s^{-1} but only 1×10^{-8} s^{-1} at 90 km for an overhead sun.

At wavelengths less than 130 nm (see Fig. 4.30), the most important contribution to oxygen photolysis comes from the solar Lyman α line ($\sigma = 10^{-20}$ cm^2), which penetrates to the mesosphere. The corresponding photolysis rate at zero optical depth is about 3×10^{-9} s^{-1}, but its value can change strongly with solar activity. From the experimental data of Carver et al. (1977), Nicolet and Peetermans (1980) indicate that the conventional expression can be used for $N \leqslant 10^{19}$ cm^{-2}:

$$J(O_2;L\alpha) = q_\infty \times 1 \times 10^{-20} \exp(-1 \times 10^{-20} N) \qquad (4.94a)$$

but for $N > 10^{19}$ cm^{-2}:

$$J(O_2;L\alpha) = q_\infty \times 4.17 \times 10^{-19} N^{-0.083} \exp(-4.17 \times 10^{-19} N^{0.917}) \qquad (4.94b)$$

corresponding to a variable cross section for $N > 10^{19}$ cm^{-2}. Finally, Figure 4.31 shows the contribution of each spectral region to the photolysis rate of O_2 for an overhead sun. Note the importance of the Herzberg continuum below 65 km (stratosphere and mesosphere), the Schumann-Runge bands from 65 to 90 km (mesosphere), and the Schumann-Runge continuum in the thermosphere. The Lyman α line plays a secondary role in oxygen photolysis, but becomes much more important for other atmospheric constituents.

- *Ozone*

The primary absorption by ozone occurs in the Hartley band, which is located from 200 to 310 nm. The absorption cross section maximizes at about 250 to 260 nm, where $\sigma = 10^{-17}$ cm^2. The absorption in this band has been studied since the beginning of the century by several investigators. For numerical applications the values of Vigroux (1952a, b; 1969), Inn and Tanaka (1953, 1959) Tanaka, Inn and Watanabe (1953) are generally employed to yield the curve shown in Figure 4.32. Vigroux (1953, 1969) observed a weak variation of the cross section with temperature, which is probably due to the superposition of strong predissociating bands on the continuum. These were observed as early as 1929 by Lambrey and Chalonge.

Around 300 nm, the Hartley band becomes weak, and from 310 to 350 nm it blends with the temperature dependent Huggins bands. Figure 4.33 shows the values of Simons et al. (1973) from 310 to 370, and demonstrates the importance of this temperature sensitivity, especially at long wavelengths. A careful calculation is required to account for the details of the spectral structure as a function of temperature for aeronomic applications.

Measurements of the ozone cross section in the wavelength range from 245-345 nm have been reported by Bass and Paur (1981). These data obtained for different temperatures are expressed relative to the peak absorption value measured by Hearn (1961) at 254 nm. The measurements performed at room temperature are, however, 12% higher in the wavelength range of 310 to 334

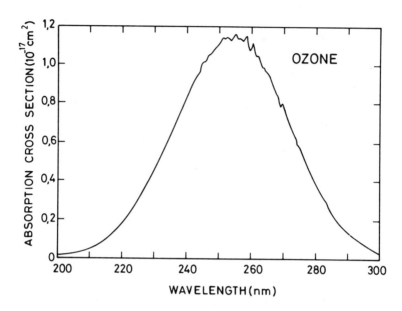

Fig. 4.32. Hartley band of ozone. Absorption cross section from 200 to 300 nm After Griggs (1968).

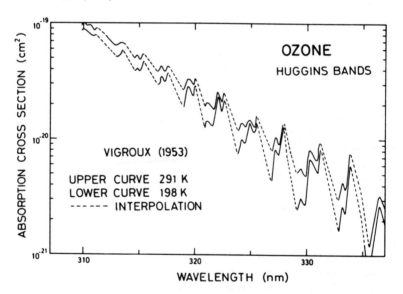

Fig. 4.33. Huggins bands of ozone. Absorption cross sections from 310 to 370 nm, showing the effects of temperature. From Nicolet (1980).

nm than the absolute values measured by Daumont et al. (1983). These discrepancies must be resolved since they lead to significant uncertainties in the calculation of the ozone photodissociation and in the value of the total ozone column derived by Dobson spectrophotometers (Dobson, 1957). Daumont et al. (1983) indicate that the use of the conventionally accepted cross sections might underestimate the ozone column by a few percent if their newly measured values are correct.

Table 4.6 Theoretical limits corresponding
to different photolysis products (nm).

	$O_2(^3\Sigma_g)$	$O_2(^1\Delta_g)$	$O_2(^1\Sigma_g+)$	$O_2(^3\Sigma_u+)$	$O_2(^3\Sigma_u^-)$
$O(^3P)$	1180	590	460	230	170
$O(^1D)$	410	310	260	167	150
$O(^1S)$	234	196	179	129	108

Ozone also absorbs in the visible region via the Chappuis bands. (Fig. 4.34). This spectral regime contributes significantly to the photodissociation of ozone and plays a dominant role in the lower stratosphere and troposphere $(z < 25$ km$)$.

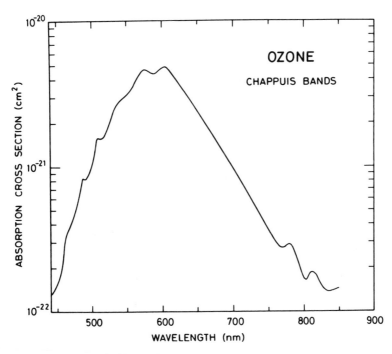

Fig. 4.34. Chappuis bands of ozone. Absorption cross section from 450 to 850 km. From Nicolet (1980).

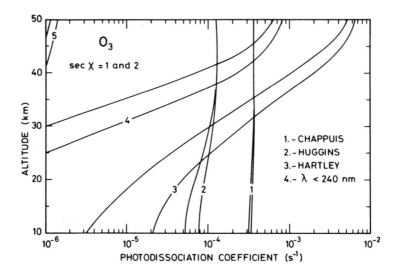

Fig. 4.35. Contribution of each spectral region to the total photodissociation of ozone as a function of altitude. From Nicolet (1980).

Finally, at wavelengths less than 200 nm, the absorption by ozone is related to the existence of large bands superimposed on a continuum. This spectral region plays a very minor role in the photochemistry of the ozonosphere because these wavelengths are absorbed by molecular oxygen at altitudes far above the ozone layer. It should be noted, however, that in the Herzberg continuum both ozone and oxygen contribute to absorption of radiation.

Photodissociation of ozone is energetically possible for wavelengths less than 1.14 μm (near infrared). The photodissociation products, O and O_2, are found in different energy states depending directly on the energy of the incident photon. Table 4.6 indicates the theoretical limits corresponding to different dissociation products. Some transitions violate the spin rules and can thus be neglected compared to the other allowed transitions in the same spectral region. Thus, for wavelengths greater than 310 nm, the following reaction occurs:

$$J_{O_3}; \quad O_3 + h\nu \rightarrow O(^3P) + O_2(^3\Sigma_g^-)$$

and for wavelengths less than 310 nm:

$$J_{O_3}*; \quad O_3 + h\nu \rightarrow O(^1D) + O_2(^1\Delta_g)$$

The formation of $O(^1S)$ is possible for $\lambda < 196$ nm. J_{O_3} corresponds partly to the Huggins bands, which yield a photodissociation coefficient of about

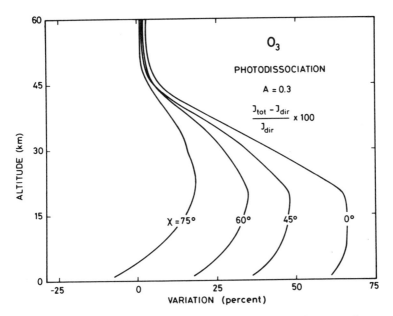

Fig. 4.36. Relative effects of scattering on the photolysis rate of ozone from the ground to 60 km for several solar zenith angles. Surface albedo = 0.3.

$$J_{\infty}(O_3; \text{Huggins bands}) = 1.2 \times 10^{-4} \text{ s}^{-1} \qquad (4.95)$$

for a typical temperature, and the contribution from the Chappuis bands is about

$$J_{\infty}(O_3; \text{Chappuis bands}) = 4.4 \times 10^{-4} \text{ s}^{-1} \qquad (4.96)$$

for zero optical depth.

J_{O_3} can be compared to the value of $J_{O_3^*}$ in the Hartley band where one obtains:

$$J_{\infty}(O_3; \text{Hartley bands}) = 9.5 \times 10^{-3} \text{ s}^{-1} \qquad (4.97)$$

such that the total ozone photolysis coefficient at zero optical depth is slightly less than 10^{-2} s^{-1}. Figure 4.35 shows the relative contribution of each band as a function of altitude for two different solar zenith angles. It should be noted that the value of the ozone photodissociation coefficient depends critically on the absorption by ozone itself, introducing a non linear coupling as a function of altitude.

The calculation of the rate of formation of the $O(^1D)$ atom and the $O_2(^1\Delta_g)$ molecule requires exact knowledge of the quantum efficiency near 310 nm. Recent work by Lin and DeMore (1973), Moortgat and Warneck (1975), Kuis et al. (1975), Kajimoto and Cvetanovic (1976) and Moortgat et al. (1977)

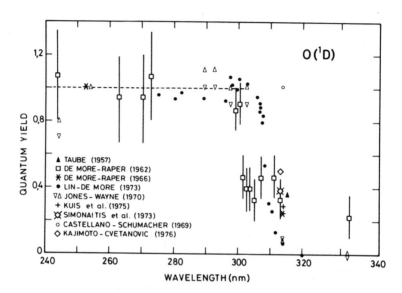

Fig. 4.37. Quantum efficiency of O(^{1}D) production by ozone from 240 to 340 nm.

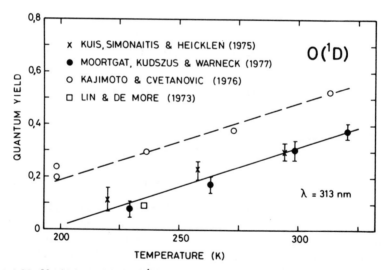

Fig. 4.38. Variation of the O(^{1}D) quantum efficiency as a function of temperature for a wavelength of 313 nm.

has led to a more precise determination of the spectral distribution of the quantum efficiency and its temperature dependence. Figure 4.37 shows some of these experimental determinations and Figure 4.38 shows the variation of the quantum efficiency with temperature for a wavelength of 313 nm. It is assumed that the proportion of $O(^1D)$ atoms produced is 100 percent below 300 nm. However, Fairchild et al. (1978) and Sparks et al. (1979) have indicated a quantum efficiency of 90 percent in this region. Finally, it should be noted that the calculation of J_{O_3} * must also include the effects of molecular scattering and albedo (see Fig 4.36).

• *Water vapor*

The absorption spectrum of water vapor consists of a continuum from 145 to 186 nm, diffuse bands from 69 to 145 nm, and a continuum below 69 nm. Photodissociation in the middle atmosphere occurs at wavelengths above about 100 nm because photoionization occurs for wavelengths less than 98 nm. In the mesosphere and thermosphere, water vapor is photodissociated essentially by the solar Lyman α line, leading to a substantial fraction of the hydrogen atom production at these altitudes. In the lower mesosphere, the continuum absorption of water vapor occurs in the domain of the Schumann Runge bands of molecular oxygen. Figure 4.39 shows the shape of the absorption cross section of water vapor above 120 nm (by Watanabe and Zelikoff, 1953). Later measurements were made by Thompson, Harbeck and Reeves (1963), Laufer and McNesby (1965), and Schurgers and Welge (1968). In particular the value of $\sigma(H_2O)$ at Lyman α has been determined to be about 1.4 $\times 10^{-17}$ cm^2. Figure 4.40 shows the contributions of the Lyman α line and the Schumann Runge bands to the vertical distribution of J_{H_2O}. Analytic formulas can be used to describe the contribution of the Lyman α line (Nicolet, private communication, 1983):

$$J(H_2O;Ly\alpha) = J_\infty(H_2O;Ly\alpha)exp\left(-4.4\times10^{-19}N^{0.917}\right) \qquad (4.98)$$

where the coefficient at zero optical depth $J_\infty(H_2O;Ly\alpha)$ varies from 4.0 to 6.5 $\times 10^{-6}$ s^{-1} according to the level of solar activity. In the Schumann Runge bands, one similarly obtains:

$$J(H_2O;SRB) = 1.2\times10^{-6}exp\left(-1\times10^{-7}N^{0.35}\right) \qquad (4.99)$$

• *Hydrogen peroxide* H_2O_2

Hydrogen peroxide was discovered more than 150 years ago. Its absorption cross section has been measured numerous times.

In the spectral domain which is important in the stratosphere, absorption cross section measurements have been made by Schurgens and Welge (1968) from 120 to 200 nm, Holt et al. (1948) and Holt and Oldenberg (1949) from 185 to 253 nm, and Urey, Dawsey and Rice (1929) from 215 to 380 nm. More

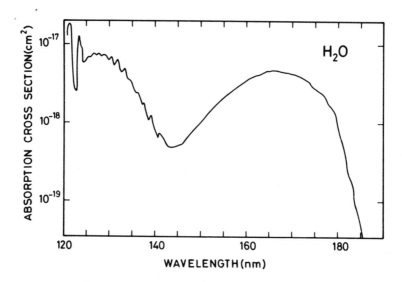

Fig. 4.39. Absorption cross section of water vapor. From Watanabe and Zelik-off (1953). (Copyright by the Optical Society of America).

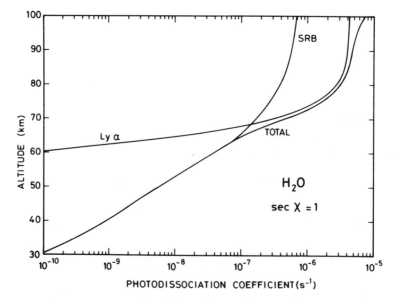

Fig. 4.40. Contributions of different spectral regions to the photodissociation of water vapor for an overhead sun.

recent measurements include those of Molina et al., (1977) and Lin et al. (1978).

• *Methane* CH_4

Methane only photodissociates in the upper part of the middle atmosphere because its absorption cross section becomes very weak at wavelengths longer than 145 nm. The most intense part of the spectrum is located below 130 nm, where the cross section is about $1.9 \times 10^{-17} cm^2$. The Lyman α line makes up almost all of the photolysis cross section with a mean value of about

$$J_\infty(CH_4;Ly\alpha) = 5.5 \times 10^{-6} s^{-1} \qquad (4.100)$$

Like water vapor, the photolysis rate varies by a factor of two over the solar cycle due to the variation in Lyman α flux.

• *Carbon dioxide* CO_2

The spectral distribution of the absorption cross section of CO_2 has been measured by several experimenters. An analysis of these studies (Nicolet, 1980) indicates that the values of Inn, Watanabe, and Zelikoff (1953) from 106 to 175 nm (see Fig. 4.41), those of Ogawa (1971) from 175 to 195 nm and

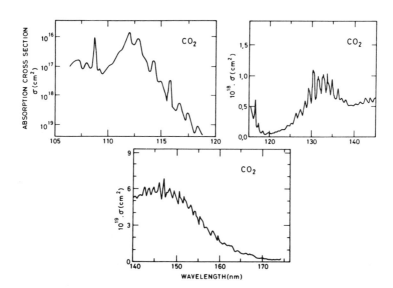

Fig. 4.41. Spectral distribution of the absorption cross section of carbon dioxide. From Inn et al. (1953). (Copyright by the Optical Society of America).

those of Shemansky (1972) above 195 nm can be adopted. The measurements of the last study show that the process of photolysis can be neglected relative to Rayleigh scattering for wavelengths greater than 210 nm. Carbon dioxide thus dissociates only at relatively high altitudes, where solar Lyman α and Schumann Runge continuum radiation is present. The corresponding photolysis rates are (Nicolet, 1980):

$$J_\infty(CO_2;Ly\alpha) = 2.20 \times 10^{-8} s^{-1} \tag{4.101}$$

$$J_\infty(CO_2;SRC) = 9.28 \times 10^{-8} s^{-1} \tag{4.102}$$

and, for T = 300K,

$$J_\infty(CO_2;SRB) = 1.90 \times 10^{-9} s^{-1} \tag{4.103}$$

These values are also subject to solar activity variations, especially the contribution of Lyman α and the Schumann Runge continuum. The complete calculation of the photolysis rate including the absorption by molecular oxygen indicates that CO_2 photolysis is due primarily to Lyman α in the upper mesosphere, and to the Schumann Runge bands in the mesosphere and stratosphere (see Fig. 4.42). Due to the strong temperature dependence of the CO_2 cross section in the Schumann Runge bands (DeMore and Patapoff, 1972), the value of $J(CO_2)$ decreases at lower temperatures (see Nicolet, 1980).

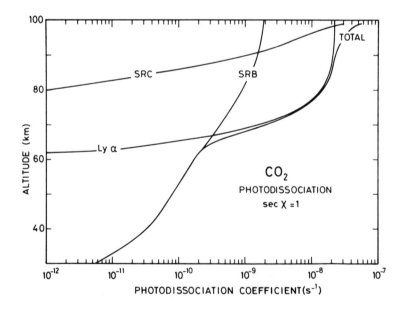

Fig. 4.42. Contribution of different spectral regions to the CO_2 photolysis rate from 30 to 100 km for an overhead sun.

• *Formaldehyde* CH_2O

The absorption cross section of formaldehyde has been studied by several workers since the experiments of Henri and Schou (1928), in particular by Bass et al. (1980). These data are shown in Figure 4.43. It has long been recognized (Herzberg, 1931; Norrish and Kirkbride, 1932) that the photolysis of CH_2O can lead to two different product paths:

$$(J_{H-CHO}); \quad CH_2O + h\nu \rightarrow H + CHO$$

$$(J_{H_2-CO}); \quad CH_2O + h\nu \rightarrow H_2 + CO$$

The relative importance of these processes has been studied by Moortgat and Warneck (1979), Cox (1979), and DeMore et al. (1979). Figure 4.44 shows the quantum efficiency associated with each process.

• *Nitric oxide NO*

Although the absorption spectrum of nitric oxide had first been studied as much as a century ago, only recently has the work of Miescher and colleagues at the University of Basel led to a detailed aeronomic study of its characteristics (Cieslik and Nicolet, 1973; Cieslik, 1977). The ultraviolet absorption spectrum of NO exhibits four band systems (Marmo, 1953) defined by the following transitions (cf. Fig. 4.45):

γ bands $A^2\Sigma \rightarrow X^2\pi$
β bands $B^2\Pi \rightarrow X^2\pi$

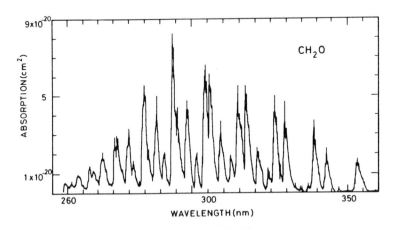

Fig. 4.43. Spectral distribution of the formaldehyde cross section. From Bass et al. (1979).

δ bands $C^2\Pi \rightarrow X^2\pi$

ϵ bands $D^2\pi \rightarrow X^2\pi$

Calculation of the photolysis rate of nitric oxide must include the predissociation process which occurs at wavelengths less than 191 nm. In particular, predissociation occurs (Fig. 4.45) in the δ bands, the β bands (v'> 6) and the ϵ bands (v'> 3). Callear and Pilling (1970 a,b) indicate that emission dominates predissociation in the ϵ bands. At zero optical depth, the contribution of the γ bands is small compared to those of the δ and β bands.

In the mesosphere and stratosphere, the β bands of quantum number v" greater than 9, are almost totally attenuated and the other β bands contribute less than 10 percent to the photolysis rate. Thus, only the two bands δ(0–0) at $\lambda \approx$ 190.9nm and δ(1–0) at $\lambda \approx$ 182.7nm need be considered in calculating the photolysis rate in the middle atmosphere. J_{NO} has been studied by Cieslik and Nicolet (1973), Park (1974), Frederick and Hudson (1979), and Nicolet and Cieslik(1980). It should be noted that the derived values are sensitive to the

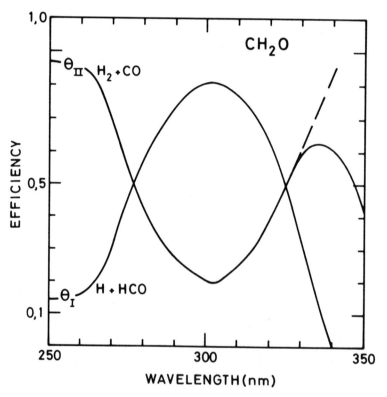

Fig. 4.44. Efficiency of each photolysis process of formaldehyde. From Nicolet (1980).

NO

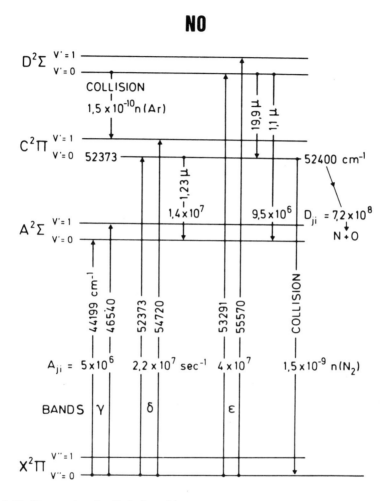

Fig. 4.45. Energy levels of nitric oxide.

different values assumed for the oscillator strength, (especially that of the $\gamma(0-0)$ band, and the solar flux. Nicolet and Cieslik (1980) indicate that for zero optical depth

$$J(NO;\delta|0-0|) = 2 \times 10^{-6} \text{ s}^{-1} \tag{4.104}$$

and

$$J(NO;\delta|1-0|) = 2.5 \times 10^{-6} \text{ s}^{-1} \tag{4.105}$$

The calculation of the photolysis rate at all atmospheric levels requires detailed knowledge of the solar spectrum. The primary attenuation of the solar flux in the $\delta(0-0)$ band is due to the (5-0) Schumann Runge band of O_2;

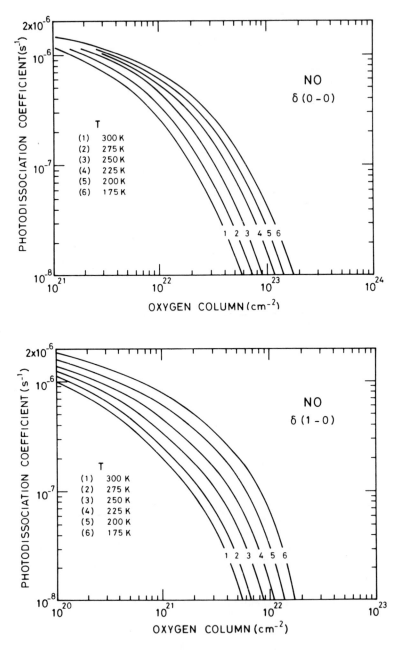

Fig. 4.46a and b. Photodissociation rate of NO as a function of the total molecular oxygen content for different temperatures. a) Contribution of the $\delta(0-0)$ band and b) contribution of the $\delta(1-0)$ band. From Nicolet and Cieslik (1980). (Copyright by Pergamon Press).

the attenuation in the $\delta(1-0)$ band is due to the $(9-0)$ and $(10-0)$ bands of O_2. Thus, the determination of J_{NO} requires consideration of the ensemble of rotational lines of each NO and O_2 band system. Further, the effect of temperature is important in the intensities of both the NO and O_2 bands. The vertical temperature structure must therefore be considered for aeronomic studies. Figure 4.46a and b show the variation of $J(NO)$ with the integrated amount of molecular oxygen (Nicolet and Cieslik, 1980). The attenuation of the $\delta(1-0)$ band is more rapid than that of the $\delta(0-0)$ band; thus the photolysis rate of NO in the stratosphere and lower mesosphere depends critically on the behavior of the $\delta(0-0)$ band and its attenuation by the $(5-0)$ band of O_2. Ozone absorption in the stratosphere must be accounted for via a correction factor multiplied into equations (4.104) and (4.105). For the $\delta(0-0)$ band, this factor is given by $\exp(-N(O_3) \times 5 \times 10^{-19})$ and $\exp(-N(O_3) \times 7.3 \times 10^{-19})$ for the $\delta(1-0)$ band (where $N(O_3)$ is the total column of O_3 along the ray path). Rough values which neglect the importance of temperature effects can be estimated from the formula given by Nicolet (1979):

$$J_{NO} = 4.5 \times 10^{-6} x \exp\left(-10^{-8}(N(O_2))^{0.38}\right) \exp\left(-5 \times 10^{-19} N(O_3)\right) s^{-1} \quad (4.106)$$

Frederick et al. (1983) have shown, however, that the inclusion of the opacity provided by nitric oxide itself reduces the calculated dissociation in the vicinity of the tropopause to 50-70% of the value obtained when this contribution is omitted. Since thermospheric NO is known to increase with increasing solar activity, a modulation of the solar penetration with the solar

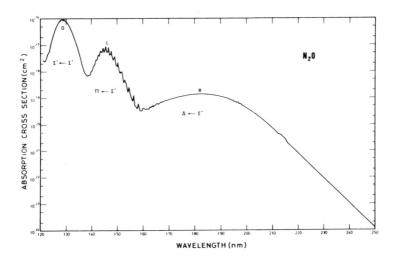

Fig. 4.47. Spectral distribution of the cross section of nitrous oxide. From Nicolet (1980).

cycle should occur at the wavelengths which photodissociate NO. The direction of this change is, however, opposite to that of the solar irradiance, which increases with solar activity. However, this mechanism provides a possible coupling between the thermosphere and lower levels in the middle atmosphere.

- *Nitrous oxide* N_2O

The absorption spectrum of N_2O has been measured by several workers since the detection of this molecule by Leifson in 1926. In contrast to previous studies (Bates and Hays, 1967), the absorption coefficient above 260 nm has been shown to be quite small (Johnston and Selwyn (1975), which implies that N_2O must be quite inert in the troposphere. The absorption spectrum shown in Figure 4.47 (Nicolet, 1980) shows the existence of three distinct spectral regions, as first indicated by Duncan in 1936.

As shown by Johnston and Selwyn (1975), the cross section of N_2O varies strongly with temperature. The atmospheric photolysis rate comes predominantly from the absorption of solar radiation in the Herzberg continuum and Schumann Runge bands. $J_\infty(N_2O) \approx 9 \times 10^{-7}$ s^{-1} for $\lambda > 175$ nm.

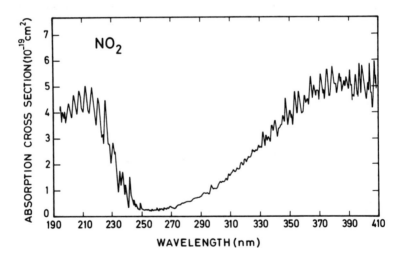

Fig. 4.48. Spectral distribution of the absorption cross section of nitrogen dioxide. From Bass et al. (1976).

• *Nitrogen dioxide* NO_2

The absorption spectrum of NO_2 has been studied in various spectral domains since its detection by Bell in 1885. It has been recently measured by Nakayama et al. (1959), Hall and Blacett (1952), Bass and Laufer (1973), Johnston and Graham (1973, 1974), Bass et al. (1976) and Harker et al. (1977). Figure 4.48 shows the shape of the absorption cross section of NO_2 from 190 to 410 nm. The photolysis quantum yield decreases with wavelength near the photolysis limit. The quantum efficiency has been measured both by Harker et al., (1977) and by Davenport (1978), and these results are about 15% lower than those of Jones and Bayes (1973). The atmospheric photolysis rate varies slightly with temperature. Thus, when the temperature changes from 300 to 235K, the photodissociation rate decreases by about 10 percent. It should be noted that this spectral region is quite sensitive to the effects of molecular scattering and albedo (Fig. 4.49a,b) so that the precise value of J_{NO_2} depends strongly on geographic conditions and on the troposphere.

• *Nitrogen trioxide* NO_3

The absorption spectrum of NO_3 exhibits several predissociating bands first observed in 1962 by Ramsay. More recent studies by Graham (1975) and Graham and Johnston (1978) provided quantitative values of the photodissociation parameters; these have been updated by the studies of Magnotta and Johnston (1980). The quantum efficiency for each of the two possible pathways

$$\text{(a)} \quad NO_3 + h\nu(\lambda < 580nm) \rightarrow NO_2 + O_2 \qquad (4.107a)$$

and

$$\text{(b)} \quad NO_3 + h\nu(\lambda > 580nm) \rightarrow NO + O_2 \qquad (4.107b)$$

is still not completely established. The work of Magnotta and Johnston (1980) indicates that the first process is dominant and that it may even occur at wavelengths longer than the limit of 580 nm at elevated temperatures. At absolute zero, J_a is only energetically possible at wavelengths less than this limit. J_b is believed to exhibit a quantum yield of no more than 25 percent from 520 to 640 nm and is very small at longer wavelengths. One can use the values determined by Magnotta and Johnston (1980) for most atmospheric applications:

$$J_a = 0.18 \pm 0.06 \text{ s}^{-1} \qquad (4.108a)$$

$$J_b = 0.022 \pm .007 \text{ s}^{-1} \qquad (4.108b)$$

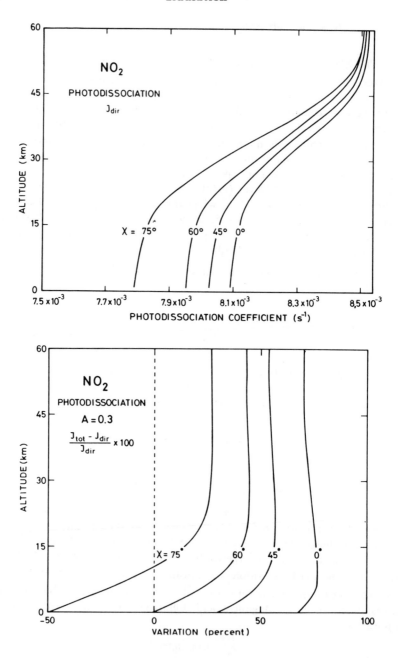

Fig. 4.49a. Vertical distribution of the NO_2 photolysis rate for several values of the solar zenith angle. b. Relative effect of scattering on the photolysis rate of NO_2. Earth albedo = 0.3.

• *Dinitrogen pentoxide* N_2O_5

The absorption cross sections of N_2O_5 were first measured by Jones and Wulf (1937). Since then, other measurements have been performed by Johnston and Graham (1974), Graham (1975), and Yao et al. (1982). The latter study examined the variation of the cross sections with temperature, and found significant temperature dependence at wavelengths longer than about 280 nm. Figure 4.50 shows the spectral distribution of the cross sections near 275K. At wavelengths longer than 280 nm, the following parameterization should be applied (Yao et al., 1982):

$$\ln \sigma = 0.432537 + (4.72848 - 0.0171263\lambda)(\frac{1000}{T})$$ (4.109)

where λ is expressed in nm and σ in $10^{-19} cm^2$.

• *Nitrous acid* HNO_2

The absorption spectrum of HNO_2 was measured by Graham (1975) and by Cox and Derwent (1976). The latter study recommends values about five time larger than the first. Recently, Stockwell and Calvert (1978) also derived absorption cross sections from 300 to 400 nm. More detailed experimental

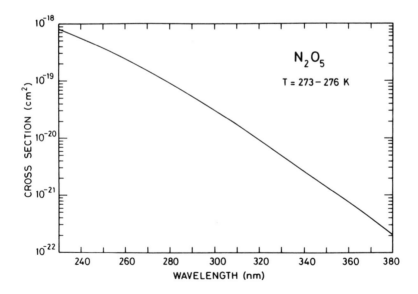

Fig. 4.50. Spectral distribution of the cross section of N_2O_5 at 273-276K (from Yao et al., 1982). (Copyright by the American Chemical Society).

work is needed, but it is clear that the photolysis of HNO_2 is fast, even at low altitudes.

• *Nitric acid* HNO_3

The absorption by HNO_3 is a continuum from 220 to 300 nm, and was observed as early as 1943 by Dalmon . Johnston and Graham (1973) and Biaume (1973) measured the absorption cross section from 180 to 330 nm. Figure 4.51 presents the data obtained by these studies, which are in good agreement with one another, at least to within the scatter of the experimental data.

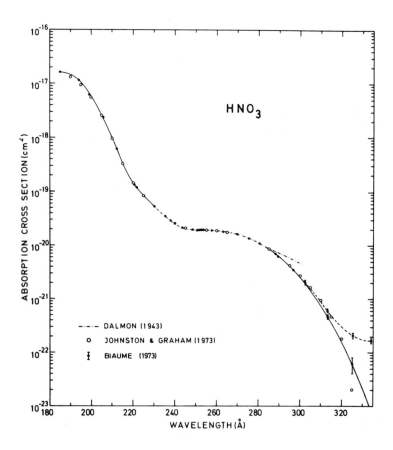

Fig. 4.51. Spectral distribution of the absorption cross section of HNO_3 from 180 to 320 nm.

• *Peroxynitric acid* HO_2NO_2 (or HNO_4)

The absorption spectrum of HO_2NO_2 was observed by Graham et al. (1978) and by Cox and Patrick (1979). These two studies are in good agreements from 205 to 260 nm, but at 195 nm the results of Graham are twice as large as those of Cox and Patrick. Recently Molina and Molina (1980) presented a spectral distribution of the absorption cross section which exhibits much smaller values than the previous studies at all wavelengths (see Fig. 4.52). The smaller cross sections affect the distribution of OH in the lower stratosphere in the direction indicated by several studies (see, Sze and Ko, 1980, and Turco et al., 1981).

• *The halocarbons*

Since the important effects of chlorocarbon compounds on atmospheric ozone were indicated by Stolarski and Cicerone (1975) and Rowland and Molina (1975), the absorption spectra of the various chlorocarbons which provide the source of chlorine to the atmosphere have been measured by several groups, including Rowland and Molina (1975), Chou et al. (1977,1978), Robbins

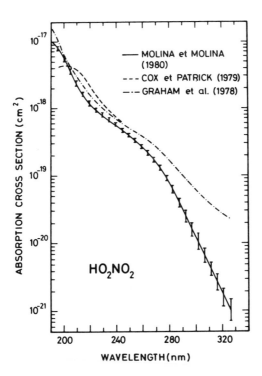

Fig. 4.52. Absorption cross section of peroxynitric acid.

(1976a and b), and Vanlaethem-Meuree, Wisemberg and Simon (1978a,b). The curves presented below (Fig. 4.53) are taken from the last study, in which the effect of temperature variations was also presented. Figure 4.54 presents the vertical distribution of the photolysis rate of several chlorocarbons.

• *Chlorine monoxide ClO*

The absorption spectrum of chlorine monoxide was studied by Johnston et al. (1969), as well as by Coxon (1976, 1977), Coxon and Ramsay (1976), Watson (1977), and Jourdain et al. (1978). The photodissociation coefficient at zero optical depth is about $7 \times 10^{-3} s^{-1}$. Theoretical studies of the ClO spectrum include those of Mandelman and Nicholls (1977), Langhoff et al., (1977) and Arnold et al. (1977). A very detailed analysis was presented by Langhoff, Jaffe and Arnold (1977). The dissociation of ClO accounts for less than 5 percent of the chemical destruction of this species, which is rapidly destroyed by reaction with atomic oxygen and nitric oxide.

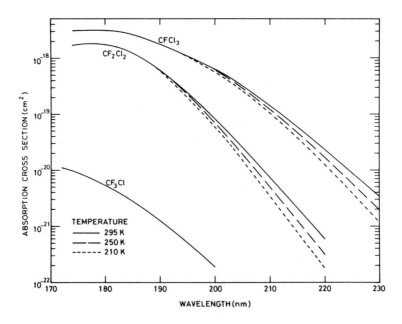

Fig. 4.53   Spectral distribution of the absorption cross sections of several halocarbons and their temperature dependence. From Vanlaethem-Meuree et al. (1978).

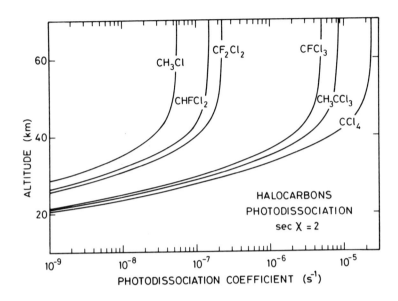

Fig. 4.54. Vertical distribution of the photolysis rate of several chlorocarbons.

Other chlorine oxides are likely to be present in the middle atmosphere: OClO, ClOO, ClO_3, Cl_2O ... The aeronomic importance of these species seems to be negligible at present. For an analysis of their cross sections, see for example, the work of Nicolet (1980) and Watson (1977).

• *Chlorine nitrate and bromine nitrate* $ClONO_2$ *and* $BrONO_2$

The absorption cross section of $ClONO_2$ was measured from 186 to 400 nm by Rowland, Spencer and Molina (1976) (Fig. 4.55). The photodissociation rate at zero optical depth is about 1×10^{-3} s^{-1}. A more recent study by Molina and Molina (1979) indicates somewhat smaller values as well as a slight temperature dependence. Due to its large photolysis rate, $ClONO_2$ is present in very small abundance in the upper stratosphere, but the photolysis rate is much slower in the lower stratosphere due to absorption by ozone. The photolysis products are not at present well established. Smith, Chou and Rowland (1977) suggest that the photolysis at 302.5 nm leads to the formation of oxygen atoms:

$$ClONO_2 + h\nu \rightarrow ClONO + O(^3P)$$

while Chang et al. (1979) find the formation of atomic chlorine most prevalent at low pressures:

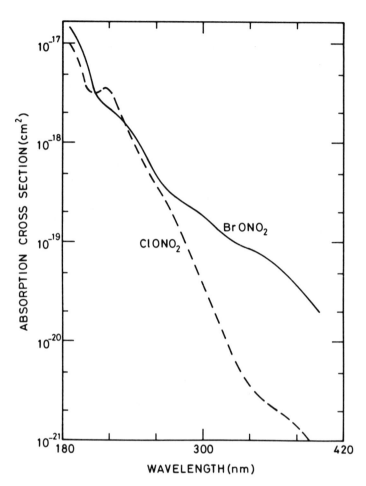

Fig. 4.55. Absorption cross section of chlorine and bromine nitrate From
Spencer and Rowland (1978). (Copyright by the American Chemical Society).

$$ClONO_2 + h\nu \rightarrow Cl + NO_3$$

Bromine nitrate possesses a similar spectrum to that of chlorine nitrate. It is,
however, shifted toward slightly longer wavelengths such that the photolysis
rate of $BrONO_2$ is faster than that of $ClONO_2$. Like chlorine nitrate, the
absorption cross section is likely to be temperature dependent. The photolysis
products have not yet been determined.

• *Hydrochloric, hydrobromic and hydrofluoric acid*
  HCl, HBr, HF

The aeronomic importance of HCl is due mostly to the fact that it constitutes a reservoir for chlorine atoms, leading to the removal of chlorine from
the stratosphere to the troposphere through washout/rainout processes. The
importance of this reservoir depends on how much chlorine it sequesters in a
form which is inert towards ozone; clearly this depends on its lifetime and
thus on the photolysis rate. The absorption cross sections (Fig. 4.56) were
measured by Vodar (1948), Romand and Vodar (1948), and Romand (1949),
and more recently by Inn (1975). These last measurements extend from 140 to
220 nm, leading to the determination of the photodissociation frequency of
about 3×10^{-6} s^{-1} at zero optical depth. To calculate J_{HCl} in the atmosphere,
the Schumann Runge bands and continuum must be considered, as well as the

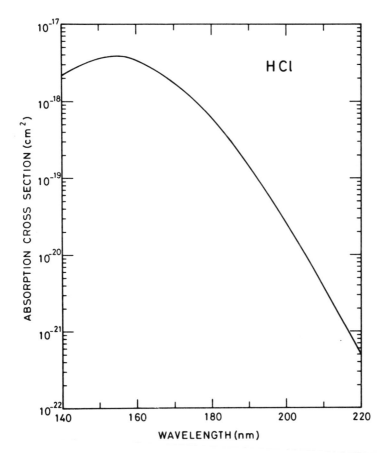

Fig. 4.56. Spectral distribution of the absorption cross section of hydrochloric
acid. From Inn (1975). (Copyright by the American Meteorological Society).

Herzberg continuum.

Measurements of the spectrum of hydrobromic acid have been presented by Goodeve and Taylor (1935), and Romand (1949). J_{HBr} at zero optical depth is about $5x10^{-6} s^{-1}$, but it is comparable to the rate of reaction with the OH radical in the stratosphere. The absorption spectrum of hydrofluoric acid indicates that its photolysis can be completely neglected in the middle atmosphere.

- ### *Hypochlorous acid HOCl*

The spectral distribution of the absorption cross section of HOCl was measured by Molina and Molina (1978) and by Knauth et al. (1979). The two studies are in good agreement. Figure 4.57 shows the values observed by Molina and Molina. The photolysis rate for zero optical depth is about $1x10^{-3}s^{-1}$.

4.6.3. Numerical calculation of photodissociation coefficients

The photodissociation frequencies for several important trace species are shown in Figure 4.58 for specific conditions (winter at 40° latitude, solar maximum). The determination of these photolysis rates J at altitude z for solar

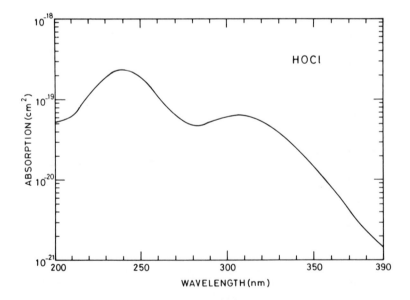

Fig. 4.57. Spectral distribution of the absorption cross section of HOCl. From Molina and Molina (1978). (Copyright by the American Chemical Society).

zenith angle χ is accomplished by the numerical methods described below.

a. When the absorption spectrum is a continuum, thus varying slowly with wavelength, the spectrum is divided into a certain number of discrete intervals and a mean value of σ_i and ϵ_i is assigned to each. Equation (4.93) can then be written as a sum of the intervals i of wavelength $\delta\lambda_i$:

$$J(z,\chi) = \sum_i \epsilon_i \sigma_i Q_i(z,\chi) \tag{4.110}$$

where

$$Q_i = \int_{\delta\lambda_i} q_\lambda d\lambda \tag{4.111}$$

is the solar flux at χ,z.

b). The absorption spectra of some molecules exhibit a structure of discrete narrow bands. Examples include NO and O_2 in the Schumann Runge bands (175-200 nm). In these cases, it is not possible to determine a mean value σ_i of the cross section for use in a broad spectral interval. The generalized expression for atmospheric absorption (4.93) must then be considered, and an integration performed over spectral intervals smaller than the width of the bands. Such a calculation is often quite expensive, and approximate methods are often used in aeronomic models in order to reduce computation time. Two components must be considered: on one hand, the rate of photolysis of O_2 must be computed, and on the other, the transmission of solar radiation including O_2 absorption must be considered in order to calculate the photodissociation coefficient in the Schumann Runge bands for constituents such as H_2O, CO_2, N_2O, HNO_3, etc. NO must be considered separately because of the very narrow width of its bands (Doppler widths of about 0.1 cm^{-1}). The Schumann Runge region can be analyzed by dividing it into 20 intervals beginning at each O_2 rotational band. The first band, (19-0) is found at 175.345 nm; while the last occurs at 204.08 nm. Other intervals can be adopted, notably the classical division in 500 cm^{-1} bands.

One method which leads to a simple treatment of the penetration of radiation in the Schumann Runge bands and the photolysis rate of O_2 involves establishing an equivalent cross section of O_2 in the interval $\delta\lambda_i$ and then using the classical methods for continuum absorption. Thus for each interval $\delta\lambda_i$ an equivalent cross section $\bar{\sigma}_i(O_2)$ is defined for use in solar penetration calculations. It can be defined by the expression:

$$e^{-\bar{\sigma}_i(O_2)N(O_2)} d\lambda \tag{4.112}$$

i. e.,

$$\bar{\sigma}_i(O_2) = \frac{-1}{N(O_2)} \ln \frac{1}{\delta\lambda_i} \int_{\delta\lambda_i} e^{-\sigma(O_2)N(O_2)} d\lambda \tag{4.113}$$

where

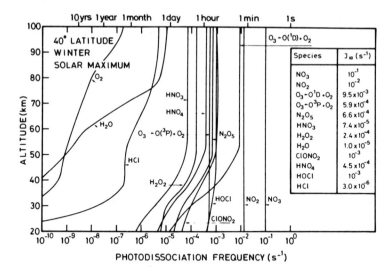

Fig. 4.58. Photodissociation rates of several important atmospheric species for winter, solar maximum, 40 ° latitude. From the model of Garcia and Solomon (1983).

$$N(O_2) = \sec\chi \int_z^\infty n(O_2)dz \qquad\qquad (4.114)$$

is the integrated O_2 column from z to the top of the atmosphere. Since the cross section in the band system varies with the temperature T, the value of $\bar{\sigma}_i(O_2)$ depends on both T and the integrated column of O_2 (that is, the altitude and solar zenith angle).

To evaluate the photodissociation of molecular oxygen, it is useful to define another equivalent cross section, $\bar{\bar{\sigma}}_i$ defined by

$$\bar{\bar{\sigma}}_i(O_2)e^{-\bar{\sigma}_i(O_2)N(O_2)}\delta\lambda = \int_{\delta\lambda_i}\sigma(O_2)e^{-\sigma(O_2)N(O_2)}d\lambda \qquad (4.115)$$

that is,

$$\bar{\bar{\sigma}}(O_2) = \frac{\displaystyle\int_{\delta\lambda_i}\sigma(O_2)e^{-\sigma(O_2)N(O_2)}d\lambda}{\displaystyle\int_{\delta\lambda_i}e^{-\sigma(O_2)N(O_2)}d\lambda} \qquad (4.116)$$

Here again, $\bar{\bar{\sigma}}_i(O_2)$ depends on both temperature and the total oxygen amount, hence the solar zenith angle.

It is often useful to introduce reduction factors defined in each spectral interval $\delta\lambda_i$ by the relations:

$$R_i = \frac{1}{\delta\lambda_i} \int_{\delta\lambda_i} e^{-\sigma(O_2)N(O_2)} d\lambda \qquad (4.117)$$

$$S_i = \frac{1}{\delta\lambda_i} \int_{\delta\lambda_i} \sigma(O_2) e^{-\sigma(O_2)N(O_2)} d\lambda \qquad (4.118)$$

These factors are related to the equivalent cross sections by the expressions

$$R_i = e^{-\overline{\sigma}_i(O_2)N(O_2)} \qquad (4.119)$$

$$S_i = \overline{\overline{\sigma}}_i R_i \qquad (4.120)$$

The solar flux Q_i and the photodissociation coefficient J_i of a minor constituent or of molecular oxygen can then be calculated:

$$Q_i(z,\chi) = Q_i(\infty)R_i(z,\chi)e^{(-\tau_i(O_3;z,\chi))} \qquad (4.121)$$

$$J_i(M;z,\chi) = Q_i(\infty)\sigma_i(M)R_i(z,\chi)e^{-\tau_i(O_3;z,\chi)} \qquad (4.122)$$

$$J_i(O_2;z,\chi) = Q_i(\infty)S_i(z,\chi)e^{-\tau_i(O_3;z,\chi)} \qquad (4.123)$$

where the factor $e^{-\tau_i(O_3;z,\chi)}$ represents the effect of ozone absorption.

To simplify the calculations it is useful to adopt analytic forms for $\overline{\sigma}_i$ and $\overline{\overline{\sigma}}_i$ which are the same as R_i and S_i as a function of the integrated amount of molecular oxygen and the temperature T. Such expressions were formulated by Hudson and Mahle (1972), Shimazaki and Ogawa (1974), Fang et al. (1974), Kockarts (1976), Nicolet and Peetermans (1980) and Allen and Frederick (1982).

For example, Nicolet and Peetermans (1980) have given the following expressions:

$$R_i = \exp\left(-\exp(d_o^i + d_1^i \ln N)\right) \qquad (4.124)$$

and

$$S_i = \overline{\overline{\sigma}}_i R_i \qquad (4.125)$$

where

$$\overline{\overline{\sigma}}_i = \sigma_{min}^i \left(\frac{\sigma_{max}}{\sigma_{min}^i}\right)^{\frac{1}{1 + \exp(P_i)}}$$

with $P_i = a_o^i + a_1^i \ln N$

The parameters d_0, d_1, a_0, a_1, σ^i_{min}, and σ^i_{max} are given by the authors in each band i and for different temperatures.

Allen and Frederick (1982) have suggested the following parameterization for an effective cross section of O_2 in each band interval and of NO in the $\delta(0-0)$ and $\delta(1-0)$ bands:

$$\log_{10}\sigma_e(z,\chi = 0) = \sum_j [a_j \log_{10}p(z)]^{j-1} \tag{4.126}$$

where p is the atmospheric pressure at altitude z. The dependence of σ_e with the solar zenith angle, χ can be simply described by

$$\sigma_e(z,\chi) = \sigma_e(z,o)(\sec\chi)^{c(z)} \tag{4.127}$$

where

$$c(z) = \sum_j b_j [\log_{10}N(z)]^{j-1} \tag{4.128}$$

The coefficients a_j and b_j have been determined by least square fits and are provided by the authors for each band.

References

Ackerman, M., Ultraviolet solar radiation released to mesospheric processes, pp. 149-159, in: Fiocco, G. (Ed.), *Mesospheric Models and Related Experiments,* D. Reidel (Dordrecht, Holland), 1971.

Ackerman, M., F. Biaume, and M. Nicolet, Absorption in the spectral range of the Schumann-Runge Bands, Canad. J. Chem., 47, 1834, 1969.

Allen, M. and J.E. Frederick, Effective photodissociation cross sections for molecular oxygen and nitric oxide in the Schumann-Runge Bands, J. Atmos. Sci., 39, 2066, 1982.

Apruzese, J.P., D.F. Strobel and M.R. Schoeberl, Parameterization of IR cooling in the middle atmosphere dynamics model, 2, Non LTE radiative transfer and the globally averaged temperature of the mesosphere and lower thermosphere, in press, J. Geophys. Res., 1984.

Arnold, J.O., E.E. Whiting and S.R. Langhoff, MCSF + CI wavefunctions and properties of the $X^2\Pi$ and $A^2\Pi$ states of ClO, J. Chem. Phys., 66, 4459, 1977.

Augustsson, T. and V. Ramanathan, A radiative-convective model study of the CO_2 climate problem, J. Atmos. Sci., 34, 448, 1977.

Banks, P., and G. Kockarts, *Aeronomy,* Academic Press, (New York), 1973.

Bass, A.M. and A.H. Laufer, Extinction coefficients of ozomethane and dimethyl mercury in the near ultra-violet, J. Photochem., 2, 465, 1974.

Bass, A.M., A. E. Ledford and A.H. Laufer, Extinction coefficients of NO_2 and N_2O_4, J. Res. NBS, 80A, 143, 1976.

Bass, A.M., L.C. Glasgow, C. Miller, J.P. Jesson and D.L. Filkin, Temperature dependent cross-sections for formaldehyde $[CH_2O]$: The effect of formaldehyde on stratospheric chlorine chemistry, Planet. Sp. Sci., 28, 675, 1980.

Bass, A.M. and R.J. Paur, UV absorption cross sections for ozone: The temperature dependence, J. Photochem., 17, 141, 1981.

Bates, D.R. and P.B. Hays, Atmospheric nitrous oxide, Planet. Sp. Sci., 15, 189, 1967.

Baum, W.A., F.S. Johnson, J.J. Obserly, C.C. Rockwood, C.V. Strain, and R. Tousey, Solar ultraviolet spectrum to 88 kilometers, Phys. Rev., 70, 781, 1946.

Bell, L., Notes on the absorption spectrum of nitrogen peroxide, Amer. Chem. J., 7, 32, 1885.

Bethke, G.W., Oscillator strengths in the far ultraviolet, I. Nitric oxide, J. Chem. Phys., 31, 662, 1959.

Biaume, F., Determination de la Valeur Absolute de l'absorption dans les bandes du systeme de Schumann-Runge de l'oxygene moleculaire, Aeronomica Acta, Brussels, A NO. 100, 1972a.

Biaume, F., Structure de rotation des bandes 0-0 a 13-0 du systeme de Schumann-Runge de la molecule d'oxygene, Acad. Roy. Belg., Mem. Cl. Sci. Coll In -8 Deg., 2E Ser., Tome 40, Fascicule 2, 66 pp., 1972b.

Biaume, F., Nitric acid vapour absorption cross-section spectrum and its photodissociation in the stratosphere, J. Photochem., 2, 139, 1973.

Blake, A.J., J.H. Carver and G.N. Haddad, Photoabsorption cross-sections of molecular oxygen between 1250 A and 2350 A, J. Quant. Spect. Rad. Transf., 6, 451, 1966.

Bossy, L. and M. Nicolet, On the variability of Lyman alpha with solar activity, Planet. Sp. Sci., 29, 907, 1981.

Bossy, L., Solar indices and solar UV irradiances, Planet. Space Sci., 31, 977, 1983.

Brasseur, G. and P.C. Simon, Stratospheric chemical and thermal response to long-term variability in solar UV irradiance, J. Geophys. Res., 86, 7343, 1981.

Brasseur, G., A. De Rudder and P.C. Simon, Implication for stratospheric composition of a reduced absorption cross section in the Herzberg continuum of molecular oxygen, Geopys. Res. Lett., 10, 20, 1983.

Brewer, A.W. and A.W. Wilson, Measurements of solar ultraviolet radiation in the stratosphere, Quart. J. Roy. Meteorol. Soc., 91, 452, 1965.

Broadfoot, A.L., The solar spectrum 2100-3200 A, Astrophys. J., 173, 681, 1972.

Brueckner, G.E., Solar radiometry: Spectral irradiance measurements, Adv. Space Res., 2, 177, 1983.

Brusa, R.W. and C. Frohlich, Recent solar constant determinations from high altitude balloons, Paper presented at the Symposium on the Solar Constant and the Spectral Distribution of Solar Irradiance, IAMAP third Scientific Assembly, Published by the Radiation Commission, Boulder, CO, USA, 1982.

Callear, A.B. and M.J. Pilling, Fluorescence of nitric oxide, 6. Predissociation and cascade quenching in NO $D^2\Sigma^+$ (v = 0) and NO $C^2\Pi$ (v = 0), and the oscillator strengths of the σ (0,0) and (0,0) bands, Trans. Faraday Soc., 66, 1886, 1970b.

Carver, J.H., H.P. Gies, T.I. Hobbs, B.R. Lewis and D.G. McCoy, Temperature dependence of the molecular oxygen photoabsorption cross section near the H Lyman alpha line, J. Geophys. Res., 82, 1955, 1977.

Castellano, E. and H.J. Schumacher, Die kinetik und der mechanismus des photochemischen ozonzerfalles im licht der wellenlange 313 nm, Z. Physik. Chem. Neue Folge, 65, 62, 1969.

Cess, R.D., Radiative transfer due to atmospheric water vapor: Global considerations of the earth's energy balance, J. Quant. Spectrosc. Radiat. Transfer, 14, 861, 1974.

Cess, R.D., Climate change: An appraisal of atmospheric feedback mechanisms employing zonal climatology, J. Atm. Sci., 33, 1831, 1976.

Cess, R.D. and V. Ramanathan, Radiative transfer in the atmosphere of Mars and that of Venus above the cloud deck, J. Quant. Spectrose Radiat. Transfer, 12, 933-945, 1972.

Chandrasekhar, S., Radiative Transfer, Oxford University Press, Oxford, 1950 (Reprinted by Dover Publ, (New York), 1960).

Chang, J.S., J.R. Barker, J.E. Davenprot and D.M. Golden, Chlorine nitrate photolysis by a new technique: Very low pressure photolysis, Chem. Phys. Letters, 60, 385, 1979.

Chapman, S., The absorption and dissociative or ionizing effect of monochromatic radiations in an atmosphere on a rotating earth, Proc. Phys. Soc., 43, 483, 1931.

Chou, C.C., W.S. Smith, H. Vera Ruiz, K. Moe, G. Crescentini, J.J. Molinar and F.S. Rowland, The temperature dependence of the ultraviolet absorption cross sections of CCl_2F_2 and CCl_3F, and their stratospheric significance, J. Phys. Chem., 81, 1977.

Chou, C.C., R.J. Milstein, W.S. Smith, H. Vera Ruiz, M.J. Molinar and F.S. Rowland, Stratospheric photodissociation of several saturated perhalo chlorofluorocarbon compounds in current technological use (Fluorocarbons -13, -113, -114 and -115), J. Phys. Chem., 82, 1, 1978.

Cieslik, S. and M. Nicolet, The aeronomic dissociation of nitric oxide, Planet. Space Sci., 21, 925, 1973.

Coulson, K.L., *Solar and Terrestrial Radiation,* Academic Press, (New York), 1975.

Cox, R.A. and R.G. Derwent, The utlraviolet absorption spectrum of gaseous nitrous acid, J. Photochem., 6, 23, 1976.

Cox, R.A. and K. Patrick, Kinetics of the reaction $HO_2 + NO_2$ (+M) → HO_2NO_2 using molecular modulation spectrometry, Int. J. Chem. Kinetics, 11, 635, 1979.

Coxon, J.A., Vibrational numbering in the $A^2\Pi$ state of ClO, J. Photochem., 5, 337, 1976.

Coxon, J.A. and D.A. Ramsay, The $A^2\Pi-X^2\Pi$ band system of ClO reinvestigation of the absorption spectrum, Canad. J. Phys., 54, 1034, 1976.

Coxon, J.A., RKR Franck-London factors and absorption cross-sections for rotational transitions in the $A^2\Pi$ - $X^2\Pi$ system of ClO, J. Photochem., 6, 439, 1977.

Crutzen, P., Comment on paper "Absorption and emission by carbon dioxide in the mesosphere", by J.T. Houghton, Quart. J. Roy. Met. Soc., 96, 767, 1970.

Curtis, A.R., Discussion of a statistical model for water vapour absorption, Quart. J. Roy. Met. Soc., 78, 638, 1952.

Curtis, A.R. and R.M. Goody, Thermal radiation in the upper atmosphere, Proc. Roy. Soc., A236, 193, 1956.

Dalmon, R., Recherches sur l'acide nitrique et ses solutions par les spectres d'absorption dans l'ultraviolet, Memoires des services de L'etat, Paris, 30, 141, 1943.

Daumont, D., J. Brion and J. Malicet, Measurement of total atmospheric ozone: Consequences entailed by new values of O_3 absorption cross sections at 223 K in the 310-350 nm spectral range, Planet. Space Sci., 31, 1229, 1983.

Davenport, J.E., Determination of NO_2 photolysis parameters for stratospheric modeling, Report No. FAA-EQ-78-14, 1978.

De More, W.B. and M. Patapoff, Temperature and pressure dependence of CO_2 extinction coefficients, J. Geophys. Res., 77, 6291, 1972.

De More, W.B. and O.F. Raper, Reaction of $O(^1D)$ with nitrogen, J. Chem. Phys., 37, 2048, 1962.

De More, W.B. and O.F. Raper, Primary processes in ozone photolysis, J. Chem. Phys., 44, 1780, 1966.

Dickinson, R.E., Method of parameterization for infrared cooling between the altitudes of 30 and 70 km, J. Geophys. Res., 78, 4451, 1973.

Dobson, G.M.B., *Observers Handbook for the Ozone Spectrophotometer*, Ann. IGU, V, 46, Pergamon Press, New York, 1957.

Donner, L. and V. Ramanathan, Methane and nitrous oxide: Their effects on the terrestrial climate, J. Atmos. Sci., 37, 119, 1980.

Dopplick, T.G., Radiative heating of the global atmosphere, J. Atmos. Sci., 29, 1278, 1972.

Ditchburn, R.W. and P.A. Young, the absorption of molecular oxygen between 1850 and 2500 A, J. Atm. Terr. Phys., 24, 127, 1962.

Duncan, A.B.F., The far ultraviolet absorption spectrum of N_2O, J. Chem. Phys., 4, 638, 1936.

Edwards, D.K., Absorption of radiation by carbon monoxide gas according to the exponential wide-band model, Appl. Optics, 4, 1351, 1965.

Ellingson, R.G. and J.C. Gille, An infrared radiative transfer model, I. model description and comparison of observations with calculations, J. Atmos. Sci., 35, 523, 1978.

Ellis, J.S., Cloudiness: The planetary radiation budget and climate, Ph.D. Thesis, Dept. of Atmos. Sci., Colorado State Univ., Fort Collins.

Elsasser, W.M., Heat transfer by infrared radiation in the atmosphere, Harvard Meteorological Studies, No. 6, Harvard Univ. Press, Cambridge, Mass., 1942.

Elsasser, W.M., Mean absorption and equivalent absorption coefficient of a band spectrum, Phys. Rev., 54, 126, 1938.

Elterman, L., UV, visible and IR attenuation for altitudes to 50 km, AFCRL Report 68-0153, Environ. Res. Papers, Bedford, MA, 1968.

Fairchild, C.E., E.J. Stone and G.M. Lawrence, Photofragment spectroscopy of ozone in the UV region 270-310 nm and 600 nm, J. Chem. Phys., 69, 3632, 1978.

Fang, T.M., S.C. Wofsy and A. Dalgarno, Capacity distribution functions and absorption in Schumann-Runge bands of molecular oxygen, Planet. Space Sci., 22, 413, 1974.

Fiocco, G., A. Mugnai and W. Forlizzi, Effects of radiation scattered by aerosols on the photodissociation of ozone, J. Atom. Terr. Phys., 40, 949. 1978.

Frederick, J.E. and R. D. Hudson, Predissociation of nitric oxide in the mesosphere and stratosphere, J. Atmos. Sci., 36, 737-745, 1979.

Frederick, J.E., R.D. Hudson and J.E. Mentall, Stratospheric observations of the attenuated solar irradiance in the Schumann-Runge band absorption region of molecular oxygen, J. Geophys. Res., 86, 9885, 1981.

Frederick, JE. and J.E. Mentall, Solar irradiance in the stratosphere: Implication for the Herzberg continuum absorption of O_2, Geophys. Res. Lett., 9, 461, 1982.

Frederick, J.E., R.B. Abrams and P.J. Crutzen, The delta band dissociation of nitric oxide: A potential mechanism for coupling thermospheric variations to the mesosphere and stratosphere, J. Geophys. Res., 88, 3829, 1983.

Frohlich, C., Contemporary measures of the solar constant, pp. 93-109, in: White, O.R. (ed.), *The Solar Output and its Variation*, University of Colorado Press, 1977.

Froidevaux, L. and Y.L. Yung, Radiation and chemistry in the stratosphere: Sensitivity to O_2 cross sections in the Herzberg continuum, Geophys. Res. Lett., 9, 854, 1982.

Ghazi, A.V., V. Ramanathan and R.E. Dickinson, Acceleration of upper stratospheric radiative damping: observational evidence, Geophys. Res. Lett., 6, 437, 1979.

Gibson, G.E. and N.S. Bayliss, Variation with temperature of the continuous absorption spectrum of diatomic molecules: Part I. Experimental absorption spectrum of chlorine, Phys. Rev., 44, 186, 1933.

Gibson, G.E., O.K. Rice and N.S. Bayliss, Variation with temperature of the continuous absorption spectrum of diatomic molecules: Part II. Theoretical, Phys. Rev., 44, 193, 1933.

Godson, W.L., The evaluation of infrared radiative fluxes due to atmosphere water vapour, Quart. J. Roy. Met. Soc., 79, 367, 1953.

Goldstein, R. and F.N. Mastrup, Absorption coefficients of the O_2 Schumann-Runge continuum from 1270 A to 1745 A using a new continuum source, 56, 765, 1966.

Goodeve, C.F. and A.C.W. Taylor, The continuous absorption spectrum of hydrogen bromide, Proc. Roy. Soc., A152, 221, 1935.

Goody, R.M., A statistical model for water-vapour absorption, Quart. J. Roy. Meteorol. Soc., 78, 165, 1952.

Goody, R.M., *Atmospheric radiation, I. Theoretical Basis*, Oxford at the Clarendon Press, 1964.

Graham, R.A. and H.S. Johnston, The photochemistry of NO_3 and the kinetics of the $N_2O_5-O_3$ system, J. Phys. Chem., 82, 254, 1978.

Graham, R.A., A.M. Wier and J.A. Pitts, Ultraviolet and infrared cross section of gas phase HO_2NO_2, Geophys. Res. Lett., 5, 909, 1978.

Grant, I.P. and G.E. Hunt, Discrete space theory of radiative transfer, I. Fundamentals, Proc. Roy. Soc. London, A313, 183, 1969.

Griggs, M., Absorption coefficients of ozone in the ultraviolet and visible regions, J. Chem. Phys., 49, 857, 1968.

Groves, K.S. and A.F. Tuck, Strospheric O_3-CO_2 coupling in a photochemical-radiative column moel, I. Without chlorine chemistry, Quart. J. Roy. Met. Soc., 106, 125, 1980.

Hall, T.C. and F.E. Blacett, Separation of the absorption spectra of NO_2 and N_2O_4 in the range of 2400-5000 A, J. Chem. Phys., 20, 1745, 1952.

Hansen, J.E., W.C. Wang and A.A. Lacis, Mount Agung eruption provides test of a global climatic perturbation, Science, 199, 1065, 1978.

Hansen, J., D. Johnson, A. Lacis, S. Lebedeff, P. Lee, D. Rind, and G. Russell, Climate impact of increasing atmospheric carbon dioxide, Science, 213, 957, 1981.

Harker, A.B., N. Ho and J.J. Ratto, Photodissociation quantum yield of NO_2 in the region 375 to 420 nm, Chem. Phys. Letters, 50, 394, 1977.

Hasson, V. and R.W. Nicholls, Absolute spectral absorption measurements on molecular oxygen from 2640-1920 A. II. Continuum measurements 2430-1920 A, J. Phys. B., 4, 1789, 1971.

Hearn, A.G., The absorption of ozone in the ultraviolet and visible region of the spectrum, Proc. Phys. Soc., 79, 932, 1961.

Heath, D.F. and M.P. Thekaekara, The solar spectrum between 1200 and 3000 A, in *The Solar Output and Its Variations*, Oran R. White, (ed.), Colorado Associated University Press, Boulder, Colorado, 193-212, 1979.

Heath, D.F., A review of observational evidence for short and long term ultraviolet flux variability of the Sun, in Proceedings of the International Conference on Sun and Climate, Centre National D'Etudes Spatiales, p. 163, France, 1980.

Henri, V. and S.A. Schou, Struktur und akitivierung der molekel des formaldehyds, eine analyse auf grund des ultrvioletten absorption-spektrums des dampfes, Zeit. Phys., 49, 774, 1928.

Hering, W.S., C.N. Touart and T.R. Borden, Ozone heating and radiative equilibrium in the lower stratosphere, J. Atmos. Sci., 29, 402, 1967.

Herman, J.R. and J.E. Mentall, The direct and scattered solar flux within the stratosphere, J. Geophys. Res., 87, 1319, 1982a.

Herman, J.R. and J.E. Mentall, O_2 absorption cross section (187-225 nm) from stratospheric solar flux measurements, J. Geophys. Res. 87, 8967, 1982b.

Heroux, L. and R.A. Swirbalus, Full-disk solar fluxes between 1230 and 1940 A, J. Geophys. Res. 81, 436, 1976.

Herzberg, G., Ultraviolet absorption spectra of acetylene and formaldehyde, Trans. Faraday. Soc., 27, 378, 1931.

Herzberg, L., in *Physics of the Earth's upper atmosphere*, Hines, C., I. Paghis, T. R. Hartz, and J. A. Fejer (eds.), Prentice Hall, (Englewood Cliffs, N. J.), 1965.

Hinteregger, H.E., Solar UV irradiance at wavelengths below 185 nanometers observed for sunspot cycle 21, EGS, Uppsala, 1981.

Holt, R.B., C.K. McLane and O. Oldenberg, Ultraviolet absorption spectrum of hydrogen peroxide, J. Chem. Phys., 16, 225-229, 1948. Erratum: J.

Chem. Phys., 16, 638, 1948.

Holt, R.B. and O. Oldenberg, Role of hydrogen peroxide in the thermal combination of hydrogen and oxygen, J. Chem. Phys., 17, 1091, 1949.

Houghton, J.T., Absorption and emission by carbon dioxide in the mesosphere, Quart. J. Roy. Met. Soc., 95, 1, 1969.

Houghton, J.T., *The Physics of Atmospheres,* Cambridge University Press (Cambridge), 1977.

Hudson, R.D., V.L. Carter and J.A. Stein, An investigation of the effect of temperature on the Schumann-Runge absorption continuum of oxygen, 1580-910 A, J. Geophys. Res., 71, 2295, 1966.

Hudson, R.D. and V.L. Carter, Absorption of oxygen at elevated temperatures (300 to 900 K) in the Schumann-Runge system, J. Opt. Soc. Amer., 58, 1621, 1968.

Hudson, R.D., V.L. Carter and E.L. Breig, Predissociation in the Schumann-Runge band system of O_2: Laboratory measurements and atmospheric effects, J. Geophys. Res., 74, 4079, 1969.

Hudson, R.D. and S.H. Mahle, Photodissociation rates of molecular oxygen in the mesosphere and lower thermosphere, J. Geophys. Res., 77, 2902, 1972.

Huffman, R.E., Y. Tanaka and J.C. Larrabee, Nitrogen and oxygen absorption cross sections in the vacuum ultraviolet, Disc. Faraday Soc., 37, 159, 1964.

Hummel, J.R. and W.R. Kuhn, An atmospheric radiative-convective moel with interactive water vapor transport and cloud development, Tellus, 33, 372, 1981.

Inn, E.C.Y. and Y. Tanaka, Absorption coefficient of ozone in the ultraviolet and visible regions, J. Opt. Soc. Amer., 43, 8760, 1953.

Inn, E.C.Y., K. Watanabe and M. Zelikoff, Absorption coefficients of gases in the vacuum ultraviolet: 3. CO_2, J. Chem. Phys., 21, 1648, 1953.

Inn, E.C.Y., Absorption coefficient of HCl in the region 1400 to 2200 A, J. Atmos. Sci., 32, 2375, 1975.

Iribarne, J. V., and H. R. Cho, *Atmospheric Physics,* D. Reidel Publishing Company, (Dordrecht, Holland), 1980.

Isaksen, I.S.A., K.H. Modtbo, J. Sunde and P.J. Crutzen, A simplified method to include the molecular scattering and reflection calculations of photon fluxes and photodissociation rates, Geophys. Norv. 31, 11, 1977.

Johnson, F.S., J.D. Porcell, R. Tousey and K. Watanabe, Direct measurements of the vertical distribution of atmospheric ozone to 70 km altitude, J. Geophys. Res., 57, 157, 1952.

Johnston, H.S., E.D. Morris, Jr and J. Van den Bogaerde, Molecular modulation kinetic spectrometry, ClOO and ClO radicals in the photolysis of chlorine in oxygen, J. Amer. Chem. Soc., 91, 7712-7727, 1969.

Johnston, H.S. and R.A. Graham, Gas-phase ultraviolet spectrum of nitric acid vapor, J. Chem. Phys., 77, 62, 1973.

Johnston, H.S. and R.A. Graham, Photochemistry of NO_x compounds, Canad. J. Chem., 52, 1415, 1974.

Johnston, H.S. and G. Selwyn, New cross sections for the absorption of near ultraviolet radiation by nitrous oxide (N_2O), Geophys. Res. Lett., 2, 549, 1975.

Jones, E.J. and O.R. Wulf, The absorption coefficient of nitrogen pentoxide in the ultraviolet and the visible absorption spectrum NO_3, J. Chem. Phys., 5, 873, 1937.

Jones, I.T.N. and K.D. Bayes, Photolysis of nitrogen dioxide, J. Chem. Phys., 59, 4836, 1973.

Jones, I.T.N. and R.P. Wayne, The photolysis of ozone by ultraviolet radiation. V. Photochemical formation of O_2 ($^1\Delta_g$), Proc. Roy. Soc. London, A321, 409, 1971.

Jourdan, J.L., G. Le Bras, G. Poulet, J. Combourieu, P. Rigaud and B. Leroy, UV absorption spectrum of ClO ($A^2\Pi-X^2\Pi$) up to the (1-0) band, Chem. Phys. Letters, 57, 109, 1978.

Junge, C.E., *Air Chemistry and Radioactivity*, Academic Press (New York), 1963.

Kajimoto, O. and R.J. Cvetanovic, Temperature dependence of $O(^1D)$ production in the photolysis of ozone at 313 nm, Chem. Phys. Letters, 37, 533, 1976.

Kiehl, J.T. and V. Ramanathan, CO_2 radiative parameterization used in climate models: Comparison with narrow band moels and with laboratory data, J. Geophys. Res., 88, 5191, 1983.

Knauth, H.-D., H. Alberti and H. Clausen, Equilibrium constant of the gas reaction $Cl_2 + H_2O$ ultraviolet spectrum of HOCl, J. Phys. Chem., 83, 1604, 1979.

Kockarts, G., Penetration of solar radiation in the Schumann-Runge bands of molecular oxygen in *Mesopheric Models and Related Experiment*, ed. G. Fiocco, Reidel Publ. Co. (Dordrecht, Holland), 160-176, 1971.

Kockarts, G., Absorption and photodissociation in the Schumann-Runge bands of molecular oxygen in the terrestrial atmosphere, Planet. Space Sci., 24, 589, 1976.

Kondratyev, K.Y., *Radiation in the Atmosphere*, Academic Press (New York, N.Y.), 1969.

Kourganoff, V., *Basic Methods in Transfer Problems*, Oxford University Press (London), 1952.

Kuhn, W.R., and J. London, Infrared radiative cooling in the middle atmosphere (30-110 km), J. Atmos. Sci., 26, 189, 1969.

Kuis, S., R. Simonaitis and J. Heicklen, Temperature dependence of the photolysis of ozone at 3130 A, J. Geophys. Res., 80, 1328, 1975.

Lacis, A.A. and J.E. Hansen, A parameterization for the absorption of solar radiation in the Earth's atmosphere, J. Atmos. Sci., 31, 118, 1974.

Lambrey, M. and D. Chalonge, Structure de la bande ultraviolette de l'ozone, Gerl. Beitr. Geophys., 24, 42, 1929.

Langhoff, J.R., J.P. Dix, J.O. Arnold, R.W. Nicholls and L.L. Danylewych, Theoretical intensity parameters for the vibration-rotation bands of ClO, J. Chem. Phys., 67, 4306, 1977.

Langhoff, S.R., R.L. Jaffe and J.O. Arnold, Effective cross sections and rate constants for predissociation of ClO in the earth's atmosphere, J. Quant. Spectrosc. Rad. Transfer, 18, 227, 1977.

Lean, J.L. and A.J. Blake, The effect of temperature on thermospheric molecular oxygen absorption in the Schumann-Runge continuum, J. Geophys. Res., 86, 211, 1981.

Lean, J.L., O.R. White, W.C. Livingston, D.F. Heath, R.F. Donnelly and A. Skumanisch, A three-component model of the variability of the solar ultraviolet flux: 145-200 nm, J. Geophys. Res., 87, 1037, 1982.

Laufer, A.M. and J.R. Mcnesby, Deuterium isotope effect in vacuum ultraviolet absorption coefficients of water and methane, Canad. J. Chem., 43, 3487, 1965.

Leifson, S.W., Absorption spectra of some gases and vapors in the Schumann region, Astrophys. J., 63, 73, 1926.

Lenoble, J., Standard procedures to compute atmospheric radiative transfer in a scattering atmosphere, I.A.M.A.P., National Center for Atmospheric Research, Boulder, Colorado 80307, USA, 1977.

Lenoble, J., Transfert radiatif, in *Physique Moleculaire - Physique de L'Atmosphere*, C. Camy-Peyret (ed.), Editions du CNRS, Paris, 1982.

Lin, C.-L. and W.B. Demore, $O(^1D)$ production in ozone photolysis near 3100 A, J. Photochem., 2, 161-164, 1973.

Lin, C.L., N.K. Rohatgi and W.B. Demore, Ultraviolet absorption cross sections of hydrogen peroxide, Geophys. Res. Letters, 5, 113, 1978.

Liou, K-N., *An Introduction to Atmospheric Radiation*, Academic Press (New York, N.Y.), 1980.

Liou, K-N. and S-C.S. Ou, Theory of equilibrium temperatures in radiative turbulent atmospheres, J. Atmos. Sci., 40, 214, 1983.

London, J., In Proceedings of the Nato Advanced Institute on Atmospheric Ozone (Portugal), U.S. Dept. of Transportation, FAA - Washington, D.C., USA - No. FAA-EE-80-20, 1980.

Luther, F.M. and R.J. Gelinas, Effect of molecular multiple scattering and surface albedo on atmosphere photodissociation rates, J. Geophys. Res., 81, 1125, 1976.

Luther, F.M., D.J. Wuebbels, W.H. Duewer and J.C. Chang, Effect of multiple scattering on species concentrations and model sensitivity, J. Geophys. Res., 83, 3563, 1978.

Magnotta, F. and H.S. Johnston, Photodissociation quantum yields for the NO_3 free radical, Geophys. Res. Letters, 7, 769, 1980.

Malkmus, W., Random Lorentz band model with exponential-tailed S-1 line intensity distribution function, J. Opt. Soc. Amer., 57, 323-329, 1967.

Manabe, S. and F. Moller, On the radiative equilibrium and heat balance of the atmosphere: Mon. Weath. Rev., 89, 503, 1961.

Manabe, S. and R.F. Strickler, Thermal equilibrium of the atmosphere with a convective adjustment, J. Atmos. Sci., 21, 361, 1964.

Manabe, S. and R.T. Wetherald, Thermal equilibrium of the atmosphere with a given distribution of relative humidity, J. Atmos. Sci., 24, 241, 1967.

Mandelman, M. and R.W. Nicholls, The absorption cross sections and F-values for the $v^*=0$ progression of bands and associated continuum for the ClO $(A^2\Pi-X^2\Pi)$ system, J. Quant. Spectrosc. Rad. Transfer, 17, 481, 1977.

Marmo, F.F., Absorption coefficients of nitrogen oxide in the vacuum ultra-violet, J. Opt. Soc. Amer., 43, 1186, 1953.

Martin, H. and R. Gareis, Die kinetik der reaktion von ClO_2 mit NO_2 in der loesungsphase, Zeit. Elektrochem., 60, 959, 1956.

McCartney, E.J., Optics of the Atmosphere: Scattering by Molecules and Particles, Wiley (New York, N.Y.), 1976.

McClatchey, R.A., et al., Optical properties of the atmosphere, 3rd ed., AFCRL-72-0497, Air Force Cambridge Research Labs, Bedford, Mass., 1972.

McClatchey, R.S., R.W. Fenn, J.E.A. Selby, F.E. Volz and J.S. Garing, Optical properties of the atmosphere, AFCRL-71-0279, Air Force Cambridge Research Laboratories, 85 pp., Cambridge, MA, 1973.

Meier, R.R., D.E. Anderson, Jr., and M. Nicolet, Radiation field in the troposphere and stratosphere from 240 to 1000 nm - I. General analysis, Planet. Space Sci., 30, 923, 1982.

Metzger, P.H. and G.R. Cook, A reinvestigation of the absorption cross sectiosn of molecular oxygen in the 1050-1800 A region, J. Quant. Spectros. Rad. Trans., 4, 107, 1964.

Mie, G., Beitrage zur optik trueber Medien, Speziell koloidaller metaloesungen, Ann. der Phys., 25, 377, 1908.

Miescher, E., Rotationsanalyse der NO^+ banden, Helv. Phys. Acata, 29, 135, 1956a.

Miescher, E., Rotationanalyse der B'-banden ($^2\Delta-X^2\Pi$) des NO-molekuls, Helv. Phys. Acta, 29, 401 1956b.

Miescher, E., Excited NO levels, J. Opt. Soc. Amer., 49, 1130, 1959.

Miescher, E., Spectrum and energy levels of the NO molecule, J. Quant. Spectros. Rad. Trans., 2, 421, 1962.

Miescher, E., Analysis of the spectrum of nitric oxide molecule, Report AFCRL-69-0268, 1968.

Miescher, E., The fine structure of the spectrum of the electronic NO laser, J. Mol. Spec., 53, 302, 1974.

Milne, E.A., *Handbuch der Astrophysik*, 3, Part I, 1930 (Reprinted in "Selected Papers on the Transfer of Radiation", Dover, 1966).

Mitchell, A.C.G. and W.M. Zemansky, *Resonance Radiation and Excited Atoms,* Harvard Univ. Press, Cambridge, MA, 1934 (Reprinted 1961).

Molina, L.T., S.D. Schinke and M.J. Molina, Ultraviolet absorption spectrum of hydrogen peroxide vapor, Geophys. Res. Letters, 4, 580, 1977.

Molina, L.T. and M.J. Molina, Ultraviolet spectrum of HOCl, J. Phys. Chem., 42, 2410, 1978.

Molina, L.T. and M.J. Molina, Chlorine nitrate ultraviolet absorption spectrum at stratospheric temperatures, J. Photochem., 11, 139-144, 1979.

Molina, L.T. and M.J. Molina, J. Photochem., 15, 97, 1981.

Moortgat, G.K., E. Kudszus and P. Warneck, Temperature dependence of $O(^1D)$ formation in the near UV photolysis of ozone, J. Chem. Soc., Faraday Trans 11, 73, 1216, 1977.

Moortgat, G.K. and P. Warneck, Relative $O(^1D)$ quantum yields in the near UV photolysis of ozone at 298 K, Naturforsch., 30A, 835, 1975.

Moortgat, G.K. and P. Warneck, CO and H_2 quantum yields in the photodecomposition of formaldehyde in air, J. Chem. Phys., 70, 3639, 1979.

Mount, G.H., G.J. Rottman, and J.G. Thimothy, The solar spectral irradiance 1200-2550 A at solar maximum, J. Geophys. Res., 85, 4271, 1980.

Mount, G.H. and G.J. Rottman, The solar spectral irradiance 1200-1284 A near solar maximum: July 15, 1980, J. Geophys. Res., 86, 9193, 1981.

Mount, G.H. and G.J. Rottman, The solar absolute spectral irradiance 1150-3173 A: 17 May 1982, J. Geophys. Res., 88, 5403, 1983.

Murgatroyd, R.J. and R.M. Goody, Sources and sinks of radiative energy from 30 to 90 km, Quart. J. Roy. Met. Soc., 84, 225, 1958.

Nakayamat, T., M.Y. Kitamura and K. Watanabe, Ionization potential and absorption coefficients of nitrogen dioxide, J. Chem. Phys., 30, 1180, 1959.

Neckel, H. and D. Labs, Improved data of solar spectral irradiance from 0.33 to 1.25 microns, Solar Phys., 74, 231, 1981.

Nicolet, M., Photodissociation of nitric oxide in the mesosphere and stratosphere: Simplified numerical relations for atmosphere model calculation, Geophys. Res. Letters, 6, 866, 1979.

Nicolet, M., The chemical equations of stratospheric and mesopheric ozone, Proceedings of Nato Advanced Study Institute on Atmospheric Ozone (Portugal), edited by US Dept. of Transportation, FAA Washington, D.C., USA, Rapport No. FAA-EE-80-20, 1980.

Nicolet, M. and S. Cieslik, The photodissociation of nitric oxide in the mesophere and stratosphere, Planet. Space Sci., 28, 105, 1980.

Nicolet, M. and W. Peetermans, Atmospheric absorption in the O_2 Schumann-Runge band spectral range and photodissociation rates in the stratosphere and mesophere, Planet. Space Sci., 28, 85, 1980.

Nicolet, M., The photodissociation of water vapor in the mesophere, J. Geophys. Res., 86, 5203, 1981.

Nicolet, M., R.R. Meier, ànd D.E. Anderson, Radiation field in the troposphere and stratosphere - II. Numerical analysis, Planet. Space Sci., 30 935, 1982.

Nicolet, M., Photodissociation of molecular oxygen in the terrestrial atmosphere: Simplifeid numerical relatiosn for the spectral range of the Schumann-Runge bands, J. Geophys. Res., in press, 1984.

Norrish, R.G.W. and F.N. Kirkbride, Primary photochemical processes, Part I. The decomposition of formaldehyde, J. Chem. Soc., pp. 1518-1530, 1932.

Ogawa, M., Absorption cross sections of O_2 and CO_2 continua in the Schumann-Runge and far-UV regions, J. Chem. Phys. Letters, 9, 603, 1971.

Ogawa, S. and M. Ogawa, Absorption cross section of O_2 $(A^1\Delta g)$ and O_2 $(X^3\Delta_g)$ in the region from 1087 to 1700 A., Canad. J. Phys., 53, 1845, 1975.

Owens, A.J., C.H. Hales, D.L. Filkin, C. Miller, A. Yokozeki, J.M. Steed and J.P. Jesson, A coupled one-dimensional radiative-convective chemistry - transport moel of the atmosphere, I. Model structure and steady state perturbation calculations, manuscript,

Park, J.H., The equivalent mean absorption cross sections for the O_2 Schumann-Runge bands: Aplication to the H_2O and NO photodissociation rates, J. Atmos. Sci., 312, 1893, 1974.

Penndorf, R., Tables of the refractive index for standard air and the Rayleigh scattering coefficient for the spectral region between 0.2 and 20.0 μm and their application to atmospheric optics, J. Opt. Soc. Am., 47, 176, l957.

Perner, D., and U. Platt, Absorption of light in the atmosphere by collision pairs of oxygen $(O_2)_2$, Geophys. Res. Lett., 7, 1053, l980.

Ramanathan, V., Radiative transfer within the Earth's troposphere and stratosphere: A simplified radiative-convective model, J. Atmos. Sci., 33, 1330, 1976.

Ramanathan, V. and J.A. Coakley, Climate modeling through radiative convective models, Rev. Geophys. Space Phys., 16, 465, 1978.

Ramanathan, V., E.J. Pitcher, R.C. Malone and M.L. Blackmon, The response of a spectral general circulation model refinements in radiative processes, J. Atmos. Sci., 40, 605, 1983.

Rasool, S.I. and S.H. Schneider, Atmospheric carbon dioxide and aerosols effects of large increases on global climate, Science, 173, 138, 1971.

Richards, P.G., D.G. Torr, and M.A. Torr, Photodissociation of N_2: A significant source for thermospheric atomic nitrogen, J. Geophys. Res., 86, 1495, 1981.

Robbins, D.E., Photodissociation of methyl chloride and methyl bromide in the atmosphere, Geophys. Res. Letters, 3, 213, 1976a, and Erratum: Ibid, 757, 1976b.

Rodgers, C.D. and C.D. Walshaw, The computation of infra-red cooling rate in planetary atmospheres, Quart. J. Roy. Met. Soc., 92, 67, 1966.

Romand, J. and B. Vodar, Spectre d'absorption de la'acide chlorhydrique gazeux dans la region de Schumann, C.R. Acad. Sci. Paris, 226, 238, 1948.

Romand, J., Absorption ultraviolette dans la region of Schumann-Runge, Etude de ClH, BrH et IH gazeux, Ann. Phys., Paris, 4, 527, 1949.

Romand, J. and Mayence, J., Spectre d'absorption de l'oxyde azoteux gazeux dans la region de Schumann, C.R. Acad. Sci. Paris, 228, 998, 1949.

Rothman, L.S., A. Goldman, J.R. Gillis, R.H. Tipping, L.R. Brown, J.S. Margolis, A.G. Maki and L.D.G. Young, AFGL trace gas compilation: 1980 version, Appl. Opt., 20, 1323, 1980.

Rottman, G., Personal communication, 1981.

Rottman, G., C. Barth, R. Thomas, G. Mount, G. Lawrence, D. Rusch, R. Saunders, G. Thomas and J. London, Solar spectral irradiance, 120 to 190 nm, October 13, 1981 - January 3, 1982, Geophys. Res. Letters, 9, 587, 1982.

Rottman, G.J., 27-day variations observed in solar ultraviolet (120-300 nm) irradiance, Planet. Space Sci., 31, 1001, 1983.

Rowland, F.S., and M.J. Molina, Chlorofluoromethanes in the enviornment, Rev. Geophys. Space Phys., 13, 1, 1975.

Rowland, F.S., J.E. Spencer and M.J. Molina, Stratospheric formation and photolysis of chlorine nitrate, J. Phys. Chem., 80, 2711, 1976.

Samain, D. and P.C. Simon, Solar flux determination in the spectral range 150-210 nm, Solar Phys., 49, 33, 1976.

Schoeberl, M.R. and D.F. Strobel, The zonally averaged circulation of the middle atmosphere, J. Atmos. Sci., 35, 577, 1978.

Schurgers, M. and K.H. Welge, Absorptionskoeffizient von H_2O_2 und N_2H_4 zwischen 1200 und 2000 A, Zeit. Naturforsch., 23A, 1508, 1968.

Seery, D.J. and D. Britton, The continuous absorption spectra of chlorine, bromine, bromide chloride, iodine chloride and iodine bromide, J. Phys. Chem., 68, 2263, 1964.

Shardanand, and A.D. Prasad-Rao, Collision-induced absorption of O_2 in the Herzberg continuum, J. Quant. Spectrosc. Radiat. Transfer, 17, 443, 1977.

Shaw, J., Solar Radiation, Ohio Journal of Science, LIII, 258, 1953.

Shemansky, D.E., CO_2 extinction coefficient 1700-3000 A, J. Chem. Phys., 56, 1582, 1972.

Simon, P.C., Solar irradiance between 120 and 400 nm and its variations, Solar Phys., 74, 273, 1982.

Simon, P.C., R. Pastiels and D. Nevejans, Balloon observations of solar ultraviolet irradiance at solar minimum, Planet. Space Sci., 30, 67, 1982a.

Simon, P.C., R. Pastiels, D. Nevejans and D. Gillotay, Balloon observatiosn of solar ultraviolet irradiance during solar cycle 21, in Proceedings of the Symposium on the Solar Constant and the Spectral Distribution of Solar Irradiance (J. London and C. Frohlich, ed.), p. 95, IAMAP, Third Scientific Assembly, Boulder, CO, USA, 1982b.

Simonaitis, R., S. Braslavsky, J. Heicklen and M. Nicolet, Photolysis of O_3 at 3130 A, Chem. Phys. Lett., 19, 601, 1973.

Simons, J.W., R.J. Paur, H.A. Webster, III and E.J. Bair, Ozone ultraviolet photolysis, IV. The ultraviolet spectrum, J. Chem. Phys., 59, 1203, 1973.

Simpson, C.J.S.M., P.D. Gait and J.M. Simmie, The vibrational deactivation of the bending moe of CO_2 by O_2 and by N_2, Chem. Phys. Lett., 47, 133, 1977.

Smith, E.V.P. and D.M. Gottlieb, Solar flux and its variations, Space Sci. Rev., 16, 771, 1974.

Smith, W.S., C.C. Chou and F.S. Rowland, The mechanism for ultraviolet photolysis of gaseous chlorine nitrate at 302.5 nm, Geophys. Res. Letters, 4, 517, 1977.

Sobolev, V.V., A Treatise of Radiative Transfer, D. Van Nostrand (Princeton, N.J.), 1963.

Spencer, J.E. and F.S. Rowland, Bromine nitrate and its stratospheric significance, J. Phys. Chem., 82, 7, 1978.

Stockwell, W.R. and J.C. Calvert, The near ultraviolet absorption spectrum of gaseous HONO and N_2O_4, J. Photochem., 8, 193, 1978.

Stolarski, R.S. and R.J. Cicerone, Stratospheric chlorine, A possible sink for ozone, Can. J. Chem., 52, 1610, 1974.

Stone, P. and J. Carlson, Atmospheric lapse rate regimes and their parameterization, J. Atmos. Sci., 36, 415, 1976.

Swider, W. and M.E. Gardner, On the accuracy of certain approximations for the Chapman function, Environmental Research Papers No. 272, Air Force Cambridge Research, Bedford, MA, USA, 1967.

Sze, N.D. and M.K.W. Ko, The effects of the rate for $OH + HNO_3$ and HO_2NO_2 photolysis on stratospheric chemistry, Atmos. Environm., 15, 1301, 1981.

Tanaka, Y., E.C.Y. Inn and K. Watanabe, Absorption coefficients of gases in the vacuum ultraviolet, Part IV. Ozone, J. Chem. Phys., 21, 1651, 1953.

Taube, H., Photochemical reactions of ozone in solution, Trans. Faraday Soc., 53, 657, 1957.

Thompson, B.A., P. Harteck and R.R. Reeves, Jr., Ultraviolet absorption coefficients of CO_2 CO, O_2, H_2O, N_2O, NH_3, NO, SO_2 and CH_4 between 1850 and 4000 A, J. Geophys. Res., 68, 6431, 1963.

Tiwari, S.N., Models for infrared atmospheric radiation, Adv. Geophys., 20, 1, 1978.

Turco, R.P., R.C. Whitten, O.B. Toon, E.C.Y. Inn and P. Hamil, Stratospheric hydroxyl radical concentrations: New limitations suggested by observations of gaseous and paticulate sulfur, J. Geophys. Res., 86, 1129, 1981.

Urey, H.C., L.C. Dawsey and F.O. Rice, The absorption spectrum and decomposition of hydrogen peroxide by light, J. Amer. Chem. Cos., 51, 1371, 1929.

Valley, S.L. (ed.), Handbook of Geophysics and Space Environment, Air Force Cambridge Research Laboratory, 1965.

Van de Hulst, H.C. Light Scattering by Small Particles, Wiley, (New York), 1957.

Van Laethem-Meuree, N., J. Wisemberg and P.C. Simon, Absorption des chloromethanes dans l'ultraviolet: Mesures des sections efficaces d'absorption en fonction de la temperature, Bull. Acad. Roy. Belgique, Cl. Sci., 64, 34, 1978a.

Van Laethem-Meuree, N., J. Wisemberg and P.C. Simon, Influence de la temperature sur les sections efficaces d'absorption des chlorofluorométhanes dans l'ultraviolet, Bull. Acad. Roy. Belgique, Cl. S i., 64, 42, 1978b.

Vernazza, J., E. H. Avrett, and R. Loeser, Structure of the solar chromosphere. II. The underlying photosphere and temperature minimum region, Astrophys. J., 30, 1, 1976.

Vidal-Madjar, A., The solar spectrum at Lyman Alpha, in the Solar Output and its Variation (edited by O. White), p. 2313, Colorado Associated University Press, Boulder, Colorado, USA, 1977.

Vigroux, E., Mesures absolues des coefficients d'absorption de l'ozone dans la region des bandes de Huggins, A 18 degres, C.R. Acad. Sci. Paris, 234, 2351, 1952a.

Vigroux, E., Absorption de l'ozone dans la region des bandes de Huggins, influence de la temperature, C.R. Acad. Sci. Paris, 234, 2439, 1952b.

Vigroux, E., Contribution experimentale de l'absorption de l'ozone, Ann. Phys., Paris, 12eme Serie, 8, 709, 1953.

Vigroux, E. Coefficients d'absorption de l'ozone dans la bande de Hartley, Ann. Geophys., 25, 169, 1969.

Vodar, M.B., Spectre d'absorption ultraviolet du gaz chlorhydrique et courbe d'energie potentielle de l'etat excite de la molecule ClH, J. Phys. Rad., 9, 166, 1948.

Wang, W.C., W.B. Rossow, M.S. YaO and M. Wolfson, Climate sensitivity of a one-dimensional radiative-convective model with cloud feedback, J. Atmos. Sci., 38, 1167, 1981.

Wang, W.C. and P.H. Stone, Effect of ice-albedo feedback on global sensitivity in a one-dimensional radiative-convective climate model, J. Atmos. Sci., 37, 545, 1980.

Watanabe, K. and F.F. Marmo, Photoionization and total absorption cross section of gases, II. O_2 and N_2 in the region 850-1500 A, J. Chem. Phys., 25, 965, 1956.

Watanabe, K. and M. Zelikoff, Absorption coefficient of water vapor in the vacuum ultraviolet, J. Opt. Soc. Amer., 43, 753, 1953.

Watanabe, K., E.C. Inn and M. Zelikoff, Absorption coefficients of oxygen in the vacuum ultraviolet, J. Chem. Phys., 21, 1026, 1953.

Watson, R.T., Rate constants for reactions of ClO of atmospheric interest, J. Phys. Chem. Ref. Data, 6, 871, 1977.

Williams, A.P., Relaxation of the 2.7 micron and 4.3 micron bands of carbon dioxide, in *Mesospheric models and related Experiments,* Reidel Publishing Company, 177, 1971.

Williams, A.P. and C.D. Rodgers, Radiative transfer by the 15 miron CO_2 band in the mesosphere, Proceedings of the International Radiation Symposium, Sendai, Japan, 253-260, 26 May-2 June 1972.

Wiscombe, W.J., Extension of the doubling method to inhomogeneous sources, J. Quant. Spectrosc. Radiat. Transfer, 16, 477, 1976a.

Wiscombe, W.J., On initialization, error and flux conservation in the doubling method, J. Quant. Spectrosc. Radiat. Transfer, 16, 637, 1976b

WMO (World Meteorological Organization), *The stratosphere 1981: theory and measurements,* Report no. 11, Geneva, Switzerland, 1982.

Wyatt, P.J., V.R. Stull and G.N. Plass, Quasi-random model of band absorption, J. Opt. Soc. Amer., 52, 1209, 1962.

Yao, F., I. Wilson and H. Johnston, Temperature dependent ultraviolet absorption spectrum for dinitrogen pentoxide, J. Phys. Chem., 86, 3611, 1982.

Zeilik, M., *Astronomy: the evolving universe,* Prentice Hall, (Englewood Cliffs, New Jersey), 1965.

Chapter 5

Composition and Chemistry

5.1 General

The composition of the atmosphere at ground level is relatively well known due to numerous measurements of the concentration and variability of many of the important gases. The local surface sources of atmospheric species include photochemical processes, biological and microbiological activity, volcanic eruptions, and human activities. Some of the species emitted in the troposphere are also destroyed at the surface by biological, chemical and physical processes. Others are transported upward, eventually reaching the stratosphere, where they undergo chemical transformations. For most constituents, natural sources are larger than those of anthropogenic origin. However, for CO_2 and some halocarbons, industrial production is not negligible compared to natural production.

Some minor constituents are produced at high altitudes by photochemical reactions. The altitude at which these processes occur depends on the penetration depth of the solar radiation which initiates the photochemistry.

In this chapter, the chemical and photochemical processes relevant to the principal constituents of the middle atmosphere will be discussed. Knowing the distribution and behavior of the various constituents, global atmospheric budgets can be derived. The budgets of atmospheric methane, carbon monoxide, and molecular hydrogen will be discussed in particular detail. This chapter will deal only with neutral species; the ions will be addressed in the following chapter.

The terrestrial atmosphere is made up of a multitude of chemical constituents, the most abundant being molecular nitrogen N_2, oxygen O_2, and to a lesser degree, argon. Certain species possess very long lifetimes, and are therefore insensitive to perturbations in chemical processes (c. f. §3.4). In particular, the noble gases fall into this category. Their mean atmospheric mixing ratios are indicated below:

Argon $(9340\pm 10)\mathrm{x}10^{-6}$

Neon $(18.18 \pm 0.04) \times 10^{-6}$

Helium $(5.239 \pm 0.002) \times 10^{-6}$

Krypton $(1.14 \pm 0.01) \times 10^{-6}$

Xenon $(0.086 \pm 0.001) \times 10^{-6}$

These represent mean values because the noble gases are geochemically produced under certain conditions, so that local gradients can sometimes be observed.

The study of atmospheric chemistry must begin with the distribution of the dominant gases, molecular oxygen and nitrogen. The relative concentrations of these species are essentially constant in the homosphere, as was mentioned in Chapter 3. Generally, the following mole fractions are assumed for dry air:

N_2 : 78.084$\pm$ 0.004 percent

O_2 : 20.946$\pm$ 0.06 percent

In the middle atmosphere, molecular nitrogen is particularly stable since it cannot be photodissociated below the mesopause. On the other hand, the photodissociation of molecular oxygen can occur at altitudes as low as 20 km. This primary process initiates a series of reactions which determine the chemistry of the "oxygen atmosphere"; these will be the subject of the next section.

The mixing ratios of virtually all other atmospheric species are variable in space and time, and are dependent on chemical production (P_i) and loss (L_i) processes, and/or on transport. Theoretical models can predict the abundances of these constituents by solving the continuity equation for each such species n:

$$\frac{\partial n}{\partial t} = \sum_i P_i - \sum_i L_i(n) - \vec{\nabla} \cdot \phi(n) \tag{5.1}$$

where $\phi(n)$ represents the flux vector of species n and t is time. In numerical models, the equations must be solved simultaneously for all of the many species under consideration. There are a variety of numerical methods which can be used to provide solutions to this coupled set of equations (see, e. g., Chang et al., 1974). The simplest method (one sometimes used in chemical studies) is the explicit forward Euler method:

$$n_{t + \Delta t} = n_t + \left[\sum_i P_i - \sum_i L_i(n_t) - \vec{\nabla} \cdot \phi(n_t) \right] \Delta t \tag{5.2}$$

This method has the advantage of being rapid and thus relatively cheap for computational purposes, which can be important because of the large number of photochemically active species which must be considered in numerical models of the atmosphere. More sophisticated methods are also used. It is beyond the scope of this volume to present a detailed analysis of numerical methods. Applications of some other methods in aeronomic problems are discussed, for example, by Turco and Whitten (1974) and Gelinas (1972). Gear (1971) describes the stability and accuracy of several numerical methods for solving these kinds of equations. For the explicit forward Euler method, the solution should be stable and mass conserving, provided that the time step used is no greater than about 1/5 of the time constant characterizing the equation (generally $1/\sum_i L_i$ if the species is chemically controlled as discussed in Chapter 2, or it can be dictated by dynamical parameters as shown in Chapter 3). This proves to be an important constraint in many atmospheric chemical models, and has led to the widespread use of chemical "families" in order to facilitate the solution of the equations, as shown in some detail below.

Most of the sections of this chapter are divided into subsections which deal with budgets and observations, discussion and definition of chemical families, and detailed chemistry. Subsections which present the detailed chemistry are denoted by the symbol ‡, the others by •. We do not wish to encourage the reader to neglect the detailed chemistry, but we recognize that these subsections may not be of great interest to non-chemists, and could be omitted if desired. On the other hand, we feel that the subsections which describe the families and their partitioning provide a simplified framework (although certainly not the only possible one) for understanding atmospheric chemistry and, in particular, its interaction with dynamics, with minimal detail.

5.2 Oxygen compounds

5.2.1 *Pure oxygen chemistry* ‡

The photodissociation of molecular oxygen by ultraviolet radiation at wavelengths less than 242.4 nm produces atomic oxygen (Chapman, 1930):

$$(J_2); \quad O_2 + h\nu \rightarrow O + O \tag{5.3}$$

These atoms may recombine directly in a three body process in the upper atmosphere:

$$(k_1); \quad O + O + M \rightarrow O_2 + M \tag{5.4}$$

or they may react with molecular oxygen to produce ozone:

$$(k_2); \quad O + O_2 + M \rightarrow O_3 + M \tag{5.5}$$

This species can also recombine with atomic oxygen according to the reaction

$$(k_3); \quad O + O_3 \rightarrow 2O_2 \tag{5.6}$$

The photodissociation of ozone leads to formation of oxygen atoms, in either their ground state:

$$(J_3); \quad O_3 + h\nu(\lambda > 310nm) \rightarrow O_2(^3\Sigma_g^-) + O(^3P) \tag{5.7}$$

or in their first excited state:

$$(J_3^*); \quad O_3 + h\nu(\lambda < 310nm) \rightarrow O_2(^1\Delta_g) + O(^1D) \tag{5.8}$$

In the upper part of the middle atmosphere, some $O(^1D)$ atoms can also be produced by molecular oxygen photolysis:

$$(J_2^*); \quad O_2 + h\nu(\lambda < 175.9nm) \rightarrow O(^1D) + O(^3P) \tag{5.9}$$

When molecular oxygen is in the excited $^1\Delta_g$ state, it can be deactivated by collision with ground state oxygen:

$$(k_6); \quad O_2(^1\Delta_g) + O_2(^3\Sigma_g^-) \rightarrow 2O_2(^3\Sigma_g^-) \tag{5.10}$$

or it can relax radiatively by emitting a photon at $\lambda = 1.27\mu m$

$$(A_{1.27}); \quad O_2(^1\Delta_g) \rightarrow O_2(^3\Sigma_g^-) + h\nu \tag{5.11}$$

The observation of this infrared emission can be used to deduce the concentration of $O_2(^1\Delta_g)$ and O_3 (see, e. g., Evans et al., 1968; Thomas et al., 1983). $O(^1D)$ can be quenched either by collision with N_2 or O_2:

$$(k_{4a}); \quad O(^1D) + N_2 \rightarrow O(^3P) + N_2 \tag{5.12}$$

$$(k_{4b}); \quad O(^1D) + O_2 \rightarrow O(^3P) + O_2(^1\Sigma_g) \tag{5.13}$$

Because it is produced almost exclusively by photolysis reactions, the concentration of $O(^1D)$ is strongly dependent on local photochemical conditions. Figure 5.1 shows the distribution of $O(^1D)$ calculated for several solar zenith angles.

5.2.2. The odd oxygen family and some observations •

These chemical reactions (see Fig. 5.2) lead to a simple theory of stratospheric ozone for a pure oxygen atmosphere, first proposed by Chapman in 1930. We shall later see that substantial modifications to this simple theory were introduced over the years after 1930 (see § 5.9), as the important interactions between oxygen, hydrogen, nitrogen, and finally chlorine, were recognized. It is, however, useful to examine the photochemistry of the pure oxygen system to provide a framework to which other processes may be added.

Considering the eleven reactions given above, and omitting transport, the continuity equations for O_3, $O(^3P)$, $O(^1D)$, and $O_2(^1\Delta_g)$ can be written as fol-

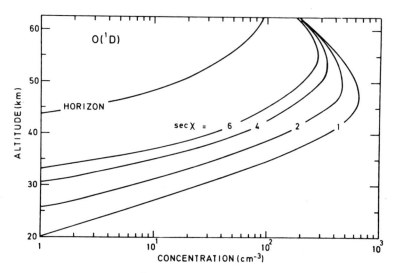

Fig. 5.1. Concentration of $O(^1D)$ atoms at various solar zenith angles. From Nicolet (1971). (Copyright by Reidel Publishing Company).

lows:

$$\frac{d(O_3)}{dt} + (J_3 + J_3^*)(O_3) + k_3(O)(O_3) = k_2(M)(O_2)(O) \qquad (5.14)$$

$$\frac{d(O)}{dt} + 2k_1(M)(O)^2 + k_2(M)(O_2)(O) + k_3(O_3)(O)$$

$$= 2J_2(O_2) + J_2^*(O_2) + J_3(O_3) + k_{4a}(N_2)(O^1D) + k_{4b}(O_2)(O^1D) \qquad (5.15)$$

$$\frac{d(O^1D)}{dt} + [k_{4a}(N_2) + k_{4b}(O_2)](O^1D) = J_3^*(O_3) + J_2^*(O_2) \qquad (5.16)$$

and

$$\frac{d(O_2{}^1\Delta_g)}{dt} + A_{1.27}(O_2{}^1\Delta_g) + k_6(O_2)(O_2{}^1\Delta_g) = J_3^*(O_3) \qquad (5.17)$$

where (O_2) represents the concentration of ground state $O_2(^3\Sigma_g^-)$ and (O) is the density of $O(^3P)$. A similar equation could be written for ground state molecular oxygen, but would not be independent since total oxygen is always conserved:

$$\frac{3d(O_3)}{dt} + \frac{2d(O_2)}{dt} + \frac{2d(O_2{}^1\Delta_g)}{dt} + \frac{d(O)}{dt} + \frac{d(O^1D)}{dt} = 0 \qquad (5.18)$$

The photochemical lifetimes of these species as determined only by the above

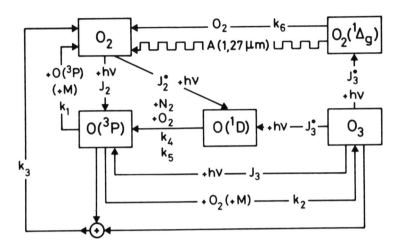

Fig. 5.2. Reaction scheme for a pure oxygen atmosphere.

reactions are:

$$\tau_{O_3} = \frac{1}{J_3 + J_3^* + k_3(O)} \approx 2000 \text{ sec at 30 km} \tag{5.19a}$$

$$\tau_O = \frac{1}{k_2(O_2)(M) + k_3(O_3) + 2k_1(M)(O)} \approx .04 \text{ sec at 30 km} \tag{5.19b}$$

$$\tau_{O^1D} = \frac{1}{k_{4a}(N_2) + k_{4b}(O_2)} \approx 10^{-8} \text{ sec at 30 km} \tag{5.19c}$$

$$\tau_{O_2(^1\Delta_g)} = \frac{1}{k_6(O_2) + A_{1.27}} \approx 1 \text{ sec at 30 km} \tag{5.19d}$$

The chemical loss of $O_2(^1\Delta_g)$ produces a ground state oxygen molecule. Since O_2 is a permanent gas always present at a constant mixing ratio in the middle atmosphere, this process is not photochemically important. Further, the lifetime of $O_2(^1\Delta_g)$ is never more than its radiative lifetime of about an hour, so that photochemical steady state (see eqn. 2.23) may be assumed for this species:

$$(O_2{}^1\Delta_g) \approx \frac{J_3^*(O_3)}{A_{1.27} + k_6(O_2)} \tag{5.20}$$

(See also, Thomas et al., 1983, for more details regarding $O_2(^1\Delta_g)$ chemistry).

On the other hand, the O_3, O, and $O(^1D)$ equations are a coupled set of partial differential equations whose time constants differ by several orders of magnitude (such a set of equations is called a "stiff" system, see, e. g., Gear, 1971). Sophisticated numerical methods are available to treat this problem, and are used, for example, in the aeronomic models of Chang (1974) and coworkers. Due to the expense involved in these methods, however, they are not widely used in aeronomic applications. If we attempt to solve these equations using the simple forward Euler method, a time step much less than one second would have to be employed. This would be prohibitively expensive if, for example, we wished to calculate the seasonal behavior of O_3, because so many time steps would be required. In view of this problem, it is very useful to define chemical families for computational purposes, whose lifetimes can be very much longer than those of the constituent members. Adding equations (5.14), (5.15) and (5.16) one obtains an equation describing the behavior of the sum of $O(^3P)$, $O(^1D)$ and O_3, which is generally referred to as the *odd oxygen family*, O_x:

$$(O_x) = (O^3P) + (O^1D) + (O_3) \tag{5.21}$$

one can then write:

$$\frac{d(O_x)}{dt} + 2k_3(O_3)(O) + 2k_1(M)(O)^2 = 2(J_2 + J_2^*)(O_2) = 2J_{O_2}(O_2) \tag{5.22}$$

where $J_{O_2} = J_2 + J_2^*$ is the total photolysis rate of O_2. The chemical lifetime of the odd oxygen family is then given by

$$\tau_{O_x} \approx \frac{(O_x)}{2k_1(M)(O)^2 + 2k_3(O)(O_3) + \text{other terms (see §5.9)}}$$

$$\approx \quad \text{weeks at 30 km} \tag{5.23}$$

Figure 5.3 presents an altitude profile of the photochemical lifetimes of the O_x family (including important reactions with hydrogen, nitrogen, and chlorine containing species, which will be introduced later), O_3, and O, as well as an estimate of the approximate time constants associated with transport by the winds and a one-dimensional vertical diffusive lifetime. In this and other figures involving lifetimes to be presented in this chapter, τ_D represents the one-dimensional vertical diffusive lifetime, assuming a vertical scale height of 5 km, and $\tau_{\bar{u}}$, $\tau_{\bar{v}^*}$, and $\tau_{\bar{w}^*}$ represent the time constants for transports by the zonal, meridional, and vertical winds at middle latitudes, assuming characteristic scales of 1000, 1000, and 5 km, respectively. Note that the adoption of this scale implies that the indicated values of $\tau_{\bar{u}}$ should be considered an lower limit, which is probably only valid when planetary waves are present, and for species with relatively large meridional gradients such as N_2O_5 (see §5.5). The vertical profiles of $\bar{u}$, $\bar{v}^*$, and $\bar{w}^*$ are taken from the model of Garcia and Solomon (1983). We will discuss the implications of the magnitudes of the

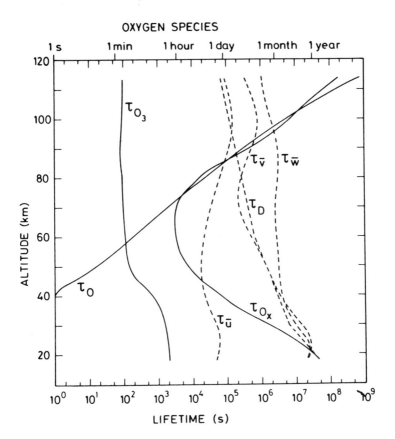

Fig. 5.3. Photochemical lifetimes of O_x, O_3, and O, and characteristic transport lifetimes.

chemical and dynamical time constants in determining the behavior of these species below.

The definition of the odd oxygen family clearly produces a substantial increment in the photochemical time constant of the equation which must be considered, enabling us to solve it more readily. The very fast reactions, such as

$$O_3 + h\nu \rightarrow O(^1D) + O_2$$

$$O(^1D) + O_2 \rightarrow O + O_2(^1\Sigma_g)$$

$$O + O_2 + M \rightarrow O_3 + M$$

now produce only an exchange, or partitioning, among members of the family, but do not appear as production or loss terms in the continuity equation for the family. For numerical stability when using the forward Euler method, the use of chemical families is valid, however, only if the time constants characterizing exchange between family members are rapid compared to the integration time step. Even if more sophisticated numerical methods are employed to avoid this constraint, the time constants characterizing exchange between family members must never exceed the dynamical time constants. Since the net transport of a chemical species depends in part upon its gradient, any species whose lifetime is sufficiently long that transport may play a role in establishing its distribution should be transported according to its own individual gradient. To some extent, the choice of integration time step and the family definitions may depend on the problem under consideration. For example, if the detailed photochemistry of the rapidly changing solar illumination found at twilight is of special interest, then the time step taken and the families used must be different from those appropriate to a study of variations on a seasonal time scale.

The use of chemical families allows a clearer distinction to be drawn between reactions which represent *net* and *gross* production and loss terms over the time scale considered. For example, the photolysis of O_3 is a gross, but not a net loss term for ozone at altitudes below about 80 km over time scales longer than a few seconds, because nearly all the oxygen atoms which are formed by this reaction rapidly reform ozone, implying no *net* loss. On the other hand, when ozone reacts with atomic oxygen to form O_2, a net loss of ozone occurs over an extended time scale. Further, since a typical time scale for meridional transport is of the order of months, it is convenient to consider odd oxygen, rather than ozone or atomic oxygen, since it is only odd oxygen which has a sufficiently long lifetime to be influenced by meridional transport processes. The importance of transport is therefore more easily seen when families are considered. Thus, the family method provides both numerical and conceptual simplifications.

This analysis allows us to obtain the concentrations of the odd oxygen family, but it must also be partitioned into its constituent parts. This can be done by examining the photochemical equilibrium expressions for certain members of the family. If the family is composed of n members, we must write photochemical equilibrium expressions for the shortest lived n-1 members of the family and use these to establish ratios between family members. Below the mesopause, the photochemical time constant for atomic oxygen is relatively short and one can assume that it is nearly in photochemical equilibrium (see eqn. 2.23). Similar conditions apply to the excited $O(^1D)$ atom. Neglecting minor terms, equations (5.15) and (5.16) then become, (assuming $J_{O_3} = J_3 + J_3^*$):

$$\frac{d(O^1D)}{dt} \approx 0 \approx J_3^*(O_3) + J_2^*(O_2) - [k_{4a}(N_2) + k_{4b}(O_2)](O^1D) \qquad (5.24)$$

$$(O^1D) \approx \frac{J_3^*(O_3) + J_2^*(O_2)}{k_{4a}(N_2) + k_{4b}(O_2)} \approx \frac{J_3^*(O_3)}{k_{4a}(N_2) + k_{4b}(O_2)} \qquad (5.25)$$

$$\frac{(O^1D)}{(O_3)} \approx \frac{J_3^*}{k_{4a}(N_2) + k_{4b}(O_2)} \qquad (5.26)$$

$$\frac{d(O^3P)}{dt} \approx 0 \approx J_3(O_3) + [k_{4a}(N_2) + k_{4b}(O_2)](O^1D) + 2J_{O_2}(O_2)$$

$$- [k_2(M)(O_2) + k_3(O_3)](O^3P) \qquad (5.27)$$

Substituting for $O(^1D)$ from the expression above,

$$(O^3P) \approx \frac{J_{O_3}(O_3) + 2J_{O_2}(O_2)}{k_2(M)(O_2) + k_3(O_3)} \approx \frac{J_{O_3}(O_3)}{k_2(M)(O_2)} \qquad (5.28)$$

and

$$\frac{(O^3P)}{(O_3)} \approx \frac{J_{O_3}}{k_2(M)(O_2)} \qquad (5.29)$$

$$\frac{(O_3)}{(O_x)} = \frac{1}{1 + \dfrac{(O^3P)}{(O_3)} + \dfrac{(O^1D)}{(O_3)}} \qquad (5.30a)$$

and

$$\frac{(O)}{(O_x)} = \frac{(O_3)}{(O_x)} \times \frac{(O)}{(O_3)} \qquad (5.30b)$$

In the stratosphere ozone is much more abundant than atomic oxygen and $O(^1D)$, (see Table 5.1) so that $(O_x) \approx (O_3)$. Since $d(O)/dt$ and $d(O^1D)/dt \approx 0$, the odd oxygen equation can be written (using eqn. 5.28):

$$\frac{d(O_x)}{dt} + \frac{2k_3 J_{O_3}(O_3)^2}{k_2(M)(O_2)} = 2J_{O_2}n(O_2) \qquad (5.31)$$

In the upper stratosphere and lower mesosphere, odd oxygen (and ozone) is very short lived, and is approximately in photochemical equilibrium, so that using (5.31) we can write

$$(O_3)_{eq} = \left[\frac{k_2}{k_3}(M)(O_2)^2 \frac{J_{O_2}}{J_{O_3}} \right]^{1/2} \qquad (5.32)$$

for pure oxygen chemistry, while below 30 km the photochemical lifetime of odd oxygen becomes much longer. The time constants associated with transport become much shorter than that for photochemistry near the tropopause, so that odd oxygen is dynamically controlled in the lower stratosphere. This will be discussed in more detail below. Table 5.1 presents typical values of the distribution of the pure oxygen species in the atmosphere (including hydrogen, nitrogen and chlorine reactions).

In the upper part of the mesosphere and in the thermosphere, atomic oxygen becomes more abundant than ozone. Ozone is short lived in this region due to a rapidly increasing photolyis rate, and is in photochemical equilibrium with atomic oxygen, whose lifetime becomes longer and longer with increasing altitude (see Fig. 5.3). It can be deduced from equation (5.14) that when $d(O_3)/dt=0$,

$$(O_3) = \frac{k_2(M)(O_2)(O)}{J_{O_3} + k_3(O)} \approx \frac{k_2(M)(O_2)(O)}{J_{O_3}} \tag{5.33}$$

and from (5.31), setting $d(O_x)/dt \approx d(O)/dt$,

Table 5.1 Example of the vertical distributions of oxygen compounds
in the middle atmosphere

Altitude	(O_2)	(O_3)	(O^3P)	(O^1D)
(km)	(cm^{-3})	(cm^{-3})	(cm^{-3})	(cm^{-3})
0	5.3(18)	8.0(11)	1.0(3)	2.7(-3)
5	3.2(18)	7.7(11)	4.6(3)	9.0(-3)
10	1.8(18)	7.5(11)	1.2(4)	1.8(-2)
15	8.8(17)	2.2(12)	1.3(5)	1.2(-1)
20	3.8(17)	4.8(12)	1.6(6)	8.9(-1)
25	1.7(17)	4.9(12)	8.6(6)	4.6(0)
30	7.8(16)	3.1(12)	3.3(7)	2.0(1)
35	3.6(16)	1.3(12)	1.2(8)	6.7(1)
40	1.6(16)	3.9(11)	3.8(8)	1.4(2)
45	8.0(15)	1.2(11)	1.1(9)	2.0(2)
50	4.2(15)	4.1(10)	2.0(9)	1.9(2)
55	2.3(15)	1.5(10)	2.7(9)	1.5(2)
60	1.2(15)	5.7(9)	3.3(9)	1.1(2)
65	6.3(14)	2.2(9)	3.8(9)	7.5(1)
70	3.2(14)	6.0(8)	3.5(9)	4.0(1)
75	1.6(14)	1.3(8)	2.8(9)	1.9(1)
80	7.1(13)	9.3(7)	1.0(10)	3.0(1)
85	4.8(13)	1.7(8)	1.1(11)	1.2(2)
90	2.3(13)	4.0(7)	4.5(11)	1.0(2)
95	1.0(13)	5.7(6)	8.6(11)	2.0(2)
100	4.6(12)	2.5(6)	1.1(12)	1.0(3)

$$\frac{d(O)}{dt} + 2\left(k_1(M) + \frac{k_3 k_2(M)(O_2)}{J_{O_3}}\right)(O)^2 = 2J_{O_2}(O_2) \qquad (5.34)$$

The photochemical time constant $\tau(O)$ for atomic oxygen is about 4 hours at 70 km, a day near 80 km and about a week at 100 km. Therefore, the distribution of atomic oxygen above the mesopause is dependent on dynamical conditions, and does not exhibit much diurnal variation above about 85 km, since its lifetime exceeds a day. At lower altitudes, on the other hand, atomic oxygen disappears rapidly after sunset, forming ozone through the recombination reaction. In the middle and lower stratosphere the abundance of atomic oxygen is very small compared to ozone, and the lifetime of ozone is greater than a day. Therefore, relatively small diurnal variations are expected to occur in stratospheric ozone concentrations. In the mesosphere, atomic oxygen densities are comparable to, and even greater than those of ozone (see Table 5.1 and Fig. 5.4), so that the recombination reaction leads to ozone increases throughout the mesosphere at the beginning of the night. Lean (1982) and Vaughn (1982) have presented measurements of the diurnal variation of mesospheric and stratospheric ozone.

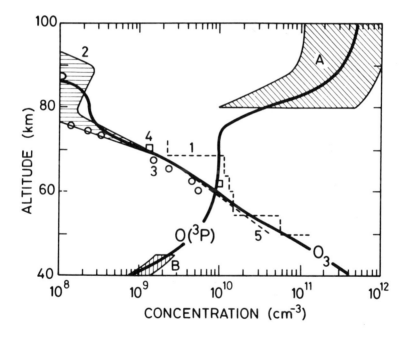

Fig. 5.4. Observed and calculated vertical distribution of ozone and atomic oxygen. 1. Carver et al. (1966); 2. Evans and Llewellyn (1970); 3. Weeks and Smith (1968); 4. de Jonckheere and Miller (1974); 5. Millier et al. (1980); A. Review of observations, Van Hemelrijck (1981); B. Anderson (1975).

Most of the total ozone molecules in a vertical column in the middle atmosphere are found at altitudes below 35 km (see Table 5.1). Measurement of the total ozone abundance can be performed from the ground using ultra-violet absorption techniques (see e.g., Dobson, 1968), and total ozone has been systematically observed for several years and at numerous locations. There-fore, the general spatial and temporal behavior of this gas in the atmosphere is very well documented (Dutsch, 1970; 1980; London, 1980). For example, it is well known that ozone is characterized by latitudinal and seasonal variations that include a maximum abundance in the region of the least production (see, for example, Dobson, 1968). Figure 5.5 shows the total ozone column versus month for several locations in the northern hemisphere. These values are expressed in Dobson units, which correspond to the height (in millicm) which the ozone column would have if all the gas were at standard temperature and pressure. It should be noted that ozone is most abundant at higher latitudes during all seasons. This increase with latitude is most pronounced in winter and spring, when ozone reaches its maximum. The seasonal variations only become large at latitudes poleward of 30 degrees north. At 80N, for example, the relative variation in total ozone is about 50 percent over six months.

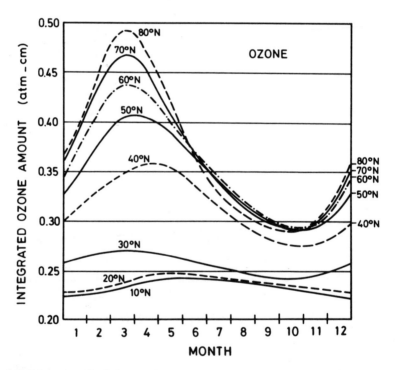

Fig. 5.5. Mean trend of the total ozone abundance during the year at different latitudes in the northern hemisphere. From Dobson (1968). (Copyright by Clarendon Press).

The global total ozone distribution versus latitude and time as derived from a network of ground based stations at different longitudes is shown in Figure 5.6. This figure shows that the peak total ozone abundance appears between 70 and 75N at the end of March and that the maximum observed in the southern hemisphere is smaller, and located at somewhat lower latitudes and earlier in the year, than that of the northern hemisphere.

Since the chemical lifetime of odd oxygen in the lower stratosphere is long compared to the time constants associated with transport by the mean meridional circulation (Fig. 5.3), transport by mean meridional advection can play an important role in determining its distribution. The hemispheric asymmetry in total ozone distributions as shown above is generally presumed to be due to differences in the stratospheric dynamics of the two hemispheres. As first pointed out based on empirical evidence by Dobson (1930), the primary mechanism accomplishing this transport can be simply understood to a first approximation in terms of the downward and poleward net transport associated with mean motion in the transformed Eulerian or diabatic framework, as discussed in Chapter 3. This implies that the slopes of the mean ozone mixing ratio surfaces will be somewhat steeper than the slopes of the isentropic surfaces represented by potential temperatures, as pointed out by Newell (1964), and discussed in more detail by Tung (1982).

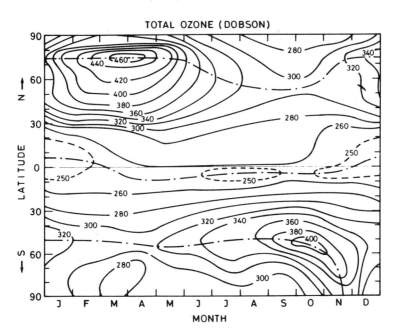

Fig. 5.6. Variation of total ozone with latitude and season. From London (1980).

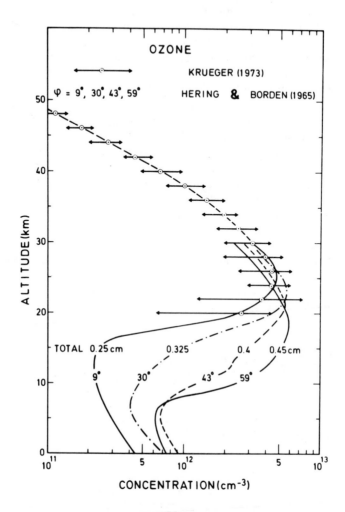

Fig. 5.7. Mean vertical distribution of the ozone concentration according to observations at different latitudes. Note the variations in total column abundance.

A more detailed understanding of the behavior of ozone requires examination of its altitude profile, so that the regions responsible for the observed total column variations can be identified. Figure 5.7 shows the vertical distribution of ozone implied by the observations of Krueger (1973) above 20 km, and those of Hering and Borden (1965) below 30 km, for different latitudes. It is clear that much of the change in total column abundance is due to differences found in the profiles below 20 km. Dobson (1968) suggested that this behavior is related to ozone transport in cyclones and anticyclones (e. g., weather systems) which propagate into the lower stratosphere. Experimental support for this

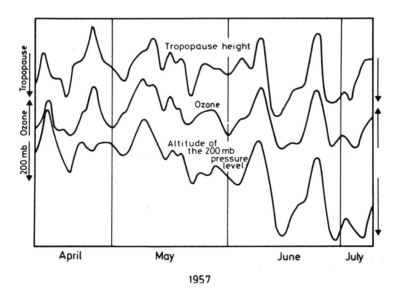

Fig. 5.8. Relationship between the variations in total ozone and the altitudes of the tropopause and the 200 mb level. From Dobson (1968). (Copyright by Clarendon Press).

view comes from the observation of close correlations between high total ozone abundances and low pressure systems, as shown, for example, in Figure 5.8. As a result of these processes, large local fluctuations in total ozone are often observed, with amplitudes as large as those related to the seasonal variations (see Fig. 5.9).

It should be noted that this behavior represents an obstacle to an accurate representation of the variability of total ozone using simple one or two-dimensional models, because the nature of transport at these lower levels is somewhat difficult to parameterize to suit the much larger physical scales resolved by such models. As pointed out, for example by Kida (1977), tracer experiments performed in his three-dimensional model imply that transport above 100 mb can be understood in terms of a residual Eulerian mean motion in the middle stratosphere (see §3.6), but that considerable dispersion related to cyclone propagation dominates the behavior for lower altitudes. Present photochemical models do not consider these processes in detail.

Satellite observations have added much to our understanding of the morphology of atmospheric ozone, both in terms of its altitude profile and total column density. For example, observations by the Backscatter Ultraviolet Spectrometer (BUV) experiment onboard Nimbus 4 have led to detailed descriptions of the monthly mean ozone abundance in the stratosphere, as discussed, for example, by Frederick et al. (1983). Figure 5.10 presents a monthly mean ozone distribution derived from that study. It is interesting to note the large abundances at the lower altitudes, particularly in the winter hemisphere,

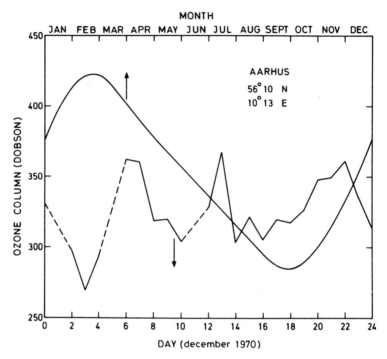

Fig. 5.9. Comparison between the amplitudes of seasonal and daily variations at Aarhus.

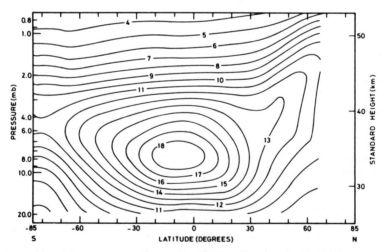

Fig. 5.10. Monthly mean stratospheric ozone distribution (μg/g) for January observed by the Nimbus 4 BUV experiment. From Frederick et al. (1983). (Copyright by the American Geophysical Union).

as indicated previously in terms of the observed total column abundances.

In the mesosphere, ozone observations are available from the Solar Mesosphere Explorer (SME) satellite. Figure 5.11 presents meridional cross sections of ozone from 50 to 90 km as observed by SME on March 14, 1982 and December 22, 1981 (Thomas et al., 1983). Latitudinal gradients and large local variations near the 90 km level are particularly noticeable. Barth et al. (1983) suggest that this may result from downward transport from the thermosphere because of the long O_x lifetime at these altitudes, as shown in Figure 5.3. Satellite studies will probably continue to improve our knowledge of the behavior of atmospheric ozone in the future.

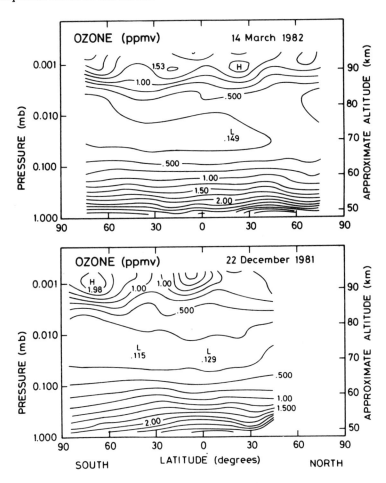

Fig. 5.11. Mesospheric ozone distributions observed by SME on March 14, 1982 and Dec. 22, 1981. From Thomas et al. (1983). (Copyright by the American Geophysical Union).

To conclude this discussion we turn to the question of long term trends in ozone. Ramanathan (1963), Angell and Korshover (1964; 1973) and Shah (1967) have noted the existence of an oscillatory component in the total ozone abundance in the tropics which is correlated with the behavior of the zonal winds in the lower stratosphere. These winds change speed and direction with a period between 24 and 33 months; this is the quasi-biennial oscillation noted in Chapter 3. Another question related to long term trends is that of a possible relationship between total ozone and the 11 year solar cycle, and this has been the subject of vigorous controversy (Willett, 1962; London and Haurwitz, 1963; Willett and Prohaska, 1965; Paetzold et al., 1972; Angell and Korshover, 1973). A theoretical study by Brasseur and Simon (1981) implies that the variation in total ozone over the solar cycle is probably less than three percent. Detection of such a small variation is difficult because of the large natural variability in total ozone due to its strong dependence on dynamical conditions, especially at middle and high latitudes (Figs. 5.5 and 5.9).

5.3 Carbon compounds

5.3.1 Methane •

Methane (CH_4) is a carbon containing species of particular importance in the atmosphere, both because it plays a role in the photochemistry of O_x, and because it is extremely useful as a tracer of atmospheric motion. This constituent is produced at ground level and is progressively transported upward until it is oxidized, providing an important source of carbon monoxide (CO).

The chemistry of atmospheric methane was first considered by Bates and Nicolet (1950). Its destruction is due to oxidation by the OH radical:

$$(c_2); \quad CH_4 + OH \rightarrow CH_3 + H_2O \tag{5.35}$$

or by reaction with excited oxygen $O(^1D)$:

$$(c_{1a}); \quad CH_4 + O(^1D) \rightarrow CH_3 + OH \tag{5.36a}$$

$$(c_{1b}); \quad CH_4 + O(^1D) \rightarrow CH_2O + H_2 \tag{5.36b}$$

or by reaction with chlorine:

$$(d_5); \quad CH_4 + Cl \rightarrow CH_3 + HCl \tag{5.37}$$

It should be noted that reaction (5.36b) is 10 times less probable than reaction (5.36a). Above the stratopause, it is also necessary to introduce the photolysis of CH_4, particularly by Lyman alpha:

$$(J_{CH_4}); \quad CH_4 + h\nu \rightarrow CH_2 + H_2(90\%) \tag{5.38a}$$

$$\rightarrow CH + H_2 + H(10\%) \qquad (5.38b)$$

with two quantum efficiencies. We note that this reaction produces molecular hydrogen. However, since the amount of methane in the upper part of the middle atmosphere is small, this process can generally be neglected in the global methane budget.

The vertical distribution of methane results from an equilibrium between its destruction by these processes and transport upward from the surface. The continuity equation can therefore be written:

$$\frac{\partial(CH_4)}{\partial t} + \vec{\nabla} \cdot \vec{\phi}(CH_4) + [c_1(O^1D) + c_2(OH) + d_5(Cl) + J_{CH_4}](CH_4) = 0 \qquad (5.39)$$

where $c_1 = c_{1a} + c_{1b}$

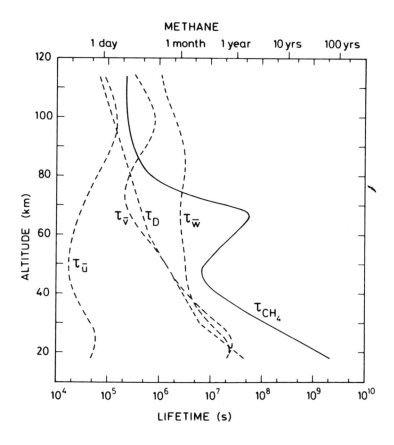

Fig. 5.12. Photochemical lifetime of CH_4, and the time constants associated with transport processes.

Figure 5.12 presents the photochemical lifetime for atmospheric methane as a function of altitude along with the time constants associated with transport by the winds and a one-dimensional diffusion profile. Since this species is not photochemically produced in the atmosphere, and its stratospheric lifetime against photochemical destruction is the same order of magnitude as transport by the mean meridional winds or vertical eddy diffusion in a one-dimensional sense, it provides an excellent tracer for transport processes (see e. g. Ackerman et al., 1978).

Migeotte (1948) first detected methane using spectroscopic methods. Since then, methane has been observed many times and its vertical distribution has been measured at various latitudes (see Ehhalt, 1974; Ehhalt et al., 1974; Ackerman et al., 1977; Bush et al., 1978; Fabian et al., 1979). Figure 5.13 summarizes

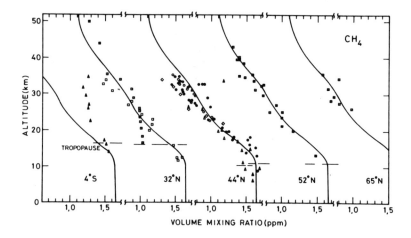

Fig. 5.13. Vertical distribution of the mole fraction of methane. Observations at various latitudes and one-dimensional model calculations (from Ehhalt and Tonnissen, 1980).

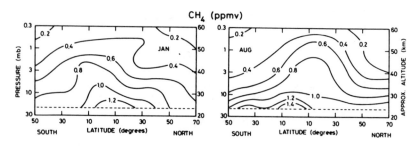

Fig. 5.14. Methane distributions observed by the SAMS satellite. From Jones and Pyle (1984). (Copyright by the American Geophysical Union).

some of these measurements, most of which have been performed by cryogenic collection and laboratory analysis, or by optical methods. The values at 32N at altitudes above 40 km were obtained by rocket, and the rest have been measured with stratospheric balloons. These data show that methane is well mixed in the troposphere with a mole fraction of 1.6×10^{-6}, while above the tropopause its concentration decreases rapidly. Recently, methane has been measured by the Stratospheric And Mesospheric Sounder (SAMS) satellite experiment, so that the global distribution of atmospheric methane can now be characterized, and used to deduce information about transport processes. Figure 5.14 presents observed methane distributions for several months as presented by Jones and Pyle (1984). The large values in the tropical lower and middle stratosphere reflect rapid upward transport from the tropospheric source region in the rising branch of the Hadley cell. The propagation of these high values into the summer hemisphere also indicates the direction of flow (see, e.g., Fig. 3.16).

As we have already noted, CH_4 is not photochemically produced in the atmosphere, because its complexity and the amount of energy required are prohibitive. Rather, the biosphere and lithosphere provide the natural sources of atmospheric methane. Most of the biological production is attributed to the effects of anaerobic bacteria such as Methanobacterium, Methanococcus and Methanosarcina. These are primarily found in alkaline regions, of relatively high temperature and enriched in organic materials. Peat bogs, swamps, ponds, rice paddies, and humid tundra produce large amounts of methane. Table 5.2 presents an estimate of the magnitude of these sources, and shows that biological production comprises between 440 and 850 MT/yr.

Human production of methane is mainly associated with industrial and mining activities. This source is probably weak on a global scale in comparison to biological sources. Comparison of the isotopic ratio of $^{14}C/^{12}C$ observed in the atmosphere versus that in organic matter can be used to estimate the

Table 5.2 Global methane production (megatons per year).
From Donahue (1980).

Natural humid zones (bogs, swamps, ponds ...)	200-300
Rice paddies	140 -280
Enteric fermentation	100 -220
Humid tundra and edges of lakes and oceans	3 - 50
Total biological sources	443- 850
Human sources related to industrial and mining activities	16-50
Methane of fossil origin	0 -160
Total of non-biological sources	16 -210
Total methane production	460 -1110

magnitude of these non-biological sources. Such studies indicate that this source must represent about 20 percent of the total production of methane, i. e., between 110 and 210 MT/yr. Measurement of the isotopic abundance must, however, be considered an upper limit due to the possibility of sample contamination. We adopt a value between 16 and 210 MT/yr for the total non-biological production of methane. A more detailed analysis of the methane budget is given by Crutzen (1983).

One may conclude that between 460 and 1110 MT of methane are produced each year, which corresponds to a mean vertical flux from 1 to 2.5x $10^{11}cm^{-2}s^{-1}$. Since the total atmospheric burden of methane is about 4600 MT, a mean lifetime is given by:

$$\frac{4600 \text{ MT}}{460 - 1110 \text{ MT produced per year}} \approx 4 - 10 \text{ years.} \tag{5.40}$$

5.3.2 *Methane oxidation chemistry* ‡

The photochemistry of the methane oxidation process will next be discussed in more detail. The oxidation of methane produces a methyl CH_3 radical, which reacts rapidly with molecular oxygen:

$$(c_4); \quad CH_3 + O_2 + M \rightarrow CH_3O_2 + M \tag{5.41}$$

to produce methyl peroxy radical. This species can be destroyed by reaction with NO, with itself, or with HO_2. The first two processes lead to the formation of the methoxy radical:

$$(c_5); \quad CH_3O_2 + NO \rightarrow CH_3O + NO_2 \tag{5.42}$$

$$(c_{14}); \quad CH_3O_2 + CH_3O_2 \rightarrow 2CH_3O + O_2 \tag{5.43}$$

Note that if reaction (5.42) is followed by photolysis of NO_2, $NO_2 + h\nu \rightarrow NO + O$, then the net effect can be to produce odd oxygen as well as CH_3O. This reaction is an important source of ozone in the lower stratosphere and upper troposphere (see Crutzen, 1971). Methoxy is in turn destroyed by reaction with molecular oxygen:

$$(c_{15}); \quad CH_3O + O_2 \rightarrow CH_2O + HO_2 \tag{5.44}$$

to produce formaldehyde. In the third case, the reaction

$$(c_7); \quad CH_3O_2 + HO_2 \rightarrow CH_3OOH + O_2 \tag{5.45}$$

produces methyl hydrogen peroxide, which either reacts rapidly with OH to reform the methyl peroxy radical:

$$(c_{17}); \quad CH_3OOH + OH \rightarrow CH_3O_2 + H_2O \tag{5.46}$$

or is photolyzed, producing a methyl peroxy and hydroxyl radical:

$$(J_{CH_3OOH}); \quad CH_3OOH + h\nu \rightarrow CH_3O + OH \tag{5.47}$$

CH_3OOH can also undergo heterogeneous reactions and precipitate in raindrops or aerosols. It seems likely however, that the primary loss process is the reaction with OH (5.46).

Therefore, regardless of the chemical path taken, the carbon atom of the methane molecule must eventually end up as formaldehyde following reaction (5.44). It should be noted that the formation of CH_2O can be accomplished directly via (5.36b). Other mechanisms which produce formaldehyde include:

$$(c_{4a}); \quad CH_3 + O_2 \rightarrow CH_2O + OH \tag{5.48}$$

$$(c_6); \quad CH_3O_2 + NO_2 \rightarrow CH_2O + HNO_3 \tag{5.49}$$

but these processes are relatively unimportant. In the mesosphere and lower thermosphere, reaction with atomic oxygen can also play a role:

$$(c_3); \quad CH_3 + O \rightarrow CH_2O + H \tag{5.50}$$

Formaldehyde has been observed (Barbe et al., 1979) in the stratosphere using infrared absorption. These measurements imply an integrated column density of 5×10^{15} cm^{-2} with a factor of two accuracy. Barbe et al. have suggested a probable vertical distribution of CH_2O based on their observations. Figure 5.15 compares these data to a calculated distribution. The calculated

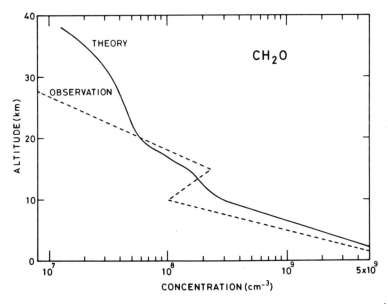

Fig. 5.15. Vertical distribution of formaldehyde. Comparison between observations by Barbe et al. (1979) and calculations by Ehhalt and Tonnisen (1980).

CH_2O profile results from an equilibrium between the production mechanisms just described and destruction by OH

$$(c_8); \quad CH_2O + OH \rightarrow CHO + H_2O \tag{5.51}$$

by atomic oxygen

$$(c_9); \quad CH_2O + O \rightarrow CHO + OH \tag{5.52}$$

by chlorine atoms

$$(c_{10}); \quad CH_2O + Cl \rightarrow HCl + CHO \tag{5.53}$$

or by photolysis. This last process can occur via two distinct pathways which lead either to the formation of the unstable CHO radical

$$(J_{H-CHO}); \quad CH_2O + h\nu \rightarrow CHO + H \tag{5.54}$$

or to the long lived species, CO and molecular hydrogen:

$$(J_{H_2-CO}); \quad CH_2O + h\nu \rightarrow CO + H_2 \tag{5.55}$$

Figure 5.16 schematically illustrates the presently understood reaction sequence for carbon compounds. We may now write the continuity equations for the carbon-containing reactive intermediates, which are all sufficiently short lived that transport may be neglected. Thus,

$$\frac{d(CH_3)}{dt} + [c_3(O) + c_4(O_2)(M)](CH_3) = [c_{1a}(O^1D) + c_2(OH) + d_5(Cl)](CH_4) \tag{5.56}$$

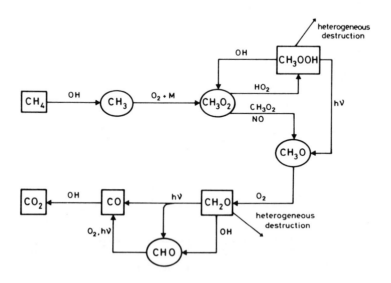

Fig. 5.16. Schematic diagram of carbon chemistry.

$$\frac{d(CH_3O_2)}{dt} + [c_5(NO) + c_6(NO_2) + c_7(HO_2) + 2c_{14}(CH_3O_2)](CH_3O_2)$$

$$= c_4(O_2)(M)(CH_3) + c_{17}(CH_3OOH)(OH) \qquad (5.57)$$

$$\frac{d(CH_3O)}{dt} + c_{15}(O_2)(CH_3O) = [c_5(NO) + 2c_{14}(CH_3O_2)](CH_3O_2)$$

$$+ J_{CH_3OOH}(CH_3OOH) \qquad (5.58)$$

$$\frac{d(CH_3OOH)}{dt} + [J_{CH_3OOH} + c_{17}(OH)](CH_3OOH) = c_7(CH_3O_2)(HO_2) \qquad (5.59)$$

$$\frac{d(CH_2O)}{dt} + [c_8(OH) + c_9(O) + J_{H-HCO} + J_{H_2-CO} + d_{10}(Cl)](CH_2O)$$

$$= c_{1b}(O^1D)(CH_4) + c_3(O)(CH_3) + c_{4a}(O_2)(CH_3)$$

$$+ c_6(NO_2)(CH_3O_2) + c_{15}(O_2)(CH_3O) \qquad (5.60)$$

$$\frac{d(CHO)}{dt} + [J_{CHO} + c_{12}(O_2)](CHO) = [J_{H-HCO} + c_8(OH)$$

$$+ c_9(O) + d_{10}(Cl)](CH_2O) \qquad (5.61)$$

Substituting expression (5.56) into (5.60), and assuming photochemical equilibrium for each species, the concentration of formaldehyde can be written as:

$$(CH_2O) = \frac{[c_1(O^1D) + c_2(OH) + d_5(Cl)](CH_4)}{J_{H-HCO} + J_{H_2-CO} + c_8(OH) + c_9(O) + d_{10}(Cl)} \qquad (5.62)$$

It should be noted that the methane oxidation chain also results in the production and destruction of the hydrogen compounds H, OH and HO_2:

$$\frac{d(H)}{dt} + \frac{d(OH)}{dt} + \frac{d(HO_2)}{dt} + [c_2(CH_4) + c_8(CH_2O)$$

$$+ c_{17}(CH_3OOH)](OH) + c_7(CH_3O_2)(HO_2) = (CH_2O)[(J_{H-HCO}) + c_9(O)]$$

$$+ (HCO)[J_{HCO} + c_{12}(O_2)] + (CH_3OOH)J_{CH_3OOH} + c_{1a}(O^1D)(CH_4) \qquad (5.63)$$

which can be rewritten as

$$\frac{d(H)}{dt} + \frac{d(OH)}{dt} + \frac{d(HO_2)}{dt} + c_{17}(CH_3OOH)(OH)$$

$$= [2c_{1a}(1 + X) + 2c_{1b}X](O^1D)(CH_4) + 2c_2X(OH)(CH_4) \qquad (5.64)$$

where

$$X = \frac{J_{H-HCO} + c_9(O) + 0.5d_{10}(Cl)}{J_{H-HCO} + J_{H_2-CO} + c_8(OH) + c_9(O) + d_{10}(Cl)} \qquad (5.65)$$

which is approximately equal to $1/2$.

5.3.3 Some end products of methane oxidation: carbon monoxide and dioxide •

The relative efficiency of the two paths of formaldehyde photolysis (reactions 5.54 and 5.55) is an important parameter because it determines the fraction of methane which ultimately produces molecular hydrogen. Reactions (5.51) and (5.52), on the other hand, produce CHO, which in turn yields hydrogen containing radicals:

$$(J_{CHO}); \quad CHO + h\nu \rightarrow CO + H \qquad (5.66)$$

$$(c_{12}); \quad CHO + O_2 \rightarrow CO + HO_2 \qquad (5.67)$$

As we will see in §5.4, these hydrogen radicals recombine to produce water vapor. Therefore, the destruction of methane by this long chain eventually produces water vapor and molecular hydrogen in the stratosphere and mesosphere, such that the sum

$$2f_{CH_4} + f_{H_2} + f_{H_2O}$$

where f is the mole fraction, is approximately constant with altitude (neglecting other much less abundant hydrogen species such as OH and H). The apparent increase in water vapor abundance in the stratosphere (see below) can be qualitatively explained by this mechanism. Each oxidized methane molecule also produces CO. Noting that the photolysis of CO_2 also plays a role in CO production, one may write the production rate for CO:

$$P(CO) = [c_1(O^1D) + c_2(OH) + c_5(Cl)](CH_4) + J_{CO_2}(CO_2) \qquad (5.68)$$

The oxidation of methane by the chemical processes just described leads to an annual production of CO of about 800 MT. This number depends directly on the mean value of the OH radical density and constitutes about 30 percent of the total CO production. Emission of CO by combustion of gasoline and natural gas must also be considered. Logan et al. (1981) suggest injection of about 450 MT through combustion, with a distribution (425 MT in the northern hemisphere and only 25 MT in the southern hemisphere) which depends on

Table 5.3. Global budget of carbon monoxide (10^{12}g CO yr^{-1})

Sources	Total	NH	SH
Fossil fuel combustion	450 (400 - 1000)	425	25
Oxidation of anthropogenic hydrocarbons	90 (0-190)	85	5
Wood for home heating	51 (17 -150)	33	17
Clearing fields†	380 (80 -1200)	260	120
Savannah burning†	200 (50 -600)	100	100
Forest fires†	25 (8-75)	22	3
Ocean	40 (20 -80)	13	27
Emission by plants	70 (21 -210)	50	20
Oxidation of isoprene and terpenes	560 (190-1700)	380	180
Methane oxidation	780± 350	400	380
TOTAL	2650 (1500-6000)	1770	880
Losses			
Uptake by soil	250	210	40
Photochemistry	3100 (2000-4000)	1850	1250
TOTAL	3350	2060	1290

† See Seiler and Crutzen (1980); Crutzen et al. (1979).

the localization of anthropogenic sources. These emissions are comprised of automobile and industrial effluents and are localized in industrial areas. This estimate of 450 MT should be compared to the estimates of Jaffe (1973) (360 MT/yr) and Seiler (1974) (640 MT/yr). The difficulty of quantitatively establishing a global source strength is apparent. In any case, the anthropogenic

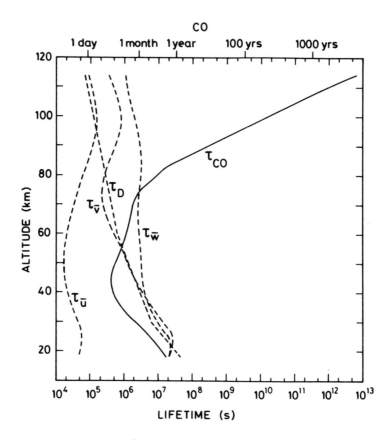

Fig. 5.17. Photochemical lifetime of carbon monoxide, and the time constants for transport processes.

production of CO seems globally smaller than the production via methane oxidation, although the amplitudes are probably comparable in the northern hemisphere. Thus it is not surprising that the observed abundance of CO varies greatly with latitude, especially in the troposphere. Seiler (1974) found that the mole fraction of CO increases from 5×10^{-8} in the southern hemisphere to about 2×10^{-7} in the more industrialized northern hemisphere.

Other sources of CO which must appear in the global budget include: domestic combustion of wood for heat by about 1.5 million persons, agricultural operations and savannah burning, principally in tropical zones, as well as forest fires and plant respiration (see, Seiler and Crutzen, 1980). Finally, the important contribution from hydrocarbon oxidation, emitted either by plants (isoprene and other terpenes) or by man, must be considered. This source was discussed by Wofsy et al. (1972), Logan et al. (1978) and by Zimmerman et al. (1978). A budget proposed by Logan (1980) and by Logan et al. (1981) is shown in Table 5.3. It should be emphasized that these numbers are quite uncertain

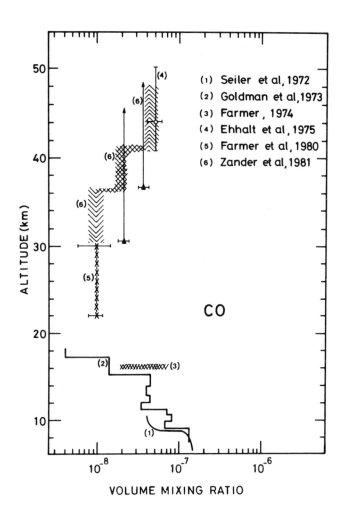

Fig. 5.18. Vertical distribution of the CO mole fraction observed in the strato-sphere. (From Zander et al., 1981, copyrighted by the American Geophysical Union).

in most cases.

Carbon monoxide is mostly destroyed in the troposphere, but some of the molecules penetrate to the stratosphere. The only important destruction process is provided by reaction with OH:

$$(c_{20} = a_{36}); \quad CO + OH \rightarrow CO_2 + H \quad (5.69)$$

and the continuity equation for CO may be written:

$$\frac{\partial (CO)}{\partial t} + \vec{\nabla} \cdot \vec{\phi}(CO) + c_{20}(OH)(CO) = P_{CO} \qquad (5.70)$$

where P_{CO} is given by expression (5.68).

Figure 5.17 presents the vertical profile of the photochemical lifetime of CO, as well as the time scale for transport by the winds and by a vertical eddy diffusion coefficient. Because of its long chemical lifetime, the distribution of CO is expected to depend strongly on meridional and vertical transport throughout the middle atmosphere. Figure 5.18 shows a number of observations of the vertical distribution of CO in the stratosphere and troposphere. The stratospheric abundances are low and a rapid decrease in mole fraction is typically observed above the tropopause. This sharp gradient suggests that transport from the troposphere to the stratosphere is not frequent at middle and high latitudes, as discussed previously based on the observed distribution of water vapor (see Chapter 3). The data obtained by Farmer et al. exhibit a constant mole fraction of about 10^{-8} from about 22 to 36 km. On the other hand, the observations by Zander et al. (1981) display an increase in the upper stratosphere, as predicted by models. CO profiles have also been measured in the mesosphere, particularly by ground based microwave methods (e. g. Clancy et al., 1982). Its usefulness as a tracer for mesospheric transport was examined, for example by Hays and Olivero (1970), and by Allen et al. (1981).

Carbon dioxide is another end product of the long chemical methane oxidation chain. Although this unreactive species is not important in the chemistry of the middle atmosphere, it plays a central role in the thermal budget of the atmosphere (see Chapter 4). CO_2 is produced at ground level by plants and

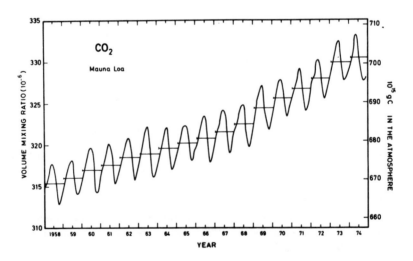

Fig. 5.19. Evolution of the relative concentration and total abundance of CO_2 in the atmosphere. From Keeling (1976). (Copyright by Tellus).

by the oceans. The combustion of fossil fuels constitutes an increasingly important anthropogenic source of CO_2, and explains its observed increase with time over recent years (see Fig. 5.19). The possible climatic effects of increasing CO_2 will be discussed briefly in Chapter 7.

The mole fraction of CO_2 is almost constant with altitude in the homosphere in spite of some photolysis in the lower thermosphere, and its 1983 value is 3.35×10^{-4}. It exhibits a weak annual variation due in large part to biological processes.

Other carbon containing species, although present in very small amounts, can also provide important information regarding the photochemistry and dynamics of the middle atmosphere. These species will not be presented here in detail, but we note, for example, that measurements of lower stratospheric ethane C_2H_6 and propane C_3H_8 have been used to deduce atomic chlorine densities there (Rudolph et al., 1981).

5.4 Hydrogen compounds

5.4.1 General •

The effects of hydrogen compounds on the behavior of other chemical species, particularly that of ozone in the mesosphere, was first examined by Bates and Nicolet (1950), and has been the subject of numerous subsequent studies (Hampson, 1966; Hunt, 1966; Hesstvedt, 1968; Crutzen, 1969; Nicolet, 1971; etc.). The high reactivity of the hydrogen free radicals, especially OH, makes these species of particular importance in atmospheric chemistry. The compounds which initiate the hydrogen radical chemistry are methane (for which the budget was discussed in §5.3), water vapor, and molecular hydrogen.

Water is quite abundant on the planet Earth, but is mostly confined to the surface and the interior. The atmosphere contains only a very small fraction of the water in the oceans. Liquid water is, however, present in small amounts compared to the vast number of molecules of H_2O in hydrated minerals which compose the solid part of the earth. Clouds contain only a few percent of the water in the atmosphere, although they cover about half of the surface of the globe, so that most of the water in the atmosphere occurs in the vapor phase.

A continuous source of atmospheric water vapor is provided by the ocean surface. For this reason, water vapor is quite abundant at low altitudes. Its ground level mole fraction is variable, changing from about 10^{-2} to about 10^{-4} in the desert. Vertical sondes taken since 1945 show that the water vapor content decreases continuously with altitude in the troposphere to reach a minimum mole fraction near the tropopause of the order of a few ppmv.

Brewer (1949) suggested that water vapor can only reach the stratosphere in the ascending branch of the Hadley cell (see Chapter 3), where the tropopause temperature is sufficiently cold (190 to 200K) to "freeze dry" the air and precipitate out much of the water. The relatively dry air penetrating to the stratosphere is then transported towards polar regions and descends into the troposphere. Other mechanisms might be examined as stratospheric water vapor sources. For example, cumulus clouds which penetrate the tropopause might be proposed as a source of stratospheric water. However, Johnston and Solomon (1979) and Danielsen (1982) have argued that such clouds would reduce the stratospheric water vapor content, particularly in the tropics. Further, several observations suggest that the minimum in tropical H_2O is located not at the tropopause, but a few kilometers above (Kley et al., 1979). This physical separation has called into question the traditional belief that the tropopause temperature limits the water vapor content of parcels of air entering the stratosphere. Danielsen (1982) suggested that transport in penetrating cumulus towers could explain this difference. This theory also implies that a substantial fraction of the tropospheric-stratospheric exchange in the tropics may be localized in such towers, rather than in a slow large scale upward transport.

In the middle and upper stratosphere, the formation of water vapor from methane oxidation occurs via the mechanisms discussed in §5.3. In the

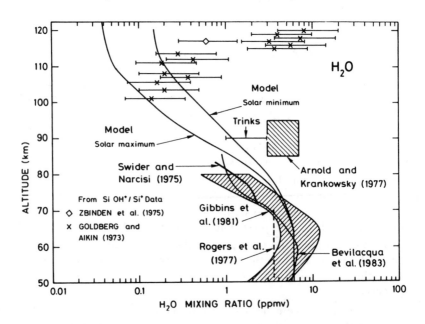

Fig. 5.20. Observations of water vapor in the mesosphere and lower thermosphere, and one-dimensional model calculations. Adapted from Solomon et al. (1982a).

mesosphere, the presence of water vapor leads to the formation of hydrated cluster ions, such as $H^+(H_2O)_n$, which have been commonly observed in the D-region by mass spectrometer (see Chapter 6). H_2O photolyzes in the thermosphere and upper mesosphere by absorption of the Lyman$-\alpha$ line:

$$(J_{H_2O}); \quad H_2O + h\nu(\lambda < 200nm) \rightarrow H + OH \tag{5.71}$$

reducing the water vapor mixing ratio and increasing those of the hydrogen free radicals. In the stratosphere and the lower mesosphere, water vapor reacts with $O(^1D)$:

$$(a_1^*); \quad H_2O + O(^1D) \rightarrow 2OH \tag{5.72}$$

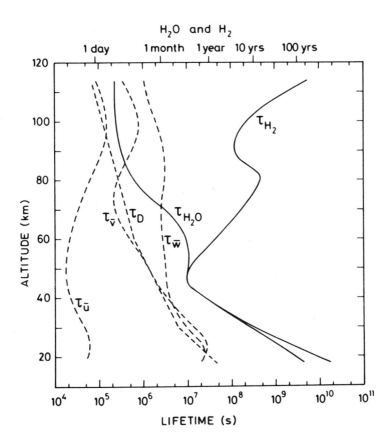

Fig. 5.21. Photochemical lifetimes of water vapor and molecular hydrogen, and the time constants for atmospheric transport processes in the middle atmosphere.

Upper stratospheric and mesospheric water vapor can be measured by microwave techniques, either from the ground (e. g., Gibbins et al., 1981; Deguchi and Muhleman, 1982; Bevilacqua et al., 1983) or from aircraft (Waters et al., 1980). In the lower thermosphere its density has been inferred from ion composition measurements (Arnold and Krankowsky, 1977; Solomon et al., 1982a). Figure 5.20 summarizes some of these observations, and shows an altitude profile of mesospheric and lower thermospheric water vapor calculated with a one-dimensional model for two different levels of solar activity.

Figure 5.21 presents the photochemical lifetime of water vapor, and the time constants characterizing atmospheric transport. The production of water vapor by methane oxidation is essentially complete by about 50 km. Because the photochemical and vertical transport lifetimes for this gas are comparable above about 50 km, and because there is no known chemical source of water vapor in this region, it provides an excellent tracer for mesospheric transport processes (Allen et al., 1981; Bevilacqua et al., 1983). In addition to its chemical properties, water vapor is radiatively active, and plays an important role in the atmospheric thermal budget through cooling by infrared emission.

Long term variations in stratospheric water vapor densities have been documented: above Washington, D. C., an increase of at least 30 percent was observed from 1964 to 1969, followed by constant values from 1970 to 1974 and then a decrease (Fig. 5.22). This modulation may be related to the 11-year cycle of solar activity through its possible influence on the tropical tropopause (see §7.3).

The vertical distribution of water vapor has been measured several times in the stratosphere (see, e.g. Fig. 5.23). It seems clear that an increase in the mixing ratio occurs with increasing altitude, as indicated by several observations (see, for example, Louisnard et al., 1980), and as expected from methane oxidation chemistry. However, it is not clear whether the observed gradient is equal to that expected theoretically (see, e. g. WMO, 1982), which may be

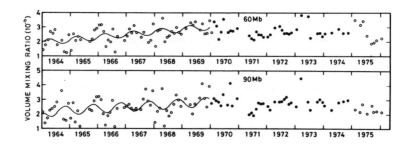

Fig. 5.22. Evolution of the water vapor mole fraction observed at altitudes from 50 to 90 mb. See Mastenbrook (1974), Mastenbrook and Oltmans (1983), and Penndorf (1978).

related to transport processes, or perhaps to uncertainties in the OH chemistry.

Because of its long chemical lifetime in the stratosphere, water vapor should be expected to exhibit variations which are related to atmospheric transport. Some observations, especially those of Kley et al. (1979, 1980) suggest that the water vapor profile is highly structured in discrete layers, with a vertical scale of these layered features of about 1000 m or less (Fig. 5.23). These variations imply a stratified structure in the lower part of the middle atmosphere, and an absence of much diffusion in the region of the measurements. This suggests that the air must be advected horizontally on a time scale much shorter than small scale diffusion.

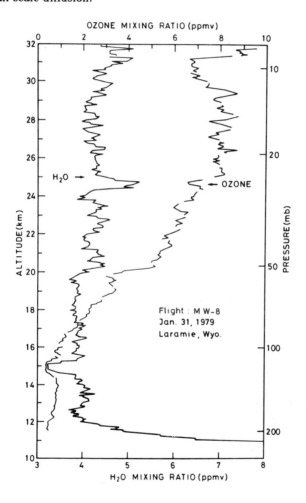

Fig. 5.23. Simultaneous observations of ozone and water vapor by Kley et al. (1980).

ZONAL MEAN WATER VAPOR CROSS SECTION (ppmv)

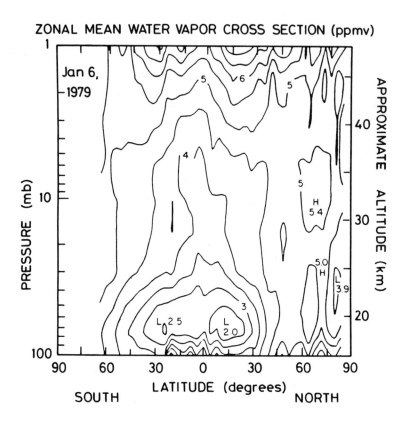

Fig. 5.24. Distribution of water vapor observed by the LIMS satellite experiment. From Gille and Russell (1984). (Copyright by the American Geophysical Union).

The Limb Infrared Monitor of the Stratosphere (LIMS) experiment onboard the NIMBUS 7 satellite measured the global distribution of stratospheric water vapor from November, 1978 to May, 1979 (Gille and Russell, 1984). A zonally averaged distribution obtained from this experiment is shown in Figure 5.24. It is interesting to note the extremely low water vapor mixing ratios observed in the tropical lower stratosphere, which are probably a result of the freeze-drying mechanism mentioned earlier. The larger values observed at higher latitudes and altitudes are at least partly due to methane oxidation. The behavior of water vapor in the atmosphere is complex because it is related to chemical, dynamical and thermodynamic conditions. The sensitivity of atmospheric water vapor to the temperature at which it is injected into the stratosphere is, however, particularly marked (see Appendix A), so that water vapor is an excellent tracer for troposphere-stratosphere exchange processes. The systematic observation of water vapor, especially by satellite, should lead to a better understanding of atmospheric transport.

The hydrogen molecule also provides a source of the free radicals H and OH, because it reacts rapidly with the $O(^1D)$ atom:

$$(a_3^*); \quad H_2 + O(^1D) \rightarrow H + OH \tag{5.73}$$

The destruction of hydrogen also occurs through

$$(a_{19}); \quad H_2 + OH \rightarrow H_2O + H \tag{5.74}$$

which plays an important role above 50 km. Mathematical models show that these processes lead to a global H_2 loss rate of 17 ± 7 MT/yr in the troposphere, and 1 ± 0.5 MT/yr in the stratosphere. In addition to this chemical loss, possible loss of hydrogen at ground level must also be considered, and this is likely to vary with the type of vegetation, temperature, etc. A few measurements of the uptake rate into various soils (Ehhalt, 1973) and an estimate of the distribution of terrestrial ecosystems suggest a global uptake rate of between 20 and 107 MT/yr. The chemical production of H_2 is due to the photolysis of formaldehyde produced by methane oxidation in the lower stratosphere. About 13 ± 3 Mt/yr are produced by this mechanism (Schmidt et al., 1980). Oxidation of isoprene and other terpenes lead to an additional production of about 10-35 MT/yr. A precise determination of these quantities requires a better understanding of the spatial distribution of formaldehyde. In the mesosphere, molecular hydrogen can be formed by recombination of the H and HO_2 radicals, which are themselves produced by water vapor photolysis. This mechanism plays a secondary role in the global hydrogen budget, but can perhaps produce large H_2 densities in the mesosphere and lower thermosphere (e. g., Liu and Donahue, 1974).

Another possible mechanism for hydrogen production must be mentioned. Formation of H_2 by anaerobic bacteria is possible but this contribution can be neglected for atmospheric purposes because these hydrogen molecules would be rapidly destroyed at ground level by other microorganisms. The oceans are however, supersaturated in hydrogen, and constitute an important source; indeed, this source is more important than that provided by volcanoes. Schmidt et al. (1980) estimate that natural ground level production is between 2 and 5 MT/yr.

Several observations have shown that hydrogen is affected by pollution sources. The hydrogen molecule is an important product of incomplete combustion in several industrial processes. The global amount of hydrogen emitted is not easy to estimate because the amount of gas produced depends on the condition of the motors. A very rough estimate of about 25 MT/yr can be derived by examining the ratio of CO and H_2 in polluted air (Schmidt, 1974) and the global CO emission rate (Logan, 1980). This contribution comes mostly from automobiles.

Formation of hydrogen through forest and savannah burning, especially for agricultural purposes in tropical regions, must also be considered. The destruction of biomass by combustion is believed to produce about 9 to 21 MT/yr according to Crutzen et al. (1979).

Table 5.4 Global molecular hydrogen budget

Production ($\times 10^{12}$ g/yr)	Global	NH	SH
Natural sources (oceans, volcanoes)	2.0-5.0	1.2-3.0	0.8-2.0
Methane oxidation	10.8-16.0	5.4-8.0	5.4-8.0
Terpene and isoprene oxidation †	10.0-35.0	6.0-20.0	4.0-15.0
Anthropogenic Sources			
Automobiles	11.5-57.1	9.8-48.5	1.7-8.6
Industry	-	-	-
Biomass burning §	9.0-21.0	4.5-10.5	4.5-10.5
Total production	43.3-134.1	26.9-90.0	16.4-44.1
Loss			
Photochemistry			
Stratosphere	0.6-1.6	0.3-0.8	0.3-0.8
Troposphere*	10.0-23.7	5.0-11.8	5.0-11.9
Destruction at surface	19.9-107.4	12.9-69.6	7.0-37.8
Total loss	30.5-132.7	18.2-82.2	12.3-50.5
Mean mixing ratio (ppmv)	0.560	0.575	0.550
Abundance (10^{12} g)	170.7	87.2	83.5

* Mean tropospheric (OH) assumed: $6.5 \pm 2.5 \times 10^6 \text{cm}^{-3}$ (Perner et al., 1976)

† Zimmerman et al. (1978)

§ Crutzen et al. (1979)

    Table 5.4 summarizes the components of the global hydrogen budget, and shows the contributions in each hemisphere. The total amount of hydrogen in the atmosphere is about 200 MT (170 MT in the troposphere and 30 MT above the tropopause). The global cycling time of hydrogen is of the order of 3 ± 2 years.

    Although the atmospheric hydrogen content at ground level was first measured many years ago (Paneth, 1937; Glueckauf and Kitt, 1957), its detection in the stratosphere occurred much later (Ehhalt and Heidt, 1973; Ehhalt et al., 1974; 1975a; Schmidt et al., 1976; Fabian et al., 1979). These observations have mostly been performed by grab sampling from balloon, and they have

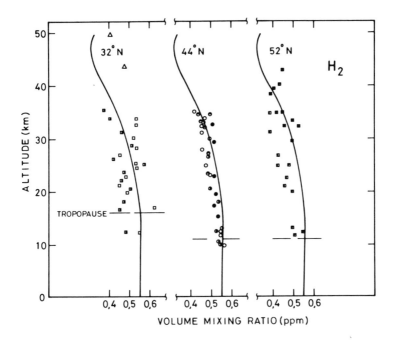

Fig. 5.25. Observations of the mole fraction of molecular hydrogen in the stratosphere, compared to a one-dimensional model. From Ehhalt and Tonnissen
(1980).

demonstrated that the H_2 mixing ratio is about 5.6×10^{-7} at the tropopause and
in the lower and middle stratosphere. Schmidt et al. (1980) reported a small
interhemispheric difference of about 5.8% at ground level and 1.5% at the tropopause. A small decrease in the observed mole fraction with increasing altitude
is also indicated by rocket measurements in the upper stratosphere, as shown in
Figure 5.25 (Scholz et al., 1970; Ehhalt et al. 1975b). This decrease in mixing
ratio is less pronounced at lower latitudes where vertical motion is more
intense.

5.4.2 Odd hydrogen chemistry ‡

The behavior of the hydrogen radicals produced by dissociation of H_2O, H_2
and CH_4 depends strongly on altitude. In the mesosphere and stratosphere, the
lifetime of atomic hydrogen is very short because the three body reaction with
oxygen rapidly produces a hydroperoxy radical:

$$(a_1); \quad H + O_2 + M \rightarrow HO_2 + M \tag{5.75}$$

but this reaction becomes slow at higher altitudes due to the low air densities found there.

Atomic hydrogen can also react with ozone, and this reaction becomes important in the upper stratosphere and mesosphere:

$$(a_2); \quad H + O_3 \rightarrow O_2 + OH\dagger(v \leqslant 9) \tag{5.76}$$

At altitudes above 40 km, the reactions of the OH and HO_2 radicals with atomic oxygen are important:

$$(a_5); \quad OH + O \rightarrow O_2 + H \tag{5.77}$$

and

$$(a_7); \quad HO_2 + O \rightarrow O_2 + OH \tag{5.78}$$

On the other hand, in the middle and lower stratosphere and in the troposphere, the reactions of the hydrogen radicals with ozone must be considered:

$$(a_6); \quad OH + O_3 \rightarrow O_2 + HO_2 \tag{5.79}$$

$$(a_{6b}); \quad HO_2 + O_3 \rightarrow 2O_2 + OH \tag{5.80}$$

These four reactions provide important loss processes for odd oxygen. Near the tropopause, the OH and HO_2 radicals react primarily with carbon and nitrogen monoxide:

$$(a_{26}); \quad HO_2 + NO \rightarrow NO_2 + OH \tag{5.81}$$

$$(c_{20} \equiv a_{36}); \quad OH + CO \rightarrow CO_2 + H \tag{5.82}$$

In the mesosphere, atomic hydrogen becomes quite abundant and the following reactions (note that molecular hydrogen can be produced) become important:

$$(a_{23a}); \quad H + HO_2 \rightarrow OH + OH \tag{5.83}$$

$$(a_{23b}); \quad H + HO_2 \rightarrow H_2 + O_2 \tag{5.83b}$$

$$(a_{23c}); \quad H + HO_2 \rightarrow H_2O + O \tag{5.83c}$$

Hydrogen peroxide also plays an important role in oxygen-hydrogen chemistry. It is produced by the HO_2 "self-reaction":

$$(a_{27}); \quad HO_2 + HO_2 \rightarrow H_2O_2 + O_2 \tag{5.84}$$

and destroyed during the day through photolysis:

$$(J_{H_2O_2}); \quad H_2O_2 + h\nu \rightarrow 2OH \tag{5.85}$$

and to a lesser extent, by reaction with OH:

$$(a_{30}); \quad H_2O_2 + OH \rightarrow H_2O + HO_2 \tag{5.86}$$

It should be noted that OH is also destroyed by reaction with HO_2:

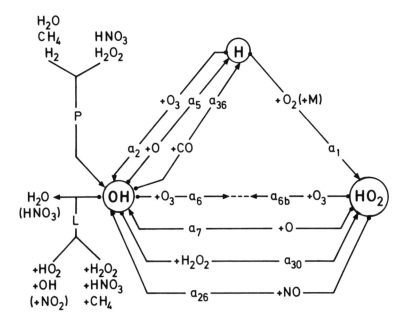

Fig. 5.26. Principal aeronomic reactions of hydrogen compounds in the stratosphere.

$$(a_{17}); \quad OH + HO_2 \rightarrow H_2O + O_2 \tag{5.87}$$

which is faster than the reaction

$$(a_{16}); \quad OH + OH \rightarrow H_2O + O \tag{5.88}$$

In the lower stratosphere, the presence of large amounts of HNO_3, HO_2NO_2 and H_2O_2 lead to an additional destruction of hydroxyl radical:

$$(b_{27}); \quad OH + HNO_3 \rightarrow H_2O + NO_3 \tag{5.89}$$

$$(b_{28}); \quad OH + HO_2NO_2 \rightarrow H_2O + O_2 + NO_2 \tag{5.90}$$

$$(a_{30}); \quad OH + H_2O_2 \rightarrow H_2O + HO_2 \tag{5.91}$$

The effect of these reactions is only important below 25 km and can only be quantitatively evaluated if the photochemistry of nitrogen compounds is considered. It should be noted that the first two reactions lead to an important loss of odd hydrogen in the lower stratosphere.

Figures 5.26 and 5.27 present schematic representations of the most important reactions in hydrogen chemistry for the lower and upper portions of the middle atmosphere, respectively. Retaining the most important of these reactions, we may write the continuity equations for the long lived oxygen-hydrogen

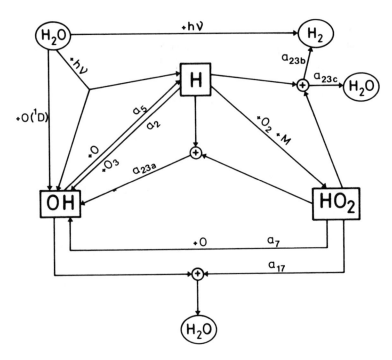

Fig. 5.27. Principal aeronomic reactions of hydrogen compounds in the meso-sphere and lower thermosphere.

species:

$$\frac{\partial(H_2O)}{\partial t} + \vec{\nabla}\cdot\vec{\phi}(H_2O) + [J_{H_2O} + a_1^*(O^1D)](H_2O)$$

$$= a_{16}(OH)^2 + a_{17}(OH)(HO_2) + a_{19}(OH)(H_2) + a_{23c}(H)(HO_2) \qquad (5.92)$$

$$\frac{\partial(H_2)}{\partial t} + \vec{\nabla}\cdot\vec{\phi}(H_2) + [a_3^*(O^1D) + a_{19}(OH)](H_2) = a_{23b}(H)(HO_2) \qquad (5.93)$$

Neglecting the formaldehyde and methyl radical reactions discussed above, as well as (b_{27}) and (b_{28}), the equations of H, OH, HO_2, and H_2O_2 can be written:

$$\frac{d(H)}{dt} + [a_1(M)(O_2) + a_2(O_3) + a_{23}(HO_2)](H) = a_5(O)(OH)$$

$$+ J_{H_2O}(H_2O) + a_3^*(O^1D)(H_2) + a_{19}(OH)(H_2) + a_{36}(CO)(OH) \qquad (5.94)$$

$$\frac{d(OH)}{dt} + [a_5(O) + a_6(O_3) + 2a_{16}(OH) + a_{17}(HO_2) + a_{19}(H_2) + a_{30}(H_2O_2)$$

$$+ a_{36}(CO)](OH) = J_{H_2O}(H_2O) + 2a_1^*(O^1D)(H_2O) + a_2(O_3)(H)$$

$$+ a_{6b}(O_3)(HO_2) + a_7(HO_2)(O) + 2a_{23a}(H)(HO_2) + a_{26}(NO)(HO_2)$$

$$+ 2J_{H_2O_2}(H_2O_2) + a_3^*(O^1D)(H_2) + a_2^*(O^1D)(CH_4) \tag{5.95}$$

$$\frac{d(HO_2)}{dt} + [a_{6b}(O_3) + a_7(O) + a_{23}(H) + a_{26}(NO) + a_{17}(OH) + 2a_{27}(HO_2)](HO_2)$$

$$= a_1(M)(O_2)(H) + a_6(OH)(O_3) + a_{30}(OH)(H_2O_2) \tag{5.96}$$

$$\frac{d(H_2O_2)}{dt} + [J_{H_2O_2} + a_{30}(OH)](H_2O_2) = a_{27}(HO_2)^2 \tag{5.97}$$

5.4.3 The odd hydrogen family and some observations •

Adding together equations (5.94) through (5.96) and introducing a transport term, we obtain an approximate equation for an *odd hydrogen family* defined by $HO_x = H + OH + HO_2$ (neglecting methyl radical and nitrogen reactions):

$$\frac{\partial(HO_x)}{\partial t} + 2a_{23b + 23c}(H)(HO_2) + 2a_{17}(OH)(HO_2) + 2a_{27}(HO_2)^2$$

$$= 2J_{H_2O}(H_2O) + 2a_3^*(O^1D)(H_2) + 2a_1^*(O^1D)(H_2O) + 2J_{H_2O_2}(H_2O_2)$$

$$+ a_2^*(O^1D)(CH_4) - \vec{\nabla} \cdot \phi(HO_x) \tag{5.98}$$

Figure 5.28 presents calculated altitude profiles of the photochemical lifetimes HO_x family and its constituent members, as well as H_2O_2, and the time constants for dynamical processes. As in the case of O_x, the lifetime of the family is several orders of magnitude greater than those of some of its members, making the solution of the continuity equation of the HO_x family relatively easy to obtain with simple numerical methods. Even so, the lifetime of the HO_x family is considerably shorter than the transport time scale up to altitudes near 80 km. Therefore, its density will not be directly dependent on transport processes except above this altitude. Throughout the middle atmosphere, the lifetimes of OH and HO_2 are quite short. Assuming photochemical equilibrium for these

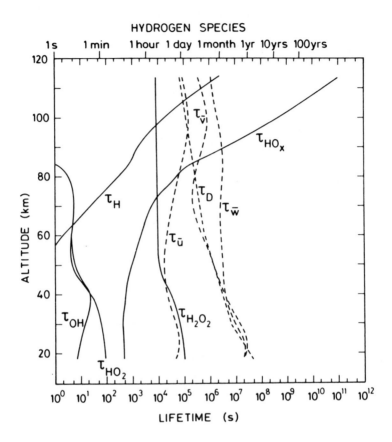

Fig. 5.28. Photochemical lifetimes of HO_x, H, OH, HO_2, and H_2O_2, and the time constants for transport.

species and keeping only the dominant terms in (5.95) and (5.96), one may write:

$$[a_5(O) + a_6(O_3) + a_{36}(CO)](OH) \approx a_2(O_3)(H) + a_7(HO_2)(O) + a_{26}(NO)(HO_2) \quad (5.99)$$

and

$$[a_{6b}(O_3) + a_7(O) + a_{26}(NO)](HO_2) \approx a_1(M)(O_2)(H) + a_6(OH)(O_3) \quad (5.100)$$

The ratio of the concentrations of OH and HO_2 can then be written:

$$\frac{(OH)}{(HO_2)} \approx \frac{a_5(O) + a_{36}(CO)}{a_7(O) + a_{26}(NO) + a_{6b}(O_3)}$$

$$\times \left| \frac{a_1(M)(O_2)}{a_1(M)(O_2) + a_2(O_3)} + \frac{a_6(O_3)}{a_5(O) + a_{36}(CO)} \right| \qquad (5.101)$$

This last expression can be further simplified; in the upper part of the middle atmosphere $(z > 40 \text{ km})$, one may assume:

$$\frac{(HO_2)}{(OH)} \approx \frac{a_5}{a_7} \times \frac{a_1(M)(O_2)}{a_1(M)(O_2) + a_2(O_3)} \qquad (5.102)$$

In the middle stratosphere, we obtain

$$\frac{(HO_2)}{(OH)} \approx \frac{a_6}{a_{6b}} \qquad (5.103)$$

and near the tropopause, it becomes

$$\frac{(HO_2)}{(OH)} \approx \frac{a_{36}(CO) + a_6(O_3)}{a_{26}(NO) + a_{6b}(O_3)} \qquad (5.104)$$

In the upper stratosphere and mesosphere,

$$(H) \approx \frac{a_5(O)(OH)}{a_1(M)(O_2) + a_2(O_3)} \qquad (5.105)$$

so that

$$\frac{(H)}{(OH)} \approx \frac{a_5(O)}{a_1(M)(O_2) + a_2(O_3)} \qquad (5.106)$$

These ratios establish the partitioning between members of the HO_x family, as was shown previously for O_x:

$$\frac{(OH)}{(HO_x)} = \frac{1}{\left| 1 + \dfrac{(H)}{(OH)} + \dfrac{(HO_2)}{(OH)} \right|} \qquad (5.107)$$

In the upper stratosphere and mesosphere, H_2O_2 is approximately in photochemical equilibrium defined by

$$(H_2O_2) = \frac{a_{27}(HO_2)^2}{J_{H_2O_2} + a_{30}(OH)} \qquad (5.108)$$

In the lower stratosphere the H_2O_2 lifetime becomes long, and the time dependent equation (5.97), must be used with a transport term $\vec{\nabla} \cdot \vec{\phi}(H_2O_2)$

added. Heterogeneous removal in clouds must also be considered in the tropo-sphere.

In the lower thermosphere, atomic hydrogen becomes the dominant odd hydrogen species. Adding expressions (5.94) through (5.96) and 2×(5.97), assuming $d(OH)/dt = d(HO_2)/dt = d(H_2O_2)/dt = 0$, omitting minor terms and adding a transport term, we find:

$$\frac{\partial(H)}{\partial t} + \vec{\nabla}\cdot\vec{\phi}(H) + 2(a_{23b} + a_{23c})(HO_2)(H) = 2J_{H_2O}(H_2O) \qquad (5.109)$$

Below 75 km, the concentration of H becomes small relative to OH and HO_2, and the addition of equations (5.94), (5.95) and (5.96) leads to the following expression, in which secondary terms are neglected but reactions with HNO_3, HO_2NO_2, and carbon compounds are added:

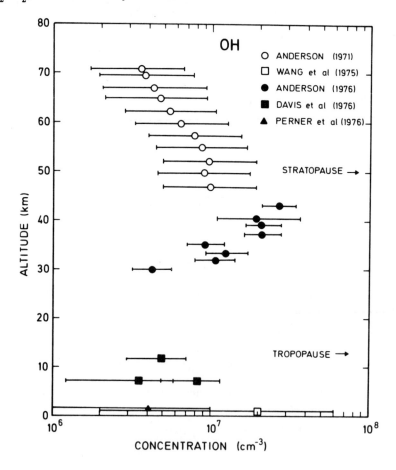

Fig. 5.29. Observations of the OH radical.

$$2a_{17}(OH)(HO_2) + 2a_{30}(OH)(H_2O_2) + 2b_{27}(OH)(HNO_3)$$

$$+ 2b_{28}(OH)(HO_2NO_2) + 2c_{17}(CH_3OOH)(OH) = 2J_{H_2O}(H_2O) + 2a_1^*(O^1D)(H_2O)$$

$$+ (O^1D)(CH_4)[2c_{1a}(1 + X) + c_{1b}X] + 2c_2X(OH)(CH_4) + 2a_3^*(O^1D)(H_2) \quad (5.110)$$

where X was defined in (5.65). This equation can be simplified in the mesosphere:

$$a_{17}(OH)(HO_2) + (a_{23b} + a_{23c})(HO_2)(H)$$

$$= [J_{H_2O} + a_1^*(O^1D)](H_2O) + a_3^*(O^1D)(H_2) \quad (5.111)$$

and in the stratosphere above 30 km:

$$a_{17}(OH)(HO_2) = a_1^*(O^1D)(H_2O) + a_3^*(O^1D)(H_2)$$

$$+ (1 + X)c_{1a}(O^1D)(CH_4) + c_2X(OH)(CH_4) \quad (5.112)$$

while below 30 km,

$$a_{17}(OH)(HO_2) + a_{30}(OH)(H_2O_2) + b_{27}(OH)(HNO_3) + b_{28}(OH)(HO_2NO_2)$$

$$= (O^1D)[a_1^*(H_2O) + a_3^*(H_2) + (1 + X)c_{1a}^*(CH_4)] + c_2X(OH)(CH_4) \quad (5.113)$$

OH has been measured by resonance fluorescence in the mesosphere (Anderson, 1971) and in the stratosphere (Anderson 1976; Heaps and McGee, 1983). Total column measurements of OH also provide important information about its stratospheric abundance and variability (Burnett and Burnett, 1981;

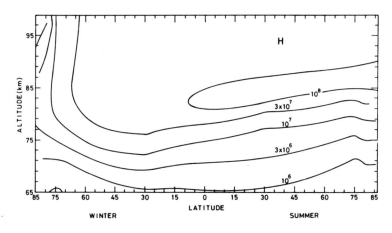

Fig. 5.30. Calculated meridional distribution of atomic hydrogen from 65 to 95 km. From Brasseur (1982).

1982). Pyle et al. (1983) have used the ratio of HNO_3 to NO_2 as observed by satellite to deduce a complete OH distribution for the middle stratosphere. Figure 5.29 presents a summary of available in-situ OH measurements. Note that OH has not yet been measured, however, in the important region between the tropopause and 30 km, where its role in the photochemistry of ozone (see §5.9) and other constituents, such as methane and HNO_3, becomes extremely important. HO_2 has also been measured using resonance fluorescence (Anderson et al., 1980) and microwave techniques (DeZafra et al., 1984). The stratospheric abundance of hydrogen peroxide has been reported by Waters et al. (1981), and is greater than those of OH and HO_2. Mesospheric atomic hydrogen has been inferred from observations of the Lyman alpha and Lyman beta emissions (Anderson et al., 1980). Figure 5.30 shows a calculated two-dimensional distribution of atomic hydrogen in the mesosphere and lower thermosphere.

Solution of the simultaneous continuity equations (including transport, chemistry, and solar insolation) yield theoretical distributions of chemical species which may be compared to available observations, provided that similar conditions apply to both calculated and measured values. Thus, for example, comparison of calculated distributions of long lived species (water vapor,

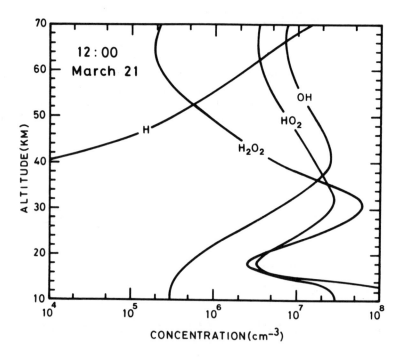

Fig. 5.31. Calculated distributions of some hydrogen compounds (noon, equinox). From the model of Brasseur et al. (1983).

methane, molecular hydrogen) to observations requires that the transport parameters adopted in the model be comparable to those present in the atmosphere. On the other hand, for short lived species such as OH, HO_2, etc. such comparisons are only valid if the solar flux and zenith angles adopted in the model are comparable to those present at the time of the observations, and if the calculated distributions of the source molecules (water vapor, methane) are close to those observed. Figure 5.31 shows calculated vertical distributions of several hydrogen compounds (noon, equinox).

5.5. Nitrogen compounds

The study of nitrogen in the middle atmosphere can be subdivided into those processes occurring in the stratosphere and those appropriate to the thermosphere. Important production processes occur in each of these regions, and these both contribute to the global budget of nitrogen species. Sections 5.5.1-5.5.3 deal with stratospheric nitrogen, while §5.5.4 and 5.5.5 focus on nitrogen compounds in the mesosphere and lower thermosphere.

5.5.1 Sources of stratospheric nitrogen oxides •

In the stratosphere, nitric oxide (NO) is produced mostly by dissociation of N_2O by reaction with an excited oxygen atom in the (^1D) state (Nicolet, 1971; Crutzen, 1971; McElroy and McConnell, 1971, etc.):

$$(b_{39}); \quad N_2O + O(^1D) \rightarrow 2NO \qquad (5.114)$$

The following products are also possible:

$$(b_{38}); \quad N_2O + O(^1D) \rightarrow N_2 + O_2 \qquad (5.115)$$

The rate of NO production by reaction (5.114) is given by:

$$P(NO) = 2b_{39}(N_2O)(O^1D) \qquad (5.116)$$

and is therefore dependent on the vertical distribution of N_2O. In the stratosphere, N_2O is destroyed primarily by photolysis:

$$(J_{N_2O}); \quad N_2O + h\nu \rightarrow N_2 + O(^1D) \qquad (5.117)$$

The equation of continuity for nitrous oxide can be written:

$$\frac{\partial (N_2O)}{\partial t} + \vec{\nabla} \cdot \vec{\phi}(N_2O) + (N_2O)[J_{N_2O} + (b_{38} + b_{39})(O^1D)] = 0 \qquad (5.118)$$

Nitrous oxide is present throughout the troposphere (and is injected into the lower stratosphere) at a mole fraction of about 3×10^{-7}. It is produced principally by bacterial processes associated with complex nitrification and

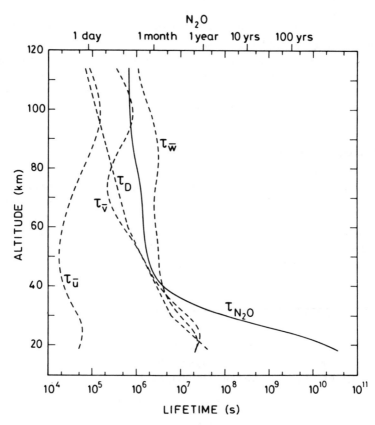

Fig. 5.32. Photochemical lifetime of N_2O, as well as the time constants for transport by the zonal and meridional winds, and a one-dimensional diffusive time constant.

denitrification mechanisms in soils (Delwiche, 1978). Thus, as first indicated by Bates and Witherspoon (1952), the biosphere plays a major role in determining the source of N_2O. In spite of numerous studies, the global budget of atmospheric N_2O is not yet well known. According to McElroy (1980), the rate of N_2O production is of the order of 10 MT/yr of nitrogen, but might be as much as 50 MT/yr. With these values, the lifetime of N_2O is about 100 years in the troposphere and lower stratosphere, and the production is balanced by present estimates of the rate of atmospheric photodissociation. Other studies (see, for example, Hahn and Junge, 1977), have derived a lifetime of only about 10 years based on analyses of the observed variability of this gas. This seems unlikely, as it would imply a ground level source which is ten times more intense that that derived above, as well as an additional destruction mechanism besides photodissociation.

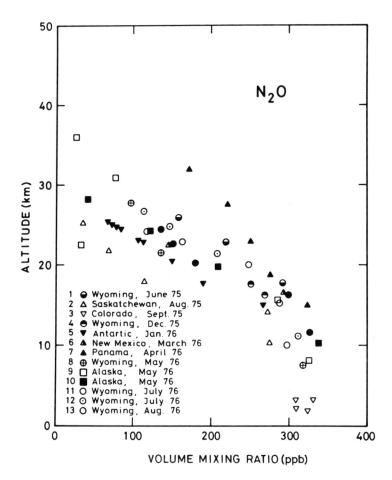

Fig. 5.33. Observations of the mole fraction of nitrous oxide by Schmeltekopf et al. (1977). (Copyright by the American Meteorological Society).

To examine the formation of N_2O by nitrification and denitrification, all of the processes which fix molecular nitrogen must be considered: living organisms (agricultural fields, forest, and uncultivated land) as well as industrial processes (fertilizers, industry). Industrial fixation of nitrogen accounts for about 25 percent of the present total and is expected to become larger in the future. Therefore, anthropogenic production of N_2O, especially through nitrogen fertilizers, combustion, and organic wastes, can perturb the aeronomic balance of the atmosphere and may influence the ozone layer.

Figure 5.32 presents the photochemical time constant of N_2O and those appropriate to transport processes. Like CH_4, N_2O is an excellent tracer for transport in the middle stratosphere, where its lifetime is comparable to those for advection by the mean meridional circulation. At higher altitudes in the upper stratosphere and lower mesosphere, the N_2O lifetime remains close to the mean meridional transport lifetimes, making it a more sensitive tracer than CH_4 in this region (because the lifetime of methane here becomes somewhat longer than transport, see Fig. 5.12).

The distribution of N_2O has been measured by cryogenic sampling and gas phase chromatography (Scholz et al., 1970; Ehhalt et al., 1975; Schmeltekopf et al., 1977) as well as by infrared absorption (Harries, 1973; Farmer, 1974; Farmer et al, 1974, 1980). Figure 5.33 shows the observations by Schmeltekopf et al. at different latitudes. In each case, a decrease in mole fraction is observed above the tropopause. As was shown previously for CH_4, this decrease is less rapid at the equator than at higher latitudes, presumably because of the rapid rate of upward transport associated with the rising branch of the Hadley cell at these latitudes. Figure 5.34 presents monthly and zonally averaged N_2O mixing ratio distributions as observed by the SAMS satellite (Jones and Pyle, 1984), and these may help elucidate important information about stratospheric transport. Like CH_4, (Fig. 5.14) large values are obtained in the tropical lower stratosphere, and a "tilt" into the summer hemisphere can be seen. Both of these features are qualitatively consistent with a circulation like the one depicted in Figure 3.16, when we recall that the source of these constituents lies exclusively in the troposphere.

Figure 5.35 shows the profile of production of nitric oxide by oxidation of N_2O (Nicolet and Peetermans, 1972). Curves are shown for two solar zenith angles, two $O(^1D)$ distributions, and two values of the vertical eddy diffusion coefficient (K_{min} and K_{max}). It should be noted that the production of NO from N_2O is directly related to atmospheric dynamics; it reaches a maximum in the mid-stratosphere where its value is about 100 molec $cm^{-3}s^{-1}$. The uncertainty due to atmospheric transport has its largest effect at high altitudes, while the uncertainty due to the formation of $O(^1D)$ is largest near the tropopause. Nicolet and Vergison (1971) derived a global stratospheric production rate of

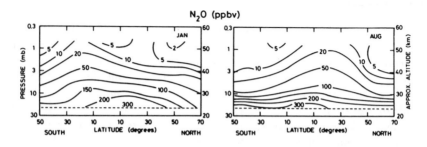

Fig. 5.34. Monthly averaged N_2O distributions observed by SAMS. From Jones and Pyle (1984). (Copyright by the American Geophysical Union).

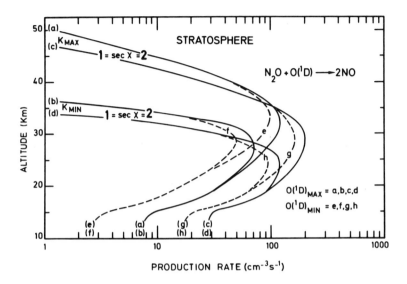

Fig. 5.35. Rate of production of NO in the stratosphere through reaction of N_2O with O^1D. Values calculated by Nicolet and Peetermans (1972) for different transport intensity and different values for the efficiency of formation of $O(^1D)$.

$1.5 \pm 1 \times 10^8$ molec NO cm^{-2} s^{-1}, assuming a tropopause height of 12 km. Crutzen and Schmailzl (1983) suggest a value between 1.1 and 1.9×10^8 molec NO cm^{-2} s^{-1}, using the N_2O distribution measured by the SAMS satellite as shown in Figure 5.34.

Stratospheric nitrogen atoms can be formed by dissociative ionization and dissociation of molecular nitrogen by galactic cosmic rays. These processes generally occur through secondary electrons ejected by heavy cosmic particles. One ion pair formed by cosmic radiation leads to the production of 1-1.6 atoms of nitrogen (see Chapter 6). Figure 5.36 shows the spatial distribution of P(N) in 1965 (from Nicolet, 1975). In the stratosphere, atomic nitrogen reacts rapidly with molecular oxygen to form nitric oxide. Stratospheric production of nitrogen oxides by cosmic radiation should not be neglected, especially in polar regions where the oxidation of N_2O is slow. Nicolet (1975) indicated that cosmic radiation leads to an integrated NO production of about $5 \pm 1 \times 10^7$ cm^{-2} s^{-1} in polar regions and 3×10^7 cm^{-2} s^{-1} in the tropics.

Crutzen et al. (1975) noted that penetration of large amounts of protons into the stratosphere during solar proton events would also lead to an intense production of atomic nitrogen. Large solar proton events occurred in November, 1960, September, 1966, and especially in August, 1972. These events can produce enough nitrogen oxide to influence the ozone content, particularly at high latitude. The chemistry of the ionization and odd nitrogen production

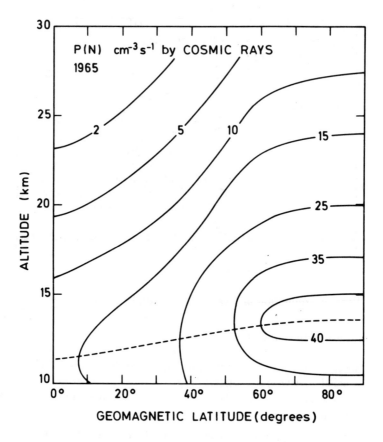

Fig. 5.36. Distribution of the rate of production of NO by cosmic rays at solar minimum. Adapted from Nicolet (1975). (Copyright by Pergamon Press).

processes is described in more detail in Chapters 6 and 7.

5.5.2 Chemistry of odd nitrogen and nitric acid in the stratosphere ‡

In the stratosphere, $N(^4S)$ is produced almost entirely by NO photolysis:

$$(J_{NO}); \quad NO + h\nu \rightarrow N(^4S) + O \qquad (5.119)$$

and rapidly destroyed by reaction with O_2:

$$(b_7); \quad N(^4S) + O_2 \rightarrow NO + O \qquad (5.120)$$

In turn, nitric oxide is converted to nitrogen dioxide by reaction with ozone:

$$(b_4); \quad NO + O_3 \rightarrow NO_2 + O_2 \qquad (5.121)$$

or, at low altitudes, by reaction with peroxy radicals (RO_2):

$$(a_{26}); \quad NO + HO_2 \rightarrow NO_2 + OH \qquad (5.122)$$

$$(c_5); \quad NO + CH_3O_2 \rightarrow NO_2 + CH_3O \qquad (5.123)$$

It can also be destroyed, but much more slowly, by the reaction with atomic nitrogen:

$$(b_6); \quad N + NO \rightarrow N_2 + O \qquad (5.124)$$

The NO_2 molecule reacts rapidly during the day to reform NO, either by reaction with atomic oxygen:

$$(b_3); \quad NO_2 + O \rightarrow NO + O_2 \qquad (5.125)$$

or by photolysis:

$$(J_{NO_2}); \quad NO_2 + h\nu(\lambda < 405nm) \rightarrow NO + O(^3P) \qquad (5.126)$$

Reaction of NO with O_3 (5.121), followed by reaction of NO_2 with O (5.125) represents a catalytic cycle which provides the major loss process for odd oxygen in the stratosphere.

Additional reactions should also be considered in the stratosphere. These lead to the formation of other nitrogen containing species such as N_2O_5, HNO_3, and HO_2NO_2, which represent nitrogen "reservoirs" (so called because, unlike NO and NO_2, these species do not destroy odd oxygen). The formation of NO_3 is an important reaction at night:

$$(b_9); \quad NO_2 + O_3 \rightarrow NO_3 + O_2 \qquad (5.127)$$

but during the day its rapid photolysis via

$$(J_{NO_3}); \quad NO_3 + h\nu \rightarrow NO_2 + O \qquad (5.128)$$

$$\rightarrow NO + O_2 \qquad (5.129)$$

leads to very small daytime densities for this species in spite of its additional possible production from HNO_3 (reaction 5.135) and $ClONO_2$ (reaction 5.198).

Dinitrogen pentoxide, formed by the reaction of NO_2 and NO_3, is produced almost entirely at night:

$$(b_{12}); \quad NO_3 + NO_2 + M \rightarrow N_2O_5 + M \qquad (5.130)$$

It is destroyed during the day through photolysis

$$(J_{N_2O_5}); \quad N_2O_5 + h\nu \rightarrow NO_3 + NO_2 \qquad (5.131)$$

and by collisional decomposition

$$(b_{32}); \quad N_2O_5 + M \rightarrow NO_3 + NO_2 + M \qquad (5.132)$$

Another important stratospheric nitrogen containing species is HNO_3. Nitric acid is readily incorporated into rain droplets, and this process leads to a flux of nitrogen species from the stratosphere to the troposphere. Nitric acid also introduces an interaction between nitrogen and hydrogen compounds because it is formed by the three body process:

$$(b_{22}); \quad NO_2 + OH + M \rightarrow HNO_3 + M \qquad (5.133)$$

The two primary loss processes for HNO_3 are:

$$(J_{HNO_3}); \quad HNO_3 + h\nu \rightarrow OH + NO_2 \qquad (5.134)$$

$$(b_{27}); \quad HNO_3 + OH \rightarrow NO_3 + H_2O \qquad (5.135)$$

The HO_2NO_2 molecule is formed by the reaction:

$$(b_{23a}); \quad NO_2 + HO_2 + M \rightarrow HO_2NO_2 + M \qquad (5.136a)$$

and destroyed by unimolecular decomposition at low altitudes:

$$(b_{23b}); \quad HO_2NO_2 + M \rightarrow HO_2 + NO_2 + M \qquad (5.136b)$$

as well as by photolysis and reaction with OH:

$$(J_{HO_2NO_2}); \quad HO_2NO_2 + h\nu \rightarrow NO_2 + HO_2 \qquad (5.137)$$

$$(b_{28}); \quad HO_2NO_2 + OH \rightarrow H_2O + O_2 + NO_2 \qquad (5.138)$$

The chemistry of HO_2NO_2 is important in both nitrogen and hydrogen chemistry.

The primary reactions of these species are depicted in Figure 5.37. The continuity equations for NO_3, N_2O_5, and HO_2NO_2 can be written:

$$\frac{d(NO_3)}{dt} + [b_{12}(M)(NO_2) + J_{NO_3}](NO_3) = b_9(O_3)(NO_2)$$

$$+ [J_{N_2O_5} + b_{32}(M)](N_2O_5) + b_{27}(OH)(HNO_3) \qquad (5.139)$$

$$\frac{d(N_2O_5)}{dt} + (N_2O_5)[J_{N_2O_5} + b_{32}(M)] = b_{12}(M)(NO_2)(NO_3) \qquad (5.140)$$

$$\frac{d(HO_2NO_2)}{dt} + (HO_2NO_2)[J_{HO_2NO_2} + b_{28}(OH) + b_{23b}(M)]$$

$$= b_{23a}(NO_2)(HO_2)(M) \qquad (5.141)$$

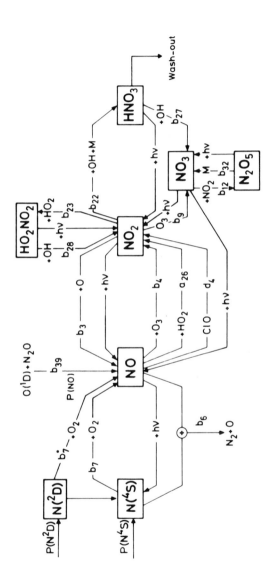

Fig. 5.37. Principal aeronomic reactions of nitrogen compounds.

5.5.3 The odd nitrogen family: lifetimes and observations •

Defining the *odd nitrogen family* (NO_x) as the sum of $N + NO + NO_2 + NO_3 + 2 \times N_2O_5 + HO_2NO_2$, the continuity equation for the family may be written as:

$$\frac{d(NO_x)}{dt} + b_{22}(NO_2)(OH)(M) + 2b_6(N)(NO)$$

$$= 2b_{39}(N_2O)(O^1D) + J_{HNO_3}(HNO_3) + b_{27}(HNO_3)(OH) \qquad (5.142)$$

The continuity equation for nitric acid is given by

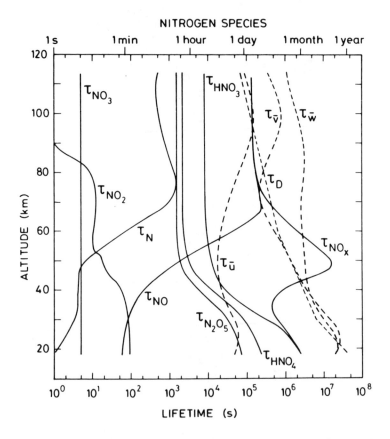

Fig. 5.38. Photochemical time constants for NO_x, N, NO, NO_2, NO_3, N_2O_5, HO_2NO_2, and HNO_3, as well as the time constants for transports by the zonal and meridional winds and a one-dimensional eddy diffusion coefficient.

$$\frac{d(HNO_3)}{dt} + (HNO_3)[J_{HNO_3} + b_{27}(OH)] = b_{22}(M)(OH)(NO_2) \qquad (5.143)$$

Figure 5.38 displays the altitude profiles of the photochemical lifetimes of the NO_x family and its constituent members, and HNO_3, as well as the time constants for transport at middle latitudes. Note that the photochemical lifetimes for NO_x and HNO_3 are both comparable to the transport lifetimes in the lower stratosphere, and they will therefore both be quite dependent on atmospheric transport. It is important to treat them separately in numerical model studies. In high latitude winter, it also becomes necessary to treat N_2O_5 separately from the other nitrogen species; this will be discussed in detail below.

The partitioning of the odd nitrogen family is somewhat more complicated than those families presented previously. The ratio between N and NO can be readily derived as follows:

$$\frac{d(N)}{dt} + b_7(N)(O_2) \approx J_{NO}(NO) \qquad (5.144)$$

and, assuming $\frac{d(N)}{dt} = 0$,

$$\frac{(N)}{(NO)} = \frac{J_{NO}}{b_7(O_2)} \qquad (5.145)$$

NO_2 is also in equilibrium with NO throughout the sunlit middle atmosphere. Because atmospheric absorption in the 400 nm region is negligibly small, J_{NO_2} is given by its optically thin value of about 1×10^{-2} even at altitudes as low as 15 km, which corresponds to a lifetime of about 2 minutes. NO_2 can therefore be considered to be in immediate photochemical equilibrium with NO during the day, and we may write:

$$\frac{d(NO_2)}{dt} = 0 = -(NO_2)[J_{NO_2} + b_3(O)] + [b_4(O_3) + a_2(HO_2) + c_5(CH_3O_2)](NO) \qquad (5.146)$$

which yields the following expression for the ratio of NO to NO_2 (with the effect of ClO added, see §5.6):

$$\frac{(NO)}{(NO_2)} = \frac{J_{NO_2} + b_3(O)}{b_4(O_3) + a_{26}(HO_2) + c_5(CH_3O_2) + d_4(ClO)} \qquad (5.147)$$

The value of this ratio is close to one over most of the stratosphere during daytime, but increases rapidly above 40 km because of increasing atomic oxygen densities. At night, NO is immediately converted to NO_2 up to altitudes of about 60 km. Figure 5.39 shows calculated one-dimensional model distributions of the various nitrogen species. The curves represent a steady state solution for average solar insolation.

Setting NO_3 in photochemical equilibrium in the sunlit stratosphere yields:

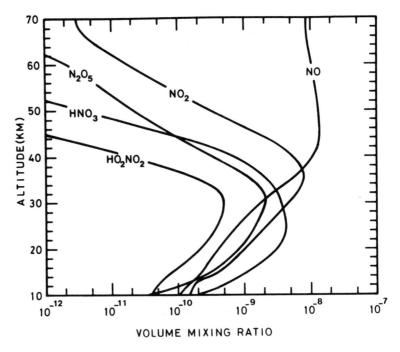

Fig. 5.39. Calculated vertical distributions of nitrogen species. From the model of Brasseur et al. (1983).

$$\frac{(NO_3)}{(NO_2)} \approx \frac{b_9(O_3)}{J_{NO_3}} \qquad (5.148)$$

while at night

$$(NO_3)_{night} = \frac{b_9(O_3)}{b_{12}(M)} \qquad (5.149)$$

For HO_2NO_2, we obtain

$$\frac{(HO_2NO_2)}{(NO_2)} = \frac{b_{23a}(HO_2)(M)}{J_{HO_2NO_2} + b_{26}(OH)} \qquad (5.150)$$

The major difficulty in partitioning the NO_x family is that a simple instantaneous ratio cannot describe the relationship between N_2O_5 and the other family members. This is because N_2O_5 is produced almost entirely at night, when concentrations of NO_3 are large. About 30-50% of the atmospheric NO_x is in the form of N_2O_5 at the end of a typical night in the cold lower stratosphere. This accumulated N_2O_5 is destroyed only during the day by photolysis, which produces a diurnal variation in NO and NO_2, see Figure 5.46. Therefore it is never really in an instantaneous stationary state; rather it achieves a diurnally chang-ing equilibrium over the 24 hour period. A diurnal average steady state ratio can be derived as follows (see also, Crutzen, 1971): First, write the appropriate continuity equation for N_2O_5:

$$\frac{d(N_2O_5)}{dt} \approx b_{12}(M)(NO_2)_{night}(NO_3)_{night}F_d - J_{N_2O_5}(N_2O_5)F_s \qquad (5.151)$$

where F_d and F_s represent the fractions of the day which are dark and sunlit, respectively. We next assume that

$$(NO_2)_{night} = (NO_2)_{day} + (NO)_{day}$$

$$= \left(1 + \frac{J_{NO_2} + b_3(O)}{b_4(O_3)}\right)(NO_2)_{day} \qquad (5.152)$$

since, as mentioned above, all the nitric oxide in the stratosphere is almost immediately converted to NO_2 at sunset (and using 5.147). Substituting for $(NO_3)_{night}$ and $(NO_2)_{night}$ with the expressions given above, and assuming steady state for N_2O_5, we obtain

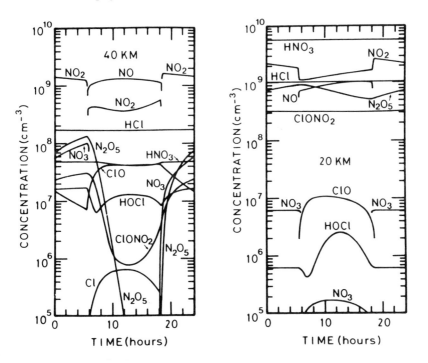

Fig. 5.40. Computed diurnal variations of nitrogen and chlorine species at 40 (left) and 20 (right) km. From the model of Brasseur et al. (1983).

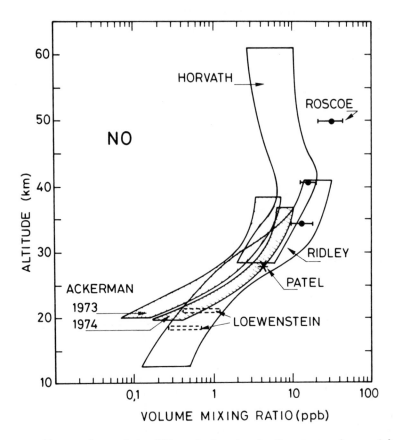

Fig. 5.41. Observations of the NO mole fraction in the stratosphere. Adapted from WMO (1982).

$$\frac{(N_2O_5)_{avg}}{(NO_2)_{day}} = \frac{b_9(O_3)\left[1 + \dfrac{J_{NO_2} + b_3(O)}{b_4(O_3)}\right]F_d}{J_{N_2O_5}F_s} \qquad (5.153)$$

The family can now be partitioned using the ratios defined above:

$$\frac{(NO_2)}{(NO_x)} = \frac{1}{1 + \dfrac{(N)}{(NO)}\times\dfrac{(NO)}{(NO_2)} + \dfrac{(NO)}{(NO_2)} + \dfrac{(NO_3)}{(NO_2)} + \dfrac{(HO_2NO_2)}{(NO_2)} + \dfrac{2\times(N_2O_5)}{(NO_2)}} \qquad (5.154)$$

and

$$\frac{(NO)}{(NO_x)} = \frac{(NO)}{(NO_2)}\times\frac{(NO_2)}{(NO_x)} \quad \text{etc.} \qquad (5.155)$$

An alternate approach is simply to integrate N_2O_5 separately from the other NO_x species. Indeed, this becomes necessary near the polar night region, as we shall see below. Figure 5.40 presents a model calculation of the diurnal variation of NO, NO_2, NO_3, N_2O_5, and HNO_3 at 20 and 40 km.

Stratospheric nitrogen species have been measured by balloon since about 1970, and more recently by satellites. Figure 5.41 shows observations of NO obtained at mid-latitude in the northern hemisphere. The mole fraction increases by a factor of about 100 from 20 to 35 km, and is about 10^{-8} from 35 to 50 km. A large degree of variability is also observed, due in part to measurement errors and calibration, but probably also due to real time dependent variations which may be related to atmospheric dynamics.

The vertical distribution of nitrogen dioxide observed by several groups is shown in Figure 5.42. An increase with increasing altitude is observed up to about 35 km, while above that altitude a progressive decrease can be seen, reflecting the changing ratio of NO_2/NO, e. g. equation (5.147).

Much of the available information on the latitudinal distribution of odd nitrogen comes from total column measurements. Numerous measurements of the NO_2 total column as a function of latitude have been reported by Noxon (1979), Noxon et al. (1979), and Coffey et al. (1981). These studies have shown that the abundance of NO_2 increases with increasing latitude in summer, while in winter the abundance decreases poleward of about 40 or 50N (Fig. 5.43). Sometimes very sharp gradients are observed in the winter hemisphere (about a factor of four in total column in only 5-10 degrees of latitude), particularly

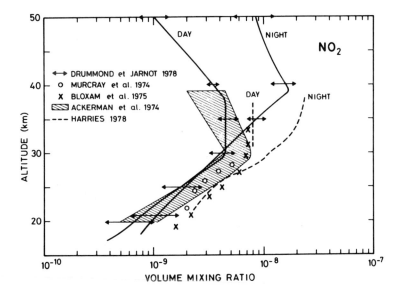

Fig. 5.42. Observations of the NO_2 mole fraction at sunset in mid-latitudes.

when the winter polar vortex is displaced by the presence of planetary waves (this is sometimes referred to as an NO_2 cliff).

The tendency for NO and NO_2 to increase at high latitudes in summer can be qualitatively understood in terms of atmospheric transport. Like ozone, NO_x is a long lived species in the lower stratosphere whose source lies in the tropical middle stratosphere. Downward poleward transport by the meridional circulation (see Chapter 3) should lead to increased total NO_x (and NO and NO_2) abundances at high latitudes, just as is found for O_3.

The observed decrease in the winter season at high latitude is probably the result of the combined effects of dynamics and chemistry. Recent studies by Miller et al. (1981), Knight et al. (1982), Evans et al. (1982) and Solomon and Garcia (1983a) have shown that N_2O_5 probably provides an important reservoir for stratospheric NO_2 in the winter hemisphere, particularly when temperature dependent ultraviolet absorption cross sections are included in the calculation of its photolysis rate. Because of the temperature dependence of the N_2O_5 absorption cross sections, its lifetime is a strong function of latitude, particularly in the lower stratosphere in winter. In the polar night region, its lifetime is more than a month, while the time scale for conversion of NO_2 to NO_3 and then to N_2O_5 is of the order of days. Therefore, nearly all of the odd nitrogen in an air parcel remaining in the polar night region for more than a few days will be in

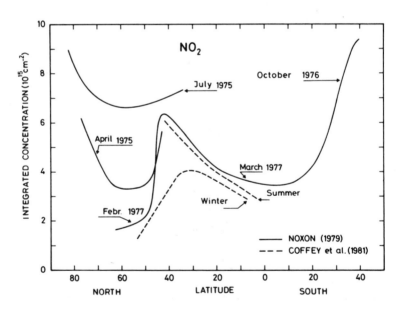

Fig. 5.43. Latitude distribution of the total NO_2 column as measured by different experimenters.

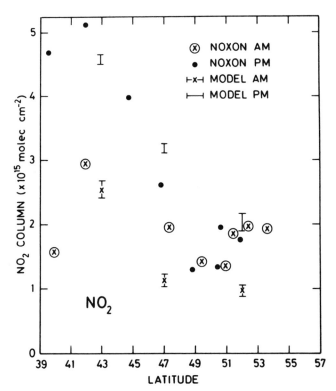

Fig. 5.44. Total column abundances as a function of latitude observed in February (1977) versus calculations. From Solomon and Garcia (1983b). (Copyright by the American Geophysical Union).

the form of N_2O_5. Its lifetime is as large as a few days in the sunlit high latitude wintertime lower stratosphere. This is comparable to the time scale associated with transport by winds of zonal strength (see Fig. 5.38). In the presence of planetary wave structure, winds of such speeds often flow across latitude lines (see, e. g., Fig. 3.6), rapidly moving N_2O_5 from regions where its lifetime is long (polar night) to regions where its lifetime is less than a day. Under these circumstances, the N_2O_5 content of a given air parcel is far from local equilibrium, and carries a "memory" of where it has been. Solomon and Garcia (1983b) have shown that in the presence of wavenumber one planetary wave structure, the photochemical-dynamical history of the air parcels can lead to sharp gradients in the sunset NO_2 column such as those observed by Noxon, and that both chemical and dynamical effects are important factors in the development of a cliff. Figure 5.44 presents calculated and observed (Noxon, 1979) NO_2 column abundances for the February, 1977 cliff. This represents perhaps one of the most striking examples of the interaction between dynamics and photochemistry.

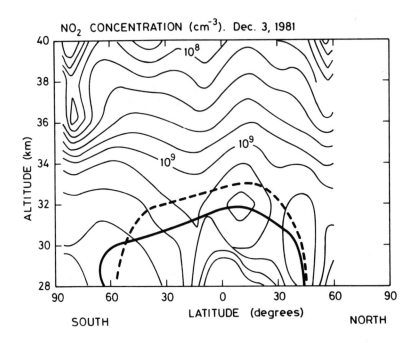

Fig. 5.45. Observed distribution of NO_2 from the SME satellite. These numbers are an average of the data for the month of December, 1982. Solid heavy line indicates the altitude of maximum NO_2 density observed by SME, while the dashed line shows that obtained with the model of Solomon and Garcia (1983a). See Mount et al. (1984) for details. (Copyright by the American Geophysical Union).

Satellite observations of NO_2 have begun to provide more detailed information, confirming many of the features inferred from ground based studies (e.g., Mount et al., 1984; Callis et al., 1983). Figure 5.45 shows a monthly averaged distribution of NO_2 measured by the SME satellite. Note the low abundances observed in the tropics, and the high values found in high latitude summer (compare, Fig. 5.43); the low densities seen in high latitude winter are probably related to the formation of N_2O_5 as discussed above. The information regarding the altitude distribution of the total column is quite important, particularly since the N_2O_5 photolysis rate and the resulting partitioning with NO and NO_2 depend sensitively on the altitude distribution of the column. Note the tendancy for the altitude of maximum abundance to decrease at higher latitudes, presumably as a result of downward transport by the mean meridional circulation.

The nighttime distribution of NO_3 has been measured by Naudet et al. (1981) using stellar occultation, and the distribution is close to that predicted by

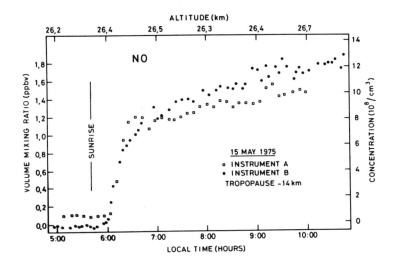

Fig. 5.46. Time variation of the NO mixing ratio at about 26.5 km by Ridley et al. (1977). (Copyrighted by the National Research Council of Canada).

theory. Its total column abundance has also been reported by Noxon et al. (1978). The difference between nighttime and daytime N_2O_5 amounts in the lower stratosphere has been reported by Roscoe (1982), and agrees well with theoretical predictions. The diurnal variation in other NO_x species as a result of the gradual formation and photolysis of N_2O_5 is reflected in several measurements. For example, Figure 5.46 shows the observed early morning increase in NO measured by Ridley et al. (1977). A diurnal variation in morning and evening twilight NO_2 column abundances has been reported by Noxon (1979), and Syed and Harrison (1981).

Nitric acid constitutes an important reservoir for odd nitrogen, particularly in the lower stratosphere where its density exceeds those of NO and NO_2. Figure 5.47 displays a series of observations of nitric acid obtained by optical methods (for example Murcray et al., 1975) and by in-situ filter sampling (Lazrus and Gandrud, 1974). Most of the nitric acid is contained in the lower stratosphere, where its photochemical lifetime is relatively long (Fig. 5.38), and its distribution is quite sensitive to atmospheric transport. As in the case of ozone, the nitric acid column abundance is greater at high latitudes (Fig 5.48), where its variability also increases.

Both NO_2 and HNO_3 mixing ratio observations were obtained by the LIMS experiment onboard NIMBUS 7. Observations of the global distribution of NO_2 at altitudes from 20 to 70 km have been derived using radiance averaging techniques by Russell et al. (1984). These observations indicate the importance of planetary waves in establishing odd nitrogen distributions in the mesosphere as well as the stratosphere. They also suggest that odd nitrogen transport from

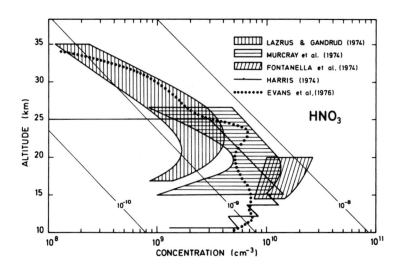

Fig. 5.47. Observations of the nitric acid concentration in the stratosphere.

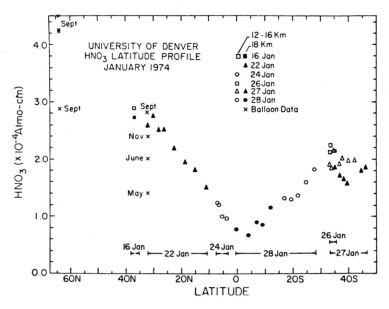

Fig. 5.48. Variation of the HNO_3 total column with season and latitude. From Murcray et al., (1975). (Copyright by the American Geophysical Union).

the thermosphere to the mesosphere, and even to the stratosphere, occurs during the high latitude polar night period. This subject will be discussed in somewhat more detail in §5.5.5.

5.5.4 Chemistry of odd nitrogen in the lower thermosphere and mesosphere ‡

In the thermosphere, atomic nitrogen can be formed by dissociation of N_2, either through the effects of solar radiation or energetic particles. The determination of the magnitude of this production (and its variability with solar activity) will be presented in Chapter 6. We note here, however, that atomic nitrogen can be produced either in the ground ^4S or the excited ^2D state by these processes. $N(^2D)$ reacts rapidly with molecular oxygen:

$$(b_7^*); \quad N(^2D) + O_2 \rightarrow NO + O \qquad (5.156)$$

While the ground $N(^4S)$ state reacts much more slowly, and the rate of its reaction depends strongly on temperature:

$$(b_7); \quad N(^4S) + O_2 \rightarrow NO + O \qquad (5.157)$$

The following reaction plays a similar role:

$$(b_{7a}); \quad N(^4S) + OH \rightarrow NO + H \qquad (5.158)$$

Photodissociation of nitric oxide initiates the primary loss process for odd nitrogen in the middle atmosphere:

$$(J_{NO}); \quad NO + h\nu \rightarrow N(^4S) + O \qquad (5.159)$$

because it leads to the formation of an $N(^4S)$ atom, which can then react with NO in the so-called "cannibalistic reaction"

$$(b_6); \quad N(^4S) + NO \rightarrow N_2 + O \qquad (5.160)$$

wherein two odd nitrogen particles are destroyed. The following kinetic equations can be written:

$$\frac{d(N^4S)}{dt} + [b_6(NO) + b_7(O_2) + b_{7a}(OH)](N^4S) = J_{NO}(NO) + P(N^4S) \qquad (5.161)$$

$$\frac{d(N^2D)}{dt} + b_7^*(O_2)(N^2D) = P(N^2D) \qquad (5.162)$$

$$\frac{d(NO)}{dt} + [b_4(O_3) + a_{26}(HO_2) + c_5(CH_3O_2) + b_6(N^4S) + J_{NO}](NO) = |J_{NO_2}$$

$$+ b_3(O)](NO_2) + b_7^*(O_2)(N^2D) + [b_7(O_2) + b_{7a}(OH)](N^4S) + P(NO) \qquad (5.163)$$

or, considering equation (5.146)

$$\frac{d(NO)}{dt} + [b_6(N^4S) + J_{NO}](NO) = b_7^*(O_2)(N^2D) + [b_7(O_2) + b_{7a}(OH)](N^4S) \qquad (5.164)$$

The rate of ionic production, $P(N^4S)$ and $P(N^2D)$ is discussed in Chapter 6 and $P(NO)$ is given in (5.116).

5.5.5 *The odd nitrogen family in the lower thermosphere and mesosphere* •

The lifetime of atomic nitrogen is short throughout the middle atmosphere, and equilibrium may therefore be assumed for $N(^2D)$ and $N(^4S)$. Thus, for example, the concentration of atomic nitrogen in the ground state (which far exceeds its concentration in the ^2D state) is given by:

$$(N^4S) = \frac{J_{NO}(NO) + P(N^4S)}{b_6(NO) + b_7(O_2) + b_{7a}(OH) + b_{7b}(O_3)} \tag{5.165}$$

This reduces approximately to

$$(N^4S) \approx \frac{J_{NO}(NO) + P(N^4S)}{b_6(NO) + b_7(O_2)} \tag{5.166}$$

If we define the "reactive nitrogen" (NO_y) family as

$$(NO_y) = (N^4S) + (N^2D) + (NO) + (NO_2) + (NO_3)$$

$$+ 2(N_2O_5) + (HNO_3) + (HO_2NO_2) \tag{5.167}$$

then in the lower stratosphere, this reduces essentially to

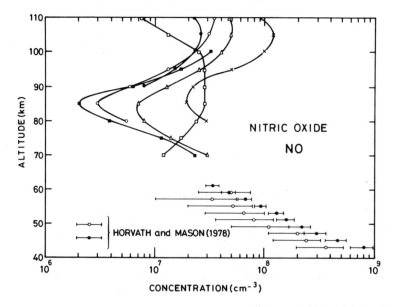

Fig. 5.49. Observations of nitric oxide from 40 to 110 km altitude. In the lower part, the values of Horvath and Mason are shown, while at higher altitudes, the observations include those of Meira (1971), Tisone (1973), Tohmatsu and Iwagami (1975, 1976), Witt et al. (1976), and Baker et al. (1977).

$$(NO_y) = (NO) + (NO_2) + (NO_3) + 2(N_2O_5) + (HNO_3) + (HO_2NO_2) \quad (5.168)$$

and in the upper stratosphere

$$(NO_y) = (NO) + (NO_2) \quad (5.169)$$

while in the mesosphere and lower thermosphere

$$(NO_y) = (NO) + (NO_2) + (N) \quad (5.170)$$

Adding the equations of continuity for all of the reactive nitrogen species, we find:

$$\frac{d(NO_y)}{dt} + 2b_6(N^4S)(NO) = P(N^2D) + P(N^4S) + P(NO) \quad (5.171)$$

In the lower thermosphere and mesosphere, we may define R as follows:

$$R = \frac{(NO_y)}{(NO)} \approx 1 + \frac{(NO_2)}{(NO)} + \frac{(N^4S)}{(NO)} \quad (5.172)$$

The value of this constant can be obtained by examination of the continuity equations for each species. Considering transport, and substituting for $N(^4S)$ with equation (5.166), we find

$$\frac{\partial(NO_y)}{\partial t} + \vec{\nabla} \cdot \vec{\phi}(NO_y) + \frac{2b_6 J_{NO}(NO_y)^2}{R(b_6(NO_y) + Rb_7(O_2))}$$

$$= P(NO) + P(N^2D) + \left| \frac{Rb_7(O_2) - b_6(NO_y)}{Rb_7(O_2) + b_6(NO_y)} \right| P(N^4S) \quad (5.173)$$

Figure 5.49 displays a series of observations of NO in the thermosphere and mesosphere. The observed variability indicates the strong sensitivity of NO to dynamic processes in the atmosphere, as one might expect based on the comparison between its photochemical lifetime and the time constant for dynamics in this region, as shown in Figure 5.38. The observed minimum near the mesopause is due to the photodissociation of NO and its subsequent recombination with N^4S. The depth of the minimum depends on the competition between downward transport and this photochemical loss. Its variability as depicted in the figure is probably related to both seasonal and short term temporal variations in mesospheric transport parameters.

In the mesosphere, NO provides the primary source of electrons during geomagnetically quiet conditions, and is responsible for the formation of the ionospheric D-region. This will be discussed in detail in Chapter 6.

Figure 5.50 shows a theoretical calculation of the distribution of NO in the mesosphere and lower thermosphere. The calculated variations with respect to latitude reflect changes in mean solar zenith angle and its influence on odd nitrogen destruction through equation (5.159), as well as spatial variability in transport processes. Note the large NO densities in the high latitude winter stratosphere, particularly near polar night, where there is virtually no

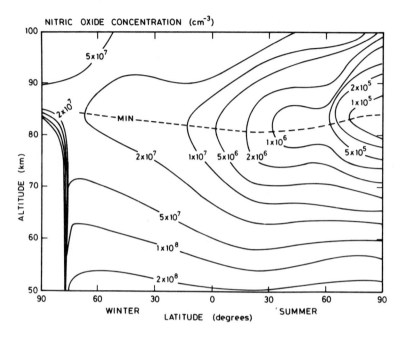

Fig. 5.50. Calculated two-dimensional NO distribution. From Brasseur (1982).

photochemical loss of odd nitrogen (because J_{NO} is zero there). As a result, models predict that NO produced at thermospheric altitudes can be transported down to the mesosphere and stratosphere at these latitudes (see, e.g., Frederick and Orsini, 1981; Solomon et al., 1982; Brasseur, 1982).

5.6 Chlorine compounds

5.6.1 General •

Numerous chlorine, fluorine, and bromine containing compounds are produced at ground level, both by natural and artificial processes. Methyl chloride (CH_3Cl), for example, is partially produced by the oceans, as well as by combustion of biomass, especially in tropical agriculture. The production is therefore subject to a seasonal cycle which should be detectable because the short lifetime of tropospheric CH_3Cl prevents uniform global mixing of this constituent on a seasonal time scale. Carbon tetrachloride (CCl_4) is also injected into the atmosphere, but it is difficult to determine the exact rate of its production because it is principally produced as an intermediate during industrial chemical manufacture. The chlorofluoromethanes, especially $CFCl_3$ (F-11), CF_2Cl_2 (F-12) and

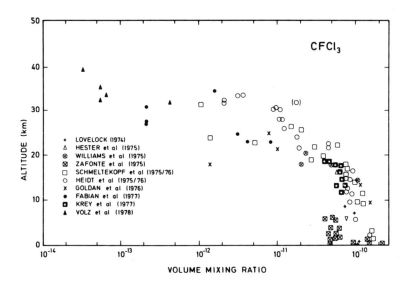

Fig. 5.51. Observed mole fraction of $CFCl_3$.

CH_3CCl_3 are produced in large quantity for use as propellants in aerosol cans and for refrigeration. Their annual production rate in 1980 was estimated at 265×10^3 tons per year for $CFCl_3$, and 393×10^3 tons per year of CF_2Cl_2. These species are generally fairly stable in the troposphere, and can therefore be transported toward the stratosphere, where they are eventually dissociated by ultraviolet radiation, or by reaction with $O(^1D)$.

The distributions of CH_3Cl and CCl_4 are not well known, particularly in the stratosphere. Tropospheric mole fractions of $5-7.5 \times 10^{-10}$ and $1-3 \times 10^{-10}$, respectively, can be assumed. For the chlorofluorocarbons, mean tropospheric mixing ratios of about 1.7×10^{-10} for $CFCl_3$, 3×10^{-10} for CF_2Cl_2, and $1-1.5 \times 10^{-10}$ for $CHCCl_3$ are appropriate for 1980. Figures 5.51 and 5.52 display some of the observations for these last two species.

Since these compounds represent the major chlorine containing constituents in the lower stratosphere and troposphere, the total amount of chlorine present in the atmosphere in 1980 should be about:

$$f_{Cl_T} \approx 4 \times f_{CCl_4} + 3 \times f_{CFCl_3} + 2 \times f_{CF_2Cl_2} + 3 \times f_{CHCCl_3} + f_{CH_3Cl} \qquad (5.174)$$

$$\approx 2.1 - 3.5 \text{ppbv}$$

A charcoal trap system followed by neutron activation detection has been used by Berg et al. (1980) to obtain measurements of f_{Cl_T}. They reported values of 2.7 ± 0.9 ppbv to 3.2 ± 0.7 ppbv at 20 km based on a series of measurements,

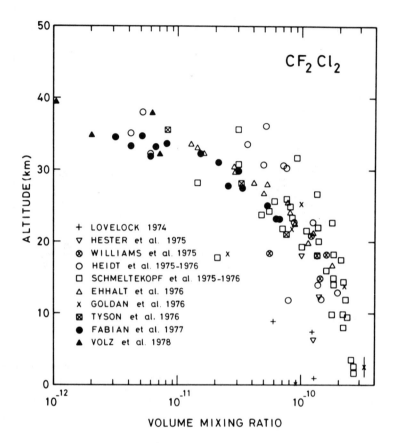

Fig. 5.52. Observed mole fraction of CF_2Cl_2.

in good agreement with the estimate derived above.

Other halogenated compounds are present in the atmosphere (e. g., $CHCl_3$, $CHCl_3$, CH_3F, CH_3Br, CF_4, etc.) but their importance in stratospheric chemistry is small. More than three thousand chlorine containing species have been detected in the troposphere. Those whose lifetime is short, however, are destroyed too rapidly to attain stratospheric heights.

5.6.2 Chlorine chemistry ‡

The natural production of chlorine atoms in the stratosphere comes from destruction of CH_3Cl, either by photolysis:

$$(J_{CH_3Cl}); \quad CH_3Cl + h\nu \rightarrow CH_3 + Cl \tag{5.175}$$

or by reaction with the OH radical:

$$(d_o); \quad CH_3Cl + OH \rightarrow CH_2Cl + H_2O \tag{5.176}$$

Among anthropogenic sources, the following mechanisms must be mentioned:

$$(J^1_{CCl_4}); \quad CCl_4 + h\nu \rightarrow CCl_3 + Cl \tag{5.177}$$

$$(J^2_{CCl_4}); \quad CCl_4 + h\nu \rightarrow CCl_2 + Cl_2 \tag{5.178}$$

$$(J_{CH_3CCl_3}); \quad CH_3CCl_3 + h\nu \rightarrow CH_3CCl_2 + Cl \tag{5.179}$$

$$(d_1); \quad CH_3CCl_3 + OH \rightarrow CH_2CCl_3 + H_2O \tag{5.180}$$

which can all be followed by other processes which eventually produce chlorine atoms. The photodissociation of chlorofluoromethanes must also be noted:

$$(J_{CFCl_3}); \quad CFCl_3 + h\nu(\lambda < 226nm) \rightarrow CFCl_2 + Cl \tag{5.181}$$

$$(J_{CF_2Cl_2}); \quad CF_2Cl_2 + h\nu(\lambda < 215nm) \rightarrow CF_2Cl + Cl \tag{5.182}$$

The importance of these processes was first noted by Molina and Rowland (1974a) and Rowland and Molina (1975), and by Stolarski and Cicerone (1974).

If the distributions of these organo-chlorine molecules were well known, the rate of production of atomic chlorine could be immediately deduced. Assuming these species are eventually completely dissociated in the stratosphere, we can write:

$$P(Cl) = 4J_{CCl_4}(CCl_4) + 3J_{CFCl_3}(CFCl_3) + 2J_{CF_2Cl_2}(CF_2Cl_2)$$

$$+ |J_{CH_3Cl} + d_o(OH)|(CH_3Cl) + 3|J_{CH_3CCl_3} + d_1(OH)|(CH_3CCl_3) \tag{5.183}$$

When a chlorine atom is produced in the stratosphere, it can react with ozone to produce chlorine monoxide:

$$(d_2); \quad Cl + O_3 \rightarrow ClO + O_2 \tag{5.184}$$

Chlorine monoxide undergoes two principal reactions in the stratosphere. Reaction with atomic oxygen destroys odd oxygen:

$$(d_3); \quad ClO + O \rightarrow Cl + O_2 \tag{5.185}$$

It can also react with NO:

$$(d_4); \quad ClO + NO \rightarrow Cl + NO_2 \tag{5.186}$$

These reactions both reform atomic chlorine. The reaction between ClO and NO represents an important coupling between the chlorine and nitrogen cycles, and the pair of reactions (5.184) and (5.185) constitutes an important catalytic cycle which destroys odd oxygen.

The photodissociation of ClO should also be mentioned:

$$(J_{ClO}); \quad ClO + h\nu \rightarrow Cl + O \tag{5.187}$$

In addition to these rapid processes (which largely determine an equilibrium between Cl and ClO), there are also slower processes which yield hydrochloric acid. Like HNO_3, HCl provides an inert reservoir which sequesters a photochemically active species and affects the rate of its catalytic reaction with odd oxygen.

$$(d_5); \quad Cl + CH_4 \rightarrow CH_3 + HCl \tag{5.188}$$

$$(d_6); \quad Cl + H_2 \rightarrow H + HCl \tag{5.189}$$

$$(d_7); \quad Cl + HO_2 \rightarrow O_2 + HCl \tag{5.190}$$

$$(d_8); \quad Cl + H_2O_2 \rightarrow HO_2 + HCl \tag{5.191}$$

$$(d_9); \quad Cl + HNO_3 \rightarrow NO_3 + HCl \tag{5.192}$$

$$(d_{10}); \quad Cl + CH_2O \rightarrow HCO + HCl \tag{5.193}$$

Reaction (5.188) is generally the most important of these, although (5.190) plays a substantial role in the upper stratosphere. HCl can be destroyed by reaction with OH and O to yield a chlorine atom:

$$(d_{11}); \quad HCl + OH \rightarrow H_2O + Cl \tag{5.194}$$

$$(d_{12}); \quad HCl + O \rightarrow OH + Cl \tag{5.195}$$

Photolysis must also be considered:

$$(J_{HCl}); \quad HCl + h\nu(\lambda < 220nm) \rightarrow H + Cl \tag{5.196}$$

HCl is soluble in water, and can therefore be removed from the atmosphere in cloud processes.

To end this brief discussion of the aeronomic reactions of chlorine species, we must consider the formation of other constituents such as ClOO, OClO, ClO_3, $ClONO_2$, HOCl, etc. Only the last two are presently believed to be important to the chemistry of the ozonosphere. Chlorine nitrate $ClONO_2$, is formed by the reaction of ClO with NO_2:

$$(d_{31}); \quad ClO + NO_2 + M \rightarrow ClONO_2 + M \tag{5.197}$$

and dissociated in the ultraviolet (Chang et al., 1979)

$$(J_{ClONO_2}); \quad ClONO_2 + h\nu \rightarrow Cl + NO_3 \tag{5.198}$$

A secondary loss with atomic oxygen should also be mentioned:

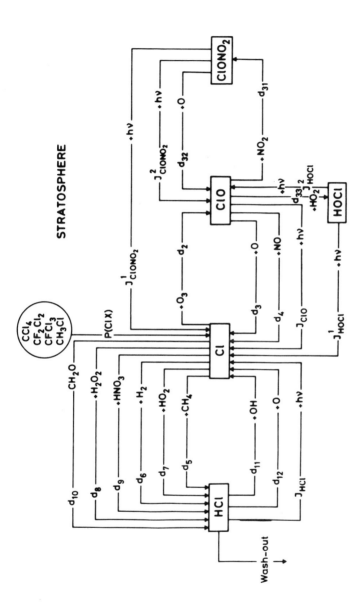

Fig. 5.53. Principal aeronomic reactions of chlorine compounds.

$$(d_{32}); \quad ClONO_2 + O \rightarrow products \tag{5.199}$$

The HOCl molecule is produced by the reaction

$$(d_{33}); \quad HO_2 + ClO \rightarrow HOCl + O_2 \tag{5.200}$$

and destroyed mostly by photolysis

$$(J_{HOCl}); \quad HOCl + h\nu(\lambda < 301nm) \rightarrow ClO + H \tag{5.201a}$$

$$HOCl + h\nu(\lambda < 503nm) \rightarrow Cl + OH \tag{5.201b}$$

and reaction with OH:

$$(d_{34}); \quad OH + HOCl \rightarrow H_2O + ClO \tag{5.202}$$

Considering the schematic diagram shown in Figure 5.53, the continuity equations for Cl, ClO, HCl, and ClONO$_2$ can be easily derived:

$$\frac{d(Cl)}{dt} + (Cl)[d_2(O_3) + d_5(CH_4) + d_6(H_2) + d_7(HO_2) + d_8(H_2O_2) + d_9(HNO_3)$$

$$+ d_{10}(CH_2O)] = P(Cl) + [d_3(O) + d_4(NO) + J_{ClO}](ClO) + [d_{12}(O)$$

$$+ d_{11}(OH) + J_{HCl}](HCl) + J_{ClONO_2}(ClONO_2) + J_{HOCl_b}(HOCl) \tag{5.203}$$

where $P(Cl)$ represents production of chlorine from organo-chlorine molecules.

$$\frac{d(ClO)}{dt} + (ClO)[d_3(O) + d_4(NO) + J_{ClO} + d_{31}(M)(NO_2)]$$

$$= d_2(O_3)(Cl) + [J_{ClONO_2} + d_{32}(O)](ClONO_2) + J_{HOCl_b}(HOCl) \tag{5.204}$$

$$\frac{d(HCl)}{dt} + (HCl)[d_{12}(O) + d_{11}(OH) + J_{HCl} + \beta_p] = (Cl)[d_5(CH_4) + d_6(H_2)$$

$$+ d_7(HO_2) + d_8(H_2O_2) + d_9(HNO_3) + d_{10}(CH_2O)] \tag{5.205}$$

where β_p represents destruction by heterogeneous processes.

$$\frac{d(HOCl)}{dt} + [J_{HOCl} + d_{34}(OH)](HOCl) = d_{33}(HO_2)(ClO) \tag{5.206}$$

$$\frac{d(ClONO_2)}{dt} + (ClONO_2)[J_{ClONO_2} + d_{32}(O)] = d_{31}(M)(ClO)(NO_2) \tag{5.207}$$

5.6.3 The odd chlorine family: lifetimes and observations •

Odd chlorine (Cl_x) is defined here as the sum of $Cl + ClO + HOCl$. The continuity equation for this family can be written as

$$\frac{\partial (Cl_x)}{\partial t} + [d_5(CH_4) + d_6(H_2) + d_7(HO_2) + d_8(H_2O_2) + d_9(HNO_3)$$

$$+ d_{10}(CH_2O)](Cl) + d_{31}(NO_2)(M)(ClO) = [d_{11}(OH) + d_{12}(O) + J_{HCl}](HCl)$$

$$+ [J_{ClONO_2} + d_{32}(O)](ClONO_2) - \vec{\nabla} \cdot \phi(Cl_x) \tag{5.208}$$

and the partitioning between family members can be readily obtained by assuming photochemical equilibrium for Cl and HOCl and retaining only the most important terms in (5.204) and (5.206):

$$\frac{(HOCl)}{(ClO)} = \frac{d_{33}(HO_2)(ClO)}{J_{HOCl} + d_{34}(OH)} \tag{5.209}$$

$$\frac{(ClO)}{(Cl)} = \frac{d_2(O_3)}{d_3(O) + d_4(NO) + J_{ClO}} \tag{5.210}$$

so that

$$\frac{(ClO)}{(Cl_x)} = \frac{1}{\left[1 + \dfrac{(HOCl)}{(ClO)} + \dfrac{(Cl)}{(ClO)} \right]} \tag{5.211}$$

and

$$\frac{(HOCl)}{(Cl_x)} = \frac{(ClO)}{(Cl_x)} \times \frac{(HOCl)}{(ClO)} \tag{5.212}$$

$$\frac{(Cl)}{(Cl_x)} = \frac{(ClO)}{(Cl_x)} \times \frac{(Cl)}{(ClO)} \tag{5.213}$$

A similar equation for "reactive chlorine", Cl_y, can be written by adding equations (5.203) through (5.207), introducing a transport term, and assuming

$$(Cl_y) = (Cl) + (ClO) + (HCl) + (ClONO_2) + (HOCl) \tag{5.214}$$

$$\frac{\partial (Cl_y)}{\partial t} + \vec{\nabla} \cdot \vec{\phi}(Cl_y) + \beta_p^*(Cl) = P(Cl) \tag{5.215}$$

where

$$\beta_p^* = \beta_p \frac{(HCl)}{(Cl_y)} \tag{5.216}$$

representing an effective heterogeneous removal coefficient of reactive chlorine unto particles. Figure 5.54 shows the photochemical lifetimes of Cl_x, Cl, ClO,

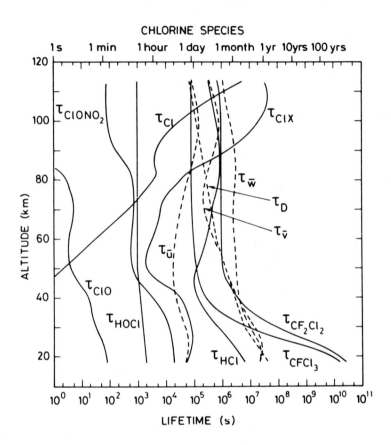

Fig. 5.54. Photochemical lifetimes of Cl_x, Cl, ClO, HOCl, ClONO$_2$, HCl, CF$_2$Cl$_2$, and CFCl$_3$, as well as the time constants appropriate to atmospheric transport.

HOCl, ClONO$_2$, HCl, CF$_2$Cl$_2$ and CFCl$_3$, as well as the transport time scales in the middle atmosphere. As we have already emphasized, photochemical equilibrium can only be assumed in aeronomic models if the species in question interchange with one another considerably more rapidly than they can be influenced by transport. This is true for ClO and Cl, for example, but not for HCl and Cl_x throughout much of the middle atmosphere. Thus it is not generally advisable to solve for reactive chlorine rather than for Cl_x for purposes of numerical modeling of the behavior of these constituents. We may, however, evaluate the equilibrium ratios between chlorine containing compounds to serve as a rough estimate of their relative abundances. For example, using equation (5.207),

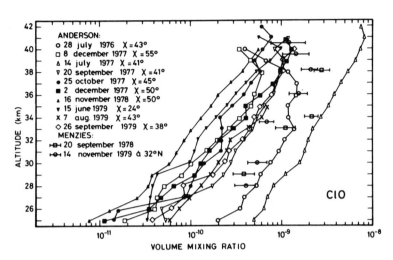

Fig. 5.55. Vertical distribution of ClO measured by Anderson (1980) and Menzies (1979).

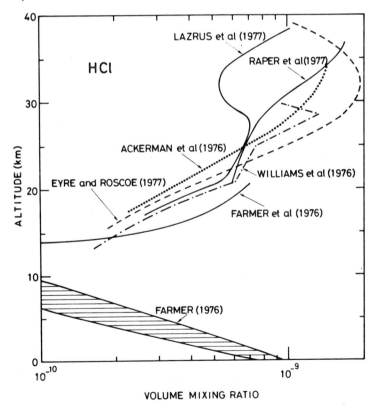

Fig. 5.56. Observations of the HCl mixing ratio.

$$\frac{(ClONO_2)}{(ClO)} = \frac{d_{31}(M)(NO_2)}{J_{ClONO_2} + d_{32}(O)} \tag{5.217}$$

and from equation (5.205),

$$\frac{(HCl)}{(Cl)} = \frac{d_5(CH_4) + d_6(H_2) + d_7(HO_2) + d_8(H_2O_2) + d_9(HNO_3)}{d_{10}(O) + d_{11}(OH) + J_{HCl}} \tag{5.218}$$

or, approximately

$$\frac{(HCl)}{(Cl)} \approx \frac{d_5(CH_4) + d_7(HO_2)}{d_{11}(OH)} \tag{5.219}$$

These ratios illustrate the importance of the interactions of chlorine chemistry with oxygen, hydrogen and nitrogen species.

We now turn to a brief discussion of available measurements of reactive chlorine species. Atomic chlorine densities have been inferred in the lower stratosphere based on measurements of ethane and propane (Rudolph et al., 1981). The ClO radical has been directly measured several times since 1976, primarily by Anderson (see Anderson, 1980), using resonance fluorescence, and by Menzies (1979), using absorption with a heterodyne laser as well as by Waters et al. (1980). These data are shown in Figure 5.55. Considerable variability is found in the observed profiles. In particular, the observation of ClO obtained in July, 1977 exhibits much larger values than the other periods. The laser

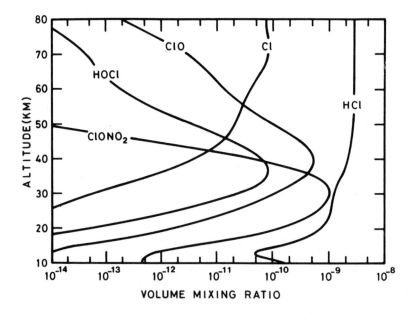

Fig. 5.57. Vertical distribution of reactive chlorine species as calculated for 1980 by the one-dimensional model of Brasseur et al., (1983).

measurements obtained after sunset in September also are relatively high. On the other hand, careful ground based infrared laser heterodyne experiments by Mumma et al. (1983) have yielded the surprising result of derived upper limits which lie far below the observations by Anderson. Systematic observations should eventually lead to an understanding of the global behavior of ClO and the amplitude of its variations. This constitutes an important element in the study of the chlorine-ozone interaction.

Hydrochloric acid has been measured by spectroscopic absorption (Ackerman et al., 1976; Farmer et al., 1976; Raper et al., 1977; Williams et al. 1976), by absorption radiometry (Eyre and Roscoe, 1977) and by in situ filter sampling and chemical analysis (Lazrus et al., 1975, 1976). Some recent measurements are shown in Figure 5.56. The optical methods are generally in good agreement, but the data obtained by Lazrus are considerably lower above 25 km. The distribution and variability of the total column abundances of HCl and HF have been reported by Mankin and Coffey (1983). Future observations should eventually provide an estimate of the HCl variability and its correlation with that of ClO.

Infrared spectra of $ClONO_2$ have been obtained by Murcray et al. (1977), and these have been reanalyzed in WMO (1982) to provide an estimate of its abundance. The derived mixing ratio is about 0.2 ppbv near 25-30 km.

One-dimensional model predictions of the theoretical distributions of chlorine compounds computed for the year 1980 are shown in Figure 5.57. The diurnal variations of these species were displayed in Figure 5.40. These densities are expected to increase progressively if the emission of industrial halocarbons continues.

5.7 Other halogens

The dissociation of chlorofluorocarbons such as $CFCl_3$ and CF_2Cl_2, as well as other species such as SF_6, represents a source of atmospheric fluorine. The fluorine atom reacts rapidly with ozone:

$$(f_2); \quad F + O_3 \rightarrow FO + O_2 \tag{5.220}$$

to produce fluorine monoxide, which reacts like chlorine monoxide, either with atomic oxygen

$$(f_3); \quad FO + O \rightarrow F + O_2 \tag{5.221}$$

or with nitric oxide:

$$(f_4); \quad FO + NO \rightarrow F + NO_2 \tag{5.222}$$

Like chlorine, the fluorine atom can also produce an acid, HF:

$$(f_5); \quad F + CH_4 \rightarrow HF + CH_3 \tag{5.223}$$

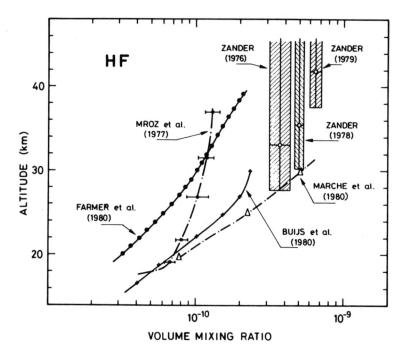

Fig. 5.58. Observations of the mole fraction of HF in the stratosphere. From Zander (1981). (Copyright by the American Geophysical Union).

$$(f_6); \quad F + H_2 \rightarrow HF + H \tag{5.224}$$

$$(f_7); \quad F + H_2O \rightarrow HF + OH \tag{5.225}$$

Unlike HCl however, HF does not react appreciably with OH, so that the fluorine atoms which react with methane, hydrogen or water vapor are effectively stabilized as HF. Since the sources of HF and HCl are similar but their loss rates are quite different as a result of the reaction of OH with HCl at low altitudes, the HF/HCl ratio can provide an indication of the abundance of the OH radical (see, e. g., Sze and Ko, 1980).

Because of the stability of HF, the atmospheric densities of F and FO are very small and the effect of fluorine on odd oxygen is insignificant. The reaction of HF with O(^{1}D) is chemically possible, but is negligible due to the low abundance of this highly excited atom. The distribution of HF is therefore largely determined by the rates of the surface emission of fluorine containing gases, of photochemical destruction of these gases, and atmospheric dynamics. Numerous measurements of HF have been reported, notably by Zander (1981) and Farmer et al. (1980). Some of these are shown in Figure 5.58.

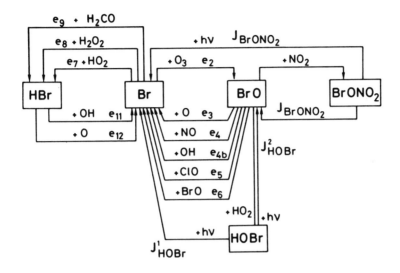

Fig. 5.59. Schematic diagram of bromine chemistry

Molecules of the type $C_xH_yBr_z$ produce bromine atoms upon photodissociation, like the other halogen atoms just discussed. Unlike fluorine, the chemistry of bromine is quite similar to that of chlorine (see Fig. 5.59) and is therefore capable of affecting the ozone budget. The production of bromine atoms can be followed by the reaction

$$(e_2); \quad Br + O_3 \rightarrow BrO + O_2 \tag{5.226}$$

which leads to a number of possible reactions for bromine monoxide:

$$(e_3); \quad BrO + O \rightarrow Br + O_2 \tag{5.227}$$

$$(e_4); \quad BrO + NO \rightarrow Br + NO_2 \tag{5.228}$$

$$(e_5); \quad BrO + ClO \rightarrow Br + OClO \tag{5.229a}$$

$$\rightarrow Br + Cl + O_2 \tag{5.229b}$$

$$(e_6); \quad BrO + BrO \rightarrow 2Br + O_2 \tag{5.230}$$

Bromine thus plays a role not only in its own catalytic ozone destruction cycle, but can also influence the chlorine cycle through the effect of reaction e_5 (see Wofsy et al., 1975). Indeed, the effect of these halogens on the ozone molecule can be significantly influenced by this "synergy" between the catalytic cycles, wherein both become more effective. Hydrobromic acid is formed almost entirely by the reaction

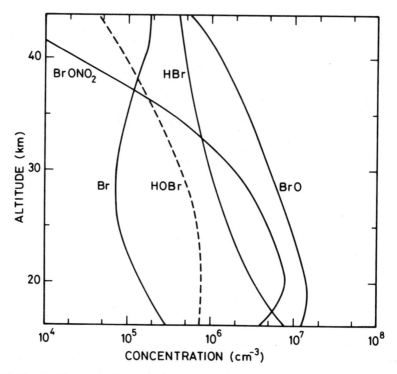

Fig. 5.60. Vertical distribution of bromine compounds in the stratosphere. From the model of Yung et al. (1980). (Copyright by the American Meteorological Society).

$$(e_7); \quad Br + HO_2 \rightarrow HBr + O_2 \tag{5.231}$$

because Br does not react with methane or molecular hydrogen, and the reaction

$$(e_8); \quad Br + H_2O_2 \rightarrow HBr + HO_2 \tag{5.232}$$

proceeds slowly at middle atmospheric temperatures. Bromine monoxide is rapidly photodissociated, in contrast to hydrobromic acid, which only photolyzes at short wavelengths. HBr reacts rapidly with OH, however:

$$(e_{11}); \quad HBr + OH \rightarrow Br + H_2O \tag{5.233}$$

The reaction with atomic oxygen

$$(e_{12}); \quad HBr + O \rightarrow Br + OH \tag{5.234}$$

is slower and thus plays only a minor role in HBr chemistry. Finally, BrO can, like ClO, associate with NO_2 and HO_2:

$$(e_{13}); \quad BrO + NO_2 + M \rightarrow BrONO_2 + M \quad\quad (5.235)$$

$$(e_{15}); \quad BrO + HO_2 \rightarrow HOBr + O_2 \quad\quad (5.236)$$

to form $BrONO_2$ and HOBr, both of which rapidly photodissociate. Figure 5.60 presents calculated vertical distributions of a number of bromine compounds (from Yung et al., 1980). Observations of the total bromine abundance in the lower stratosphere in 1980 by Berg et al. (1980) suggest that its mixing ratio is between 7 and 40 pptv, but that it may be quite variable.

5.8 Sulfur compounds and formation of aerosols

Atmospheric sulfur is produced at ground level by a number of sources. Some are natural: hydrogen sulfide, H_2S, for example, is a by product of decomposition of organic matter; dimethyl sulfide $(CH_3)_2S$ and CS_2 are produced by the oceans. Other sources are related to anthropogenic activities. Industry produces large amounts of sulfur dioxide, particularly through combustion of coal (esp. coal fired electric power plants), as well as smaller quantities of H_2S, CS_2 and carbonyl sulfide, COS. Sulfur oxides and sulfuric acid vapors are produced in combustion engines, especially in automobiles. Finally, volcanic eruptions are responsible for the injection of appreciable amounts of sulfur into the atmosphere (SO_2, H_2S).

Thus, although sulfur is always present in the atmosphere, its density can vary considerably as a result of geophysical or anthropogenic perturbations. Urban regions exhibit large densities of sulfur containing compounds, and the research related to its natural sources must generally be performed in deserts or on the ocean. In clean air, the mole fraction of SO_2 at low altitudes is about 10^{-9} or 5×10^{-10} (Friend, 1973; Kellogg et al., 1972; Breeding et al., 1973; Graedel et al., 1974). The density of SO_2 in slightly polluted air is quite variable, and depends on meteorological conditions as well as the source of the air masses, because the tropospheric lifetime of SO_2 is several days. Near the tropopause abundances of about 5×10^{-11} are characteristically observed (Jaeschke et al., 1976), indicating that less than 10% of the molecules reach the lower stratosphere. Other sulfur compounds such as H_2S and $(CH_3)_2S$ are short lived and are therefore rapidly destroyed in the troposphere. On the other hand, COS is very stable and its mole fraction is of the order of 1×10^{-10} throughout the troposphere (Hanst et al., 1975; Sandalls and Penkett; 1977; Maroulis et al., 1977), even at tropopause levels (Inn et al., 1979; Mankin et al., 1979). The lifetime of tropospheric COS is long, of the order of a year, and only above about 15 km does the mole fraction begin to decrease, reaching about 1.5×10^{-11} at 31 km. CS_2 possesses an intermediate lifetime, as evidenced by an observed tropospheric variability of about 50 % (Sandalis and Penkett, 1977). Its surface mole fraction is about 1×10^{-10}.

Therefore, middle atmospheric sulfur is mostly due to transport of COS across the tropopause as first suggested by Crutzen in 1976. COS is primarily produced by biologic, volcanic and anthropogenic sources. Thus an increase in COS, particularly as a result of the use of large quantities of coal, can alter the composition of the atmosphere, increasing the concentration of sulfur aerosols and modifying the terrestrial radiative balance (see Crutzen, 1976; Turco et al., 1980).

When COS reaches the stratosphere, photolysis occurs:

$$COS + h\nu \rightarrow CO + S \qquad (5.237)$$

producing carbon monoxide and atomic sulfur. Sulfur reacts immediately with molecular oxygen:

$$S + O_2 \rightarrow SO + O \qquad (5.238)$$

to produce sulfur monoxide. This species can also be formed by reaction of COS and atomic oxygen:

$$COS + O \rightarrow CO + SO \qquad (5.239)$$

which, like the reaction

$$COS + OH \rightarrow products \qquad (5.240)$$

represents an additional destruction mechanism for COS, but these are considerably less important than the direct photolysis process. Sulfur monoxide is in rapid equilibrium with SO_2. Its destruction is dominated by

$$SO + O_2 \rightarrow SO_2 + O \qquad (5.241)$$

which is more important (Black et al., 1982a,b) than

$$SO + O_3 \rightarrow SO_2 + O_2 \qquad (5.242)$$

and

$$SO + NO_2 \rightarrow SO_2 + NO \qquad (5.243)$$

Sulfur dioxide is relatively stable and is thus present in fairly large abundance in the stratosphere. It can be photodissociated by radiation in the 200-230 nm region, where its absorption cross section is of the order of 1×10^{-18} 1×10^{-17} cm^2.

$$SO_2 + h\nu \rightarrow SO + O \qquad (5.244)$$

or converted to HSO_3 by the following process:

$$SO_2 + OH + M \rightarrow HSO_3 + M \qquad (5.245)$$

The HSO_3 intermediate is probably destroyed by (Calvert and Stockwell, 1983; McKeen et al., 1984):

$$HSO_3 + O_2 \rightarrow HO_2 + SO_3 \qquad (5.246)$$

although the conversion process suggested by Turco (1979, 1982) cannot be ruled

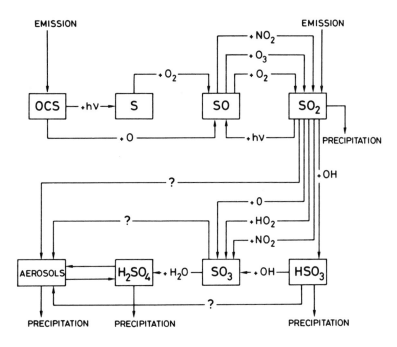

Fig. 5.61. Primary sulfur reactions in the middle atmosphere.

out:

$$HSO_3 + OH \rightarrow H_2O + SO_3 \qquad (5.247)$$

Direct conversion of SO_2 to SO_3 by reactions such as

$$SO_2 + HO_2 \rightarrow SO_3 + OH \qquad (5.248a)$$

$$SO_2 + O + M \rightarrow SO_3 + M \qquad (5.48b)$$

can be neglected in stratospheric studies.

Sulfur trioxide can combine with water vapor to form sulfuric acid:

$$SO_3 + H_2O \rightarrow H_2SO_4 \qquad (5.249)$$

The acid can be photodissociated:

$$H_2SO_4 + h\nu \rightarrow SO_2 + products \qquad (5.250)$$

or it can precipitate in water droplets, (as can SO_2 and HSO_3) yielding sulfate aerosols. A complete analysis of sulfur compounds must therefore include heterogeneous removal of these species.

It should be noted that the reaction sequence which converts SO_2 to sulfate does not modify the HO_x budget if reaction (5.247) can be neglected

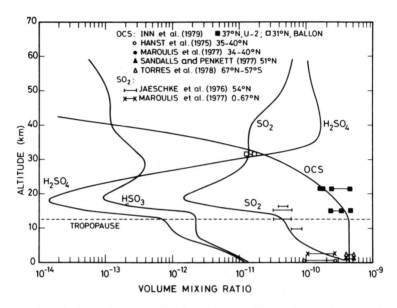

Fig. 5.62. Vertical distribution of the primary sulfur compounds, as calculated by Whitten et al. (1980). (Copyright by Birkhauser-Verlag).

compared to (5.246) (McKeen et al, 1984). If process (5.247) is dominant, then the sequence leads to one net OH loss. An additional loss of odd hydrogen is obtained if aerosol nucleation occurs directly from HSO_3 hydrates instead of H_2SO_4. The intrusion of large amounts of SO_2 in the stratosphere during volcanic eruptions could modify the amount of ozone in the middle atmosphere either by direct chemical effects (Crutzen and Schmailzl, 1983), or by the influence of aerosols (heterogeneous reactions, alteration of the ultraviolet or infrared radiation fields, etc.).

Figure 5.61 shows a schematic diagram of the chemistry of sulfur and Figure 5.62 presents calculated distributions of the gaseous sulfur compounds from the model by Whitten et al. (1980). Some observations are also shown for comparison. This model shows that below 30 km sulfur is mostly in the form of COS, while at higher altitude it is mostly sulfuric acid.

Analysis of observations of ion densities can be used to derive the abundance of $H_2SO_4 + HSO_y$ in the stratosphere (HSO_y is mostly HSO_3), as shown, for example, by Arnold and Fabian (1980) and Arijs et al. (1981). These observations suggest that the gaseous stratospheric H_2SO_4 concentration is largely determined by the equilibrium between the aerosol and vapor phases. Thus, the density of sulfuric acid in the stratosphere depends strongly on aerosol chemistry. The gas and aerosol chemistry of sulfur are discussed in detail in the excellent review by Turco (1982).

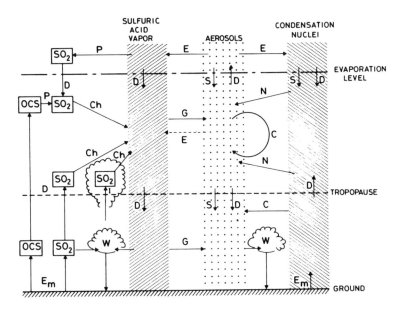

Fig. 5.63. Schematic diagram of the physical processes related to the formation, growth, transport, and destruction of atmospheric aerosols (from Whitten et al., 1980; copyright by Birkhauser-Verlag)

The H_2SO_4 molecule plays a major role in the microphysics of nucleation and condensation, and is therefore essential to the formation and growth of aerosols. These particles can also coagulate, diffuse, or sediment out under the force of gravity. The exact mechanism of nucleation, especially the initial processes, are poorly known, but it is presently thought that heterogeneous and heteromolecular nucleation of the $H_2SO_4-H_2O$ system around fine Aitken particles (condensation nuclei of radius less than 0.1 μm) is predominant. These Aitken particles are probably transported from the troposphere by dynamic processes and perhaps also produced by aircraft engines. Other processes may also be important in aerosol formation, such as homogeneous nucleation (nucleation of polar molecules such as H_2O, HNO_3, H_2SO_4 around ions). Evidence for such a process is implied by the formation of sulfate radicals and complex neutral species during ion-ion recombination.

After nucleation, the droplets grow through heteromolecular condensation, provided that the partial pressure of H_2SO_4 in the surrounding air is greater than the vapor pressure above the droplet. Otherwise, the H_2SO_4 will evaporate and the aerosol will shrink. Figure 5.63 presents a schematic diagram of the physical and chemical properties pertaining to the stratospheric aerosol layer (also called the Junge layer). It should be noted that although sulfur enters the stratosphere in gaseous form (essentially COS, CS_2, and SO_2) it leaves the middle atmosphere in the form of particles (sulfate) by sedimentation

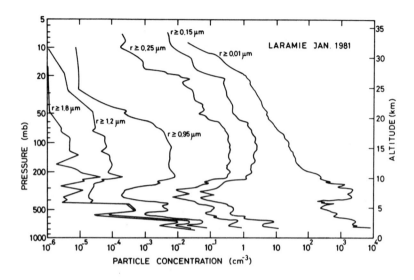

Fig. 5.64. Observed particle size distribution with altitude. From Hofmann and Rosen (1981).

and diffusion followed by precipitation in clouds, and these processes are poorly understood (see Turco, 1982).

The most fundamental aerosol property is the size distribution, which is usually expressed in an empirical form (power law, exponential, or log normal function) whose parameters are derived from observations. This size distribution depends on the nature of the particles and varies with altitude. As shown in Figure 5.64, for example, the concentration of condensation nuclei decreases abruptly with increasing height above the tropopause, while the concentration of the largest particles (radius greater than 0.15 μm) exhibit a distinct peak in the lower stratosphere. The existence of this concentration maximum, approximately 5-7 km above the tropopause, indicates that these particles nucleate in the stratosphere and grow in-situ into larger particles by sulfuric acid vapor condensation and coalescence. Figure 5.65 shows the meridional distribution of aerosols (r > 0.15 μm) observed in November, 1973, by Rosen et al. (1975), together with a distribution of ozone mixing ratios. The morphology of these two distributions is generally similar, although some divergence can be seen above 25 km, where gravitational sedimentation can have a strong influence on the transport of the particles, and where ozone starts to be photochemically controlled. Nevertheless, observed distributions of aerosols in the lower stratosphere provide useful indicators of dynamics, especially in the region where exchange through the tropopause can occur.

The seasonal variation in the tropopause height seems to modulate the total content of large aerosols in the stratosphere, according to continuous observations made over Wyoming by Hoffmann et al. (1975). These

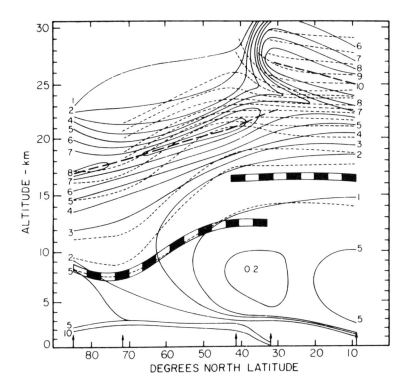

Fig. 5.65. Observed distribution of aerosols of radius greater than 0.15 μm (solid), and an observed ozone mixing ratio distribution (dashed). Heavy dashed line indicates the location of the meteorological tropopause. From Rosen et al. (1975). (Copyright by the American Meteorological Society).

investigators indicate that the aerosol layer is considerably more stable in summer than in winter, and that rapid fluctuations are observed in winter which seem to be related to the propagation of large planetary waves. These variations also appear in the aerosol extinction data derived from satellite data (McCormick et al., 1981). This behavior is consistent with the similarity between ozone and aerosol content as shown in Figure 5.65.

5.9 Generalized ozone balance

The analysis presented above permits us to establish a generalized ozone and atomic oxygen balance, and to correct the equations derived previously by considering only a pure oxygen atmosphere. Detailed reviews of the ozone balance are provided by Johnston and Podolske (1978), Cariolle (1983), and Crutzen and Schmailzl (1983). Above 40 km, the catalytic cycles involving hydrogen free radicals represent a very rapid loss process for odd oxygen

(Nicolet, l970a):

$$OH + O \rightarrow H + O_2$$

$$H + O_2 + M \rightarrow HO_2 + M$$

$$HO_2 + O \rightarrow OH + O_2$$

$$\boxed{\text{Net: } O + O \rightarrow O_2}$$

or

$$OH + O \rightarrow H + O_2$$

$$H + O_3 \rightarrow OH + O_2$$

$$\boxed{\text{Net: } O + O_3 \rightarrow 2O_2}$$

With some algebraic manipulation, the chemistry of these hydrogen catalyzed reactions can be included in the equation for the photochemical equilibrium ozone abundance between about 50 and 80 km to yield the following simple formulas at various altitudes (Allen et al., 1984):

$$(O_3)_{50 \text{ km}} \approx \frac{(J_{O_2})^{2/3}(k_{4a}a_{17})^{1/3}[k_2(M)]^{2/3}(N_2)^{1/3}(O_2)^{4/3}}{(J_{O_3})^{2/3}(J_3^* a_1^* a_5 a_7)^{1/3}(H_2O)^{1/3}} \tag{5.251}$$

$$(O_3)_{70 \text{ km}} \approx \frac{J_{O_2}(a_{17})^{1/2}k_2(M)(O_2)^2}{J_{O_3}(J_{H_2O} \, a_5 a_7)^{1/2}(H_2O)^{1/2}} \tag{5.252}$$

$$(O_3)_{80 \text{ km}} \approx \frac{(J_{O_2})^2 a_{23} k_2(M)(O_2)^2}{J_{H_2O} a_7 a_1 (H_2O)(M)\left[J_{O_3} + \dfrac{a_2}{a_1}\dfrac{J_{O_2}}{(M)}\right]} \tag{5.253}$$

The important role of odd hydrogen catalyzed destruction of odd oxygen is illustrated by comparing these expressions to equation (5.32), which presents the analytic expression for the equilibrium ozone density assuming only pure oxygen chemistry.

The hydrogen radicals also participate in the following cycle:

$$OH + O_3 \rightarrow HO_2 + O_2$$

$$HO_2 + O_3 \rightarrow OH + 2O_2$$

$$\boxed{\text{Net: } 2\,O_3 \rightarrow 3O_2}$$

This process dominates the odd oxygen destruction near the tropopause because it is the most effective catalytic cycle involving only ozone as the reactive odd oxygen species. Most of the other HO_x, NO_x, and ClO_x cycles (see below) also require reaction with atomic oxygen, which is present only in very small amounts at low altitudes. For example, the following cycles are also important in the middle and upper stratosphere:

$$OH + O_3 \rightarrow HO_2 + O_2$$

$$HO_2 + O \rightarrow OH + O_2$$

$$\boxed{\text{Net: } O + O_3 \rightarrow 2O_2}$$

or

$$OH + O \rightarrow H + O_2$$

$$H + O_2 + M \rightarrow HO_2 + M$$

$$HO_2 + O_3 \rightarrow OH + 2O_2$$

$$\boxed{\text{Net: } O + O_3 \rightarrow 2O_2}$$

The nitrogen compounds can also catalyze the destruction of odd oxygen. This cycle is most efficient near 35-45 km (Crutzen, 1970; Johnston, 1971):

$$NO + O_3 \rightarrow NO_2 + O_2$$

$$NO_2 + O \rightarrow NO + O_2$$

$$\boxed{\text{Net: } O + O_3 \rightarrow 2O_2}$$

But this cycle competes with another which has no effect on odd oxygen:

$$NO + O_3 \rightarrow NO_2 + O_2$$

$$NO_2 + h\nu \rightarrow NO + O$$

$$O + O_2 + M \rightarrow O_3 + M$$

Net: Null

Thus, the efficiency of the NO_x catalyzed destruction of odd oxygen depends on the competition between photolysis and reaction with atomic oxygen for the NO_2 radical. Other secondary cycles should also be noted:

$$NO + O_3 \rightarrow NO_2 + O_2$$

$$NO_2 + O_3 \rightarrow NO_3 + O_2$$

$$NO_3 + h\nu \rightarrow NO + O_2$$

Net: $2\ O_3 \rightarrow 3O_2$

$$NO + O_3 \rightarrow NO_2 + O_2$$

$$NO_2 + O_3 \rightarrow NO_3 + O_2$$

$$NO_3 + NO_2 + M \rightarrow N_2O_5 + M$$

$$N_2O_5 + h\nu \rightarrow NO_3 + NO_2$$

$$NO_3 + h\nu \rightarrow NO + O_2$$

Net: $2\ O_3 \rightarrow 3O_2$

Note that these cycles only result in ozone destruction when NO_3 photolysis leads to $NO + O_2$ rather than $NO_2 + O$. In the latter case, the effect of these cycles is also null.

The catalytic destruction of odd oxygen by chlorine species must also be considered, particularly in the upper stratosphere where this cycle is quite effective (Stolarski and Cicerone, 1974; Molina and Rowland, 1974):

$$Cl + O_3 \rightarrow ClO + O_2$$

$$ClO + O \rightarrow Cl + O_2$$

Net: $O + O_3 \rightarrow 2O_2$

Below 30 km the densities of Cl and ClO are much reduced due to formation of other compounds such as HOCl and $ClONO_2$, and thus the effect of the direct chlorine catalyzed destruction on odd oxygen is greatly reduced. However, these other compounds can still catalytically destroy odd oxygen. The following cycle can be important near 20 km:

$$Cl + O_3 \rightarrow ClO + O_2$$

$$ClO + HO_2 \rightarrow HOCl + O_2$$

$$HOCl + h\nu \rightarrow Cl + OH$$

$$OH + O_3 \rightarrow HO_2 + O_2$$

Net: $2\,O_3 \rightarrow 3O_2$

The following cycle may also be significant if the photolysis products of $ClONO_2$ are indeed Cl and NO_3, as indicated by Chang et al. (1979):

$$ClO + NO_2 \rightarrow ClONO_2$$

$$ClONO_2 + h\nu \rightarrow Cl + NO_3$$

$$Cl + O_3 \rightarrow ClO + O_2$$

$$NO_3 + h\nu \rightarrow NO + O_2$$

$$NO + O_3 \rightarrow NO_2 + O_2$$

Net: $2\,O_3 \rightarrow 3O_2$

If the last reaction is replaced either by

$$NO + HO_2 \rightarrow OH + NO_2$$

$$OH + O_3 \rightarrow HO_2 + O_2$$

or by

$$NO + ClO \rightarrow Cl + NO_2$$

$$Cl + O_3 \rightarrow ClO + O_2$$

the effect on ozone is still the same. These cycles illustrate the importance of interaction between the various families.

It is important to note that certain cycles can lead to ozone formation; this phenomenon is readily observed in urban atmospheres which are rich in hydrocarbons and nitrogen oxides. In the lower stratosphere, a similar cycle is initiated by the reaction of OH with methane and carbon monoxide, followed by conversion of NO to NO_2 by the intermediate products. The photodissociation of NO_2 then leads to the formation of O_3. The complete chain, which is also important in the troposphere, is as follows (Crutzen, 1974):

$$CH_4 + OH \rightarrow CH_3 + H_2O$$

$$CH_3 + O_2 + M \rightarrow CH_3O_2 + M$$

$$CH_3O_2 + NO \rightarrow CH_3O + NO_2$$

$$NO_2 + h\nu \rightarrow NO + O$$

$$O + O_2 + M \rightarrow O_3 + M$$

$$CH_3O + O_2 \rightarrow CH_2O + HO_2$$

$$HO_2 + NO \rightarrow NO_2 + OH$$

$$NO_2 + h\nu \rightarrow NO + O$$

$$O + O_2 + M \rightarrow O_3 + M$$

$$CH_2O + h\nu \rightarrow CO + H_2$$

$$\boxed{CH_4 + 4O_2 + h\nu \rightarrow H_2O + CO + H_2 + 2O_3}$$

and

$$CO + OH \rightarrow CO_2 + H$$

$$H + O_2 + M \rightarrow HO_2 + M$$

$$HO_2 + NO \rightarrow OH + NO_2$$

$$NO_2 + h\nu \rightarrow NO + O$$

$$O + O_2 + M \rightarrow O_3 + M$$

$$\boxed{\text{Net: } CO + 2O_2 + h\nu \rightarrow CO_2 + O_3}$$

The complete ozone equation assumes a complex form as a result of these processes. Considering only the most important aeronomic processes, the general behavior of ozone can be described:

$$\frac{d(O_3)}{dt} + J_{O_3}(O_3) + k_3(O)(O_3) + a_2(H)(O_3) + a_6(OH)(O_3)$$

$$+ a_{6b}(HO_2)(O_3) + b_4(NO_2)(O_3) + d_2(Cl)(O_3) = k_2(M)(O_2)(O) \qquad (5.254)$$

$$\frac{d(O)}{dt} + 2k_1(M)(O)^2 + k_2(M)(O_2)(O) + k_3(O_3)(O)$$

$$+ a_5(OH)(O) + a_7(HO_2)(O) + b_3(NO_2)(O) + d_3(ClO)(O)$$

$$= 2J_{O_2}(O_2) + J_{O_3}(O_3) + J_{NO_2}(NO_2) \qquad (5.255)$$

where equilibrium has been assumed for $O(^1D)$. Adding equations (5.254) and (5.255),

$$\frac{d(O_x)}{dt} + 2k_1(M)(O)^2 + 2k_3(O)(O_3) + a_2(H)(O_3) + |a_5(OH)$$

$$+ a_7(HO_2)|(O) + |a_6(OH) + a_{6b}(HO_2)|(O_3) + b_3(NO_2)(O) + b_4(NO)(O_3)$$

$$+ d_3(ClO)(O) + d_2(Cl)(O_3) = 2J_{O_2}(O_2) + J_{NO_2}(NO_2) \qquad (5.256)$$

Let us consider the approximate equation which expresses the photochemical equilibrium between NO_2 and NO (eqn. 5.146, with the effect of ClO added):

$$(NO)|b_4(O_3) + a_{26}(HO_2) + c_5(CH_3O_2) + d_4(ClO)| = (NO_2)|J_{NO_2} + b_3(O)| \qquad (5.257)$$

and the equation expressing the equilibrium between Cl and ClO (eqn. 5.210, neglecting J_{ClO}):

$$(ClO)|d_3(O) + d_4(NO)| = d_2(Cl)(O_3) \qquad (5.258)$$

Substituting these in equation (5.256) and adding the effect of transport:

$$\frac{\partial(O_x)}{\partial t} + \vec{\nabla} \cdot \vec{\phi}(O_x) + 2k_1(M)(O)^2 + 2k_3(O)(O_3) + a_2(H)(O_3)$$

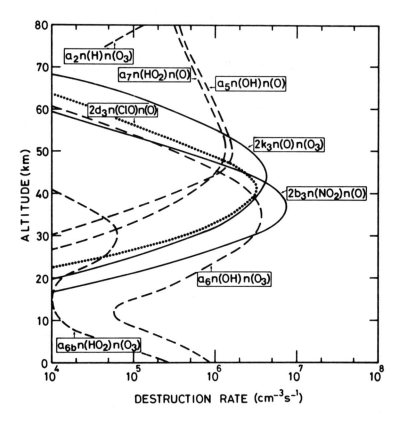

Fig. 5.66. Contribution of different reaction to the atmospheric odd oxygen balance. Nitrogen, oxygen, hydrogen and chlorine reactions are indicated. (From the model of Brasseur et al., 1983).

$$+ [a_5(OH) + a_7(HO_2)](O) + [a_6(OH) + a_{6b}(HO_2)](O_3) + 2b_3(NO_2)(O)$$

$$+ 2d_3(ClO)(O) = 2J_{O_2}(O_2) + a_{26}(HO_2)(NO) + c_5(CH_3O_2)(NO) \qquad (5.259)$$

where

$$(O_x) = (O) + (O_3) \approx (O_3) \ (z < 50km)$$

$$\approx (O) \ (z > 70km)$$

Figure 5.66 displays the effect of each of these terms on the ozone balance. These represent mean distributions of the loss rate of odd oxygen calculated with present chemistry. According to the importance of each term in the total loss as a function of altitude, one may simplify the above equation by neglecting minor terms.

In closing, we remark that although the complete chemistry of the middle atmosphere involves several hundred reactions, most of the important processes can be identified by examining the magnitudes of the photochemical terms in the continuity equations for each species, and the resulting lifetimes. When this information is used to define chemical families, a simple framework is obtained in which the species subject to the influence of transport can be readily identified, and the net versus gross production and loss processes are clearly distinguished from one another. Indeed, the situation is somewhat similar to the difficulties associated with understanding the significance of the net and gross mean meridional and eddy transports in the stratosphere in the classical and residual Eulerian representations, as was described in Chapter 3. Thus, in the photochemistry as well as the dynamics of the middle atmosphere, the occurrence of many simultaneous processes, some of which merely cancel with one another to a great degree, may require careful study to recognize the roles played by each of them. The particular frameworks chosen to accomplish that task in this volume represent one possible choice which may facilitate that study.

References

Ackerman, M., J.C. Fontanella, D. Frimout, A. Girard, N. Louisnard and C. Muller, Simultaneous measurements of NO and NO_2 in the stratosphere, Planet. Space Sci., 23, 651, 1975.

Ackerman, M., D. Frimout and C. Muller, Stratospheric CH_4, HCl and ClO and the chlorine ozone cycle, Nature, 269, 226, 1977.

Ackerman, M., D. Frimout, A. Girard, M. Gottignies and C. Muller, Stratospheric HCl from infrared spectra, Geophys. Res. Letters, 3, 81, 1976.

Ackerman, M., D. Frimout, C. Muller and D.J. Wuebbles, stratospheric methane measurements and predictions, Pure Appl. Geophys., 117, 367, 1978.

Allen, M., Y.L. Yung and J.W. Waters, Vertical transport and photochemistry in the terrestrial mesosphere and lower thermosphere (50-170 km), J. Geophys. Res., 86, 3617, 1981.

Allen, M., J.I. Lunine and Y.L. Yung, The vertical distribution of ozone in the mesosphere and lower thermosphere, in press, J. Geophys. Res., 1984.

Anderson, J.G., Rocket measurement of OH in the mesosphere, J. Geophys. Res., 76, 7820, 1971.

Anderson, J.G., The absolute concentration of O(3P) in the earth's stratosphere, Geophys. Res. Letters, 2, 231, 1975.

Anderson, J.G., The absolute concentration of OH in the earth's stratosphere, Geophys. Res. Letters, 3, 165, 1976.

Anderson, J.G., Free radicals in the earth's stratosphere: A review of recent results, pp. 233-251, in: Nicolet, M. and Aikin, A.C. (eds.), Proceedings of the Nato Advanced Study Institute on Atmospheric Ozone: Its Variation and Human Influences, U.S. Department of Transportation, Wasington, D.C. 20591, 1980.

Angell, J.K. and J. Korshover, Quasi-Biennal variations in temperature, total ozone and tropopause height, J. Atmos. Sci., 21, 5, 479, 1964.

Angell, J.K. and J. Korshover, Quasi-Biennal and long-term fluctuation in total ozone, Mon. Weather Rev., 101, 426, 1973.

Arnold, F. and D. Krankowsky, Water vapour concentrations at the mesopause, Nature, 268, 218, 1977.

Baker, K.D., A.F. Nagy, R.O. Olsen, E.S. Oran, J. Randhawa, D.F. Strobel, and T. Tohmatsu, Measurement of the nitric oxide altitude distribution in the mid-latitude mesosphere, J. Geophys. Res., 82, 3281, 1977.

Barbe, A., P. Mache, C. Secroun and P. Jouve, Measurements of tropospheric and stratospheric H_2CO by an infrared high resolution technique, Geophys. Res. Letters, 6, 463, 1979.

Barrett, E.W., P.M. Kuhn and A. Shlanta, Recent measurements of the injections of water vapor and ozone into the stratosphere by thunderstorms, Proc. Second Conf. on the Climatic Impact Assessment Program Rep. No. DOT-TSC-OST-73-4, US Dept. of Transportation, 5-7, 1973.

Barth, C.A., D.W. Rusch, R.J. Thomas, G.H. Mount, G.J. Rottman, G.E. Thomas, R.W. Sanders and G.M. Lawrence, Solar mesosphere explorer: Scientific objectives and results, Geophys. Res. Letters, 10, 237, 1983.

Bates, D.R. and M. Nicolet, Atmospheric hydrogen, Publ. Astron. Soc. Pacific, 62, 106, 1950A.

Bates, D.R., and A.E. Witherspoon, The photochemisry of some minor constituents of the earth's atmosphere (CO_2, CO, CH_4, N_2O), Mon. Notic. Roy. Astron. Soc., 112, 101, 1952.

Berg, W.W., P.J. Crutzen, F.E. Grabels, S.N. Gitlin and W.A. Sedlacek, First measurements of total chlorine in the lower stratosphere, Geophys. Res. Letters, 7, 937, 1980.

Bevilacqua, R.M., J.J. Olivero, P.R. Schwartz, C.J. Gibbins, J.M. Bologna and D.J. Thacker, An observational study of water vapor in the mid-latitude mesosphere using ground-based microwave technique, J. Geophys. Res., 88, 8523, 1983.

Brasseur, G. and P.C. Simon, Stratospheric chemical and thermal response to long-term variability in solar UV irradiance, J. Geophys. Res., 86, 7343, 1981.

Brasseur, G., Physique et chimie de l'atmosphere moyenne, Masson, Paris, 1982.

Breeding, R.J., J.P. Lodge, Jr., J.B. Pate, D.C. Sheesley, H.B. Klonis, B. Fogle, J.A. Anderson, T.R. Englert, P. Haagenson, R.B. McBeth, A.L. Morris, R. Pogue and A.F. Wartburg, Background trace gas concentrations in the central United States, J. Geophys. Res., 78, 7057, 1973.

Brewer, A.W., Evidence for a world circulation provided by the measurements of helium and water vapour distributions in the stratosphere, Quart. J. Roy. Met. Soc., 75, 351, 1949.

Burnett, C.R. and E.B. Burnett, Spectroscopic measurement of the vertical column abundance of hydroxyl (OH) in the earth's atmosphere, J. Geophys. Res., 86, 5185, 1981.

Burnett, C.R. and E.B. Burnett, Vertical column abundance of atmospheric OH at solar maximum from Fritz Peak, Colorado, Geophys. Res. Letters, 9, 708, 1982.

Buijs, H.L., G.L. Vail, G. Tremblay and D.J.W. Kendall, Simultaneous measurements of the volume mixing ratios of HCl and HF in the stratosphere, Geophys. Res. Letters,, 7, 205, 1980.

Bush, J.A., A.#L. Schmeltekopf, F.C. Fehsenfeld, D.L. Albritton, J.R. McAffe, P.D. Goldan and E.E. Ferguson, Stratospheric measurements of methane at several latitudes, Geophys. Res. Letters, 5, 1027-1029, 1978.

Carver, J.H., B.H. Horton and F.G. Burger, Nocturnal ozone distribution in the upper atmosphere, J. Geophys. Res., 71, 4189, 1966.

Chang, J.S., A.C. Hindmarsch and N.K. Madsen, Simulation of chemical kinetics transport in the stratosphere, in *Stiff Diffrential Systems* (R.A. Willoughly ed.), Plenum Publ. Corp. (New York, N.Y.), 1974.

Chang, J.S., R.R. Barker, J.E. Davenport and D.M. Goldan, Chlorine nitrate photolysis by a new technique: Very low pressure photolysis, Chem. Phys. Letters, 60, 385, 1979.

Chapman, S., On ozone and atomic oxygen in the upper atmosphere, Phil. Mag. 10, 369, 1930.

Clancy, R.T., D.O. Muhlemon and G.L. Berge, Microwave spectra of terrestrial mesosphere CO, J. Geophys. Res., 87, 5009, 1982.

Coffey, M.T., W.G. Mankin and A. Goldman, Simultaneous spectroscopic determination of the latitudinal, seasonal and diurnal variability of stratosphere N_2O, NO, NO_2 and HNO_3, J. Geophys. Res., 86, 7731, 1981.

Crutzen, P.J., Determination of parameters appearing in the "dry" and the "wet" photochemical theories for ozone in the stratosphere, Tellus, 21, 368, 1969.

Crutzen, P.J., The influence of nitrogen oxide on the atmospheric ozone content, Quart. J. Roy. Met. Soc., 96, 320, 1970.

Crutzen, P.J., Ozone production rates in oxygen-hydrogen-nitrogen oxide atmosphere, J. Geophys. Res., 76, 7311, 1971.

Crutzen, P.J., Photochemical reactions initiated by the influencing ozone in unpolluted tropospheric air, Tellus, 26, 47, 1974.

Crutzen, P.J., The possible importance of CSO for the sulfate layer of the stratosphere, Geophys. Res. Letters, 3, 73, 1976.

Crutzen, P.J., I.S.A. Isaksen and G.C. Reid, Solar Protons events: Stratospheric sources of nitric oxide, Science, 189, 457-458, 1975.

Crutzen, P.J., L.E. Heidt, J.P. Krasnec, W.H. Pollock and W. Seiler, Biomass burning as a source of the atmosphere gases CO, H_2, N_2O, NO, CH_3Cl and COS, Nature, 282, 253, 1979.

Crutzen, P.J. and U. Schmaizl, Chemical budgets of the stratosphere, Planet. Space Sci., 31, 1009, 1983.

Cunnold, D.M., F.N. Alyea and R.G. Prinn, Preliminary calculations concerning the maintenance of the zonal mean ozone distribution in the northern hemisphere, PAGEOPH, 118, 329, 1980.

Danielsen, E.F., The Laminar structure of the atmosphere and its relation to the concept of a tropopause, Arch. F. Meteor., Geophys. Bioklimat., 11, Ser. A, 293, 1959.

Danielsen, E.F., The Laminar structure of the atmosphere and its relation to the concept of a tropopause, Arch. F. Meteor. Geophys., Klim. All., 293, 1960.

Danielsen, E.F., Stratospheric-trophospheric exchange based on radioactivity, ozone, and potential vorticity, J. Atmos. Sci., 25, 501, 1968.

Danielsen, E.F., E.R. Bleck, J. Shedlozsky, A. Westburg and P. Hagenson, Observed distribution of radioactivity, ozone, and potential vorticity associated with tropopause folding, J. Geophys. Res., 75, 2353, 1970.

Danielsen, E.F., A dehydration mechanism for the stratosphere, Geophys. Res. Letters, 9, 605, 1982.

Davis, D.D., W. Heaps and T.McGee, Direct measurements of natural tropospheric levels of OH via an air-borne tunable dye laser, Geophys. Res. Letters, 3, 331, 1976.

Deguchi, S. and O.O. Muhleman, Mesopheric water vapor, J. Geophys. Res., 87, 1343, 1982.

De Jonckheere, C.G., and D.E. Miller, A measurement of the ozone concentration from 65 to 75 km at night, Planet. Space Sci., 22, 497, 1974.

Delwiche, C.C., Biological production and utilization of N_2O, Pageoph, 116, 414, 1978.

De Zafra, R.L., A. Parrish, P.M. Solomon and J.W. Barrett, A measurement of stratospheric HO_2 by ground-based nm-wave spectroscopy, J. Geophys. Res., 89, 1321, 1984.

Dickinson, P.H., W.C. Bain, L. Thomas, E.R. Williams, D.B. Jenkins, and N.D. Twiddy, The determination of the atomic oxygen concentration and associated parameters in the lower ionosphere, Proc. R. Soc. London, Ser. A, 369, 379, 1980.

Dobson, G.M.B., Observations of the amount of ozone in the earth's atmosphere and its relation to other geophysical conditions, Proc. Roy. Soc. London, Sec. A, 129, 411, 1930.

Dobson, G.M.B., *Exploring the Atmosphere,* Clarendon Press (Oxford), 1963.

Donahue, T.M., The atmosphere CH_4 budget, pp. 301-305, in: Nicolet M. and Aikin, A.C. (eds.), Proceedings of the Nato Advanced Study Institute on Atmospheric Ozone: Its Variation and Human Influences, US Dept. of Transportation, Washington, D.C. 20591, 1980.

Drummond, J.R. and R.F. Jarnot, Infrared measurements of stratospheric composition, II. Simultaneous NO and NO_2 measurements, Proc. Roy. Soc. London, A. 364, 237, 1978.

Dutsch, H.U., Atmospheric ozone: A short review, J. Geophys. Res., 75, 1707, 1970.

Dutsch, H.U., Regular ozone soundings at the aerological station of the Swiss Meteorological Office at Payenne, Switzerland, 1968-1972, Rep. 10, Lab. Atmosphaerenphys. ETH, Swiss Meteorol. Office, 1974.

Dutsch, H.U., Vertical ozone distribution and tropospheric ozone, pp. 7-30, in: Nicolet M. and Aikin, A.C. (eds.), Proceedings of the Nato Advanced Study Institute on Atmospheric Ozone: Its Variation and Human Influences, US Dept. of Transportation, Washington, D.C. 20591, 1980.

Ehhalt, D.H., On the uptake of tritium by soil water and groundwater, Water Resources Res., 9, 1073, 1973.

Ehhalt, D.H., The atmospheric cycle of methane, Tellus, 26, (1-2), 58, 1974.

Ehhalt, D.H. and L.E. Heidt, The concentration of molecular H_2 and CH_3 in the stratosphere, pageophys., 106-108, 1352, 1973.

Ehhalt, D.H. and A. Tonnissen, Hydrogen and carbon compounds in the stratosphere, pp. 129-151: Nicolet M. and Aikin, A.C. (eds.), Proceedings of the Nato Advanced Study Institute on Atmospheric Ozone: Its Variation and Human Influences, US Dept. of Transportation, Washington, D.C. 20591, 1980.

Ehhalt, D.H., L.E. Heidt, R. H. Lueb, and N. Roper, Vertical profiles of CH_4, H_2, CO, N_2O and CO_2 in the stratosphere, pp. 153-160, in: Proceedings of the Third Conference on CIAP, US Dept. of Transportation, Washington, D.C., 1974.

Ehhalt, D.H., L.E. Heidt, R.H. Lueb and W. Pollock, The vertical distribution of trace gases in the stratosphere, Pageoph, 113, 389, 1975a.

Ehhalt, D.H., L.E. Heidt, R.H. Lueb and E.A. Martell, Concentrations of CH_4, CO, CO_2, H_2, H_2O and N_2O in the upper stratosphere, J. Atmos. Sci., 32, 163, 1975b.

Ehhalt, D.H., A. Volz, H. Cosatto, and L.E. Heidt, The vertical distribution of chlorofluoromethanes in the stratosphere, pp. 252-257, in: Proceedings of the Joint Symposium on Atmospheric Ozone, V. II, Dresden, GDR, 1976.

Evans, W.J.J., O.M. Hunter, E.J. Lewellyn and A. Vallance-Jones, Altitude profile of the infrared system of oxygen in the dayglow, J. Geophys. Res.,

73, 2885, 1968.

Evans, W.F.J. and E.J. Llewellyn, Molecular oxygen emissions in the airglow, Ann. Geophys., 26, 167, 1970.

Evans, W.F.J., J.B. Kerr, D.I. Wardle, J.C. McConnell, B.A. Ridley and H.I. Schiff, Intercomparison of NO, NO_2 and HNO_3 measurements with photochemical theory, Atmosphere, 14, 189, 1976.

Evans, W.F.J., J.B. Kerr, C.T. McElroy, R.S. O'Brien and J.C. McConnell, Measurement of NO, NO_2 and HNO_3 during a stratospheric warming at 54°N in February 1979, Geophys. Res. Lett., 9, 493, 1982.

Eyre, J.R. and H.K. Roscoe, Radiometric measurements of stratospheric HCl, Nature, 266, 243, 1977.

Fabian, P., R. Borchers, K.H. Weiler, U. Schmidt, L.A. Volz, P.H. Ehhalt, W. Seiler and F. Muller, Simultaneously measured vertical profiles of H_2, CH_4, CO, N_2O, $CFCl_3$, and CF_2Cl_2 in the mid-latitude stratosphere and troposphere, J. Geophys. Res., 84, 3149, 1979.

Farmer, C.B., Infrared measurements of stratospheric composition, Can. J. Chem., 52, 1544, 1974.

Farmer, C. B., O.F. Raper, R.A. Toth and R.A. Schindler, Recent results of aircraft infrared observations of the stratosphere, Proc. Third Conf. on the Climatic Impact Assessment Program, Rep. No. DOT-TSC-OST-74-15, US Dept. of Transportation, 234-245, 1974.

Farmer, C.B., O.F. Raper and R.H. Norton, Spectroscopic detection and vertical distribution of HCl in the troposphere and stratosphere, Geophys. Res. Lett., 3 (1), 13, 1976.

Farmer, C.B., O.F. Raper, B.D. Robbins, R.A. Toth, and C. Muller, Simultaneous spectroscopic measurements of stratospheric species: O_3, CH_4, CO, CO_2, N_2O, H_2O, HCl and HF at northern and southern mid-latitudes, J. Geophys. Res., 85, 1621, 1980.

Fontanella, J.-C., A. Girard, L. Gramont and N. Louisnard, Vertical distribution of NO, NO_2 and HNO_3 as derived from stratospheric absorption infrared spectra, Appl. Opt., 14, 825, 1975.

Frederick, J.E. and N. Orsini, The distribution and variability of mesospheric odd nitrogen: A theoretical investigation, J. Atm. Terr. Phys., 44, 4798, 1982.

Frederick, J.E., F.T. Huang, A.R. Douglass and C.A. Reber, The distribution and annual cycle of ozone in the upper stratosphere, J. Geophys. Res., 88, 3819, 1983.

Friend, J.P., The global sulfur cycle, pp. 177-201, in: Rasool, I. (ed.), *Chemistry of the Lower Atmosphere,* Plenum Press (New York), 1973.

Froideveaux, L. and Y.L. Yung, Radiation and chemistry in the stratosphere: Sensitivity to O_2 absorption cross section in the Herzberg continuum,

Geophys. Res. Letters, 9, 854, 1982.

Glueckauf, E. and G.P. Kitt, The hydrogen content of air at ground level, Quart. J. Roy. Met. Soc., 83, 522, 1957.

Goldan, P.D., Data given by N. Sundararaman, in NASA Report, FAA-EQ-77-2, 1976.

Goldman, A., R.G. Murcray, F.H. Murcray, W.J. Williams, J.N. Brooks and C.M. Bradford, Vertical distribution of CO in the atmosphere, J. Geophys. Res., 78, 5273, 1973.

Garcia, R.R. and S. Solomon, A numerical model of the zonally averaged dynamical and chemical structure of the middle atmosphere, J. Geophys. Res., 88, 1379, 1983.

Gear, C. W., *Numerical initial value problems in ordinary differential equations*, Prentice Hall, (Englewood Cliffs, N. J.), 1971.

Gelinas, R.J., Stiff systems of kinetic equations - A practitioner's view, J. Comp. Phys., 9, 777, 1972.

Gibbins, C.J., P.R. Schmitz, D.L. Thacher, and R.M. Bevilacqua, The variability of mesospheric water vapor, Geophys. Res. Lett., 8, 1059, 1981.

Gille, J.C., P.L. Bailey and J.M. Russell, III, Temperature and composition measurements from LRIR and LIMS experiments on Nimbus 6 and 7, Phil. Trans. Roy. Soc. London, 296, 205, 1980.

Gille, J. C., and J. M. Russell, The limb infrared monitor of the stratosphere (LIMS): experiment description, performance and results, in press, J. Geophys. Res., 1984.

Goldan, P.D., W.C. Kuster, D.L. Albritton and A.L. Schmeltekopf, Stratospheric $CFCl_3$, CF_2Cl_2 and N_2O height profile measurements at several latitudes, J. Geophys. Res., 85, 413, 1980.

Graedel, T.E., B. Kleiner and G.C. Patterson, Measurements of extreme concentrations of tropospheric hydrogen sulfide, J. Geophys. Res., 79, 30, 4467, 1974.

Hack, W., A.W. Preuss, H.G.G. Wagner and K. Hoyermann, Reactions of H with H_2O, II. Determination of the overall rate constant, Ber. Bunsenges. Phys. Chem., 83, 212, 1979.

Hahn, J. and C. Junge, Atmospheric nitrous oxide: A critical review, 7, Nature, 32A, 190, 1977.

Hampson, J. Chemiluminescent emissions observed in the stratosphere and mesosphere, pp. 393-440, in: Nicolet, M. (ed)., Les Problemes Meteorologiques de la stratosphere et de la mesosphere, Presses Universitaries de France, Paris, 1966.

Hanst, P.L., L.L. Speller, D. M. Watts, J.W. Spence and F.M. Miller, Infrared measurements of fluorocarbons, carbon tetrachloride, carbonyl sulfide, and other atmospheric trace gases, Air Pollution Control Assn. J., 24, 12, 1220, 1975.

Harries, J.E., Measurements of stratospheric water vapor using infrared techniques, J. Atmos. Sci., 30, 1691, 1973.

Harries, J.E., D.G. Moss and N.R. Swann, H_2O, O_3, N_2O and HNO_3 in the arctic stratosphere, Nature, 250, 475, 1974.

Harries, J.E., Ratio of HNO_3 to NO_2 concentrations in the daytime stratosphere, Nature, 274, 235, 1978.

Hays, P.B. and J.J. Olivero, Carbon dioxide and monoxide above the troposphere, Planet. Space Sci., 18, 1729, 1970.

Heaps, W.S. and J.J. McGee, Balloon borne lidar measurements of stratospheric hydroxyl radical, J. Geophys. Res., 86, 5281, 1983.

Heidt, L.E., R. Lueb, W. Pollock, and D.H. Ehhalt, Stratospheric profiles of CCl_3F and CCl_2F_2, Geophys. Res. Lett., 2, 445, 1975.

Hering, W.S. and T.R. Borden, Ozone sonde observations over North America, vol. 3, Rep. AFCRL-64-30(3), Air Force Cambridge Res. Labs., Bedford, Mass., 1965.

Hesstvedt, E., On the photochemistry of ozone in the ozone layer, Geophys. Publ., 27, 4, 1968.

Hester, N.E., E.R. Stephens and C. Taylor, Fluorocarbon air pollutants, measurements in lower stratosphere, Environ. Sci. and Technol., 9, 875, 1975.

Hoffmann, D.J., J.M. Rosen, T.J. Pepin and R.G. Pinnick, Stratospheric aerosols measurements: Time variations at northern mid-latitudes, J. Atmos. Sci., 32, 1446, 1975.

Hofmann, D.J. and J. M. Rosen, Balloon-borne observations of stratospheric aerosol and condensation nuclei during the year following the Mt. St. Helens eruption, University of Wyoming Report AP-63, 1981.

Horvath, J.J. and C.J. Mason, Nitric oxide mixing ratios near the stratopause measured by a rocket-borne chemiluminescent detector, Geophys. Res. Lett., 5, 1023, 1978.

Hunt, B.G., Photochemistry of ozone in a moist atmosphere, J. Geophys. Res., 71, 1385, 1966.

Inn, E.C.Y., J.F. Vedder, B.J. Tyson and D. O'Hara, COS in the stratosphere, Geophys. Res. Lett., 6, 191, 1979.

Jaeschke, W., R. Schmitt and H.W. Georgi, Preliminary results of stratospheric SO_2 measurements, Geophys. Res. Lett., 3, 517, 1976.

Jaffe, L.S., Carbon monoxide in the biosphere: Sources, distribution and concentration, J. Geophys. Res., 78, 5293-, 1973.

Johnston, H., Reduction of stratospheric ozone by nitrogen oxide catalysts from supersonic transport exhaust, Science, 173, 517, 1971.

Johnston, H.S. and J. Podolske, Interpretation of stratospheric photochemistry, Rev. Geophys. Space Phys., 16, 491, 1978.

Johnston, H.S. and S. Solomon, Thunderstorms as possible micrometeorological sink for stratospheric water, J. Geophys. Res., 84, 3155, 1979.

Jones, R.L. and J.A. Pyle, Observations of CH_4 and N_2O by the nimbus 7 SAMS: A comparison with in-situ data and two dimensional numerical model calculations, in press, J. Geophys. Res., 1984.

Keeling, C.D., R.B. Bacastow, A.E. Bainbridge, C.A. Ekdahl, Jr., P.R. Guenther, L.S. Waterman and J.F.S. Chin, Atmospheric carbon dioxide variations at Mauna Loa observatory, Hawaii, Tellus, 28, 538, 1976.

Kellogg, W.W., R.D. Cadle, E.R. Allen, A.L. Lazrus and E.A. Martell, The sulfur cycle, Science, 175, 4022, 587, 1972.

Kida, H., A numerical investigation of the atmospheric exchange, II. Lagrangian motion of the atmosphere, J. Met. Soc., Japan, 55, 71, 1977.

Kley, D., E.J. Stone, W.R. Henderson, J.W. Drummond, W.J. Harrop, A.L. Schmeltekopf, T.L. Thompson and R.H. Winkler, in situ measurements of the mixing ratio of water vapor in the stratosphere, J. Atmos. Sci., 36, 2513, 1979.

Kley, D., J.W. Drummond and A.L. Schmeltekopf, On the structure and microstructure of stratospheric water vapor, in *Atmospheric water vapor*, A. Deepak, T.D. Wilkerson and L.H. Ruhnke (eds.), Academic Press (New York, N.Y.), 315, 1980.

Knight, W., D.R. Hastie and B. Ridley, Measurement of nitric oxide during a stratospheric warming, Geophys. Res. Lett., 9, 489, 1982.

Krey, P.W., R.J. Lagomarsino, L.E. Toonkel, Gaseous halogens in the atmosphere in 1975, J. Geophys. Res., 82, 1753, 1977.

Krueger, A.J., The mean ozone distributions from several series of rocket soundings to 52 km at latitudes from 58 °S to 64 °N, Pure Appl. Geophys., 106-108, 1271, 1973.

Krueger, A.J., B. Guenther, A.J. Fleig, D.F. Heath, E. Hilsenrath, R. McPeters and C. Prabakhara, Satellite ozone measurements, Phil. Trans. R. Soc. London, Ser. A, 196, 191, 1980.

Lazrus, A.L., B.W. Gandrud, Distribution of stratospheric nitric acid vapor, J. Atmos. Sci., 31,1102, 1974.

Lazrus, A.L., B.W. Gandrud, R.N. Woodard, and W.A. Sedlacek, Stratospheric halogen measurements, Geophys. Res. Lett., 2, 439, 1975.

Lazrus, A.L., B.W. Gandrud, R.N. Woodard and W.A. Sedlacek, Direct measurements of stratospheric chlorine and bromine, J. Geophys. Res., 81, 1067,1976.

Lazrus, A.L., B.W. Gandrud, J. Greenberg, E. Mroz and W.A. Sedlacek, Midlatitude seasonal measurements of stratospheric acid chloride vapor, Geophys. Res. Lett., 4, 587, 1977.

Lean, J.L., Observation of the diurnal variation of atmospheric ozone, J. Geophys. Res., 87, 4973, 1982.

Liu, S.C. and T.M. Donahue, The aeronomy of hydrogen in the atmosphere of the earth, J. Atmos. Sci., 31, 1118, 1974.

Loewenstein, M. and H.F. Savage, Latitudinal measurements of NO and O_3 in the lower stratospheric from 5° to 82°N, Geophys. Res. Lett., 2, 448, 1975.

Loewenstein, M., W.J. Borucki, H.F. Savage, J.G. Borucki and R.C. Whitten, Geographical variations of NO and O_3 in the lower stratosphere, J. Geophys. Res., 83, 1875, 1978.

Logan, J.A., M. Prather, S. Wofsy and M.B. McElroy, Atmospheric chemistry: Response to Human Influence, Phil. Trans. Roy. Soc. London, Ser. A, 290, 187-234, 1978.

Logan, J.A., Sources and sinks for carbon monoxide, pp. 323-343, in: Nicolet, M. and Aikin, A.C. (eds), Proceedings of the Nato Advanced Study Institute on Atmospheric Ozone: Its Variation and Human Influences, U.S. Department of Transportation, Washington, D.C. 20591, 1980.

Logan, J.A., M.J. Prather, S.C. Wofsy and M.B. Mcelroy, Tropospheric chemistry: A global perspective, J. Geophys. Res., 86, 7210-7254, 1981.

London, J., Radiative energy sources and sinks in the stratosphere and mesosphere, pp. 703-721, in: Nicolet, M. and Aikin, A.C. (eds), Proceedings of the Nato Advanced Study Institute on Atmospheric Ozone: Its Variation and Human Influences, U.S. Department of Transportation, Washington, D.C. 20591, 1980.

London, J. and G.M.W. Haurwitz, Ozone and sunspots, J. Geophys. Res., 795, 1963.

Louisnard, N., A. Girard and G. Eichen, Physique de l'atmosphère, C.R. Acad. Sci. Paris, 290(B), 385-388, 1980.

Lovelock, J.E., R.J. Maggs and R.J. Wade, Halogenated hydrocarbons in and over the Atlantic, Nature, 241, 194, 1973.

Lovelock, J.E., Atmospheric halocarbons and stratospheric ozone, Nature, 252, 292, 1974.

Mankin, W.G., M.T. Coffey, D.W.T. Griffith and S.R. Drayson, Spectroscopic measurement of carbonyl sulfide (OCS) in the stratosphere, Geophys. Res. Lett., 6, 853, 1979.

Mankin, W.G. and M.T. Coffey, Latitudinal distributions and temporal changes of stratosphere HCl and HF, J. Geophys. Res., 88, 10776, 1983.

Marche, P., A. Barbe, C. Secroun, J. Corr and P. Jouve, Mesure de HF et de HCl dans l'atmosphère par spectroscopie infrarouge a partir du sol, C.R. Acad. Sci. Paris, T. 240, Serie B, 369-371, 1980.

Maroulis, P..J., A.L. Torres and A.R. Bandy, Atmospheric concentrations of carbonyl sulfide in the south-western and eastern United States, Geophys. Res. Lett., 4, 510, 1977.

Mastenbrook, H.J., Water vapor measurements in the lower stratosphere, Can. J. Chem., 52, 1527, 1974.

Mastenbrook, H.J. and S. J. Oltmans, Stratospheric water vapor variability for Washington DC/Boulder, CO: 1964-82, J. Atmos. Sci., 40, 2157, 1983.

McCormick, M.P., W.P. Chu, G.W. Grams, P. Hamill, B.M. Herman, L.R. McMaster, T.J. Pepin, P.B. Russell, H.M. Steel and T.J. Swissler, High latitude stratospheric aerosols measured by the SAM II satellite system in 1978-1979, Science, 214, 328, 1981.

McElroy, M.B. and J.C. McConnell, Nitrous oxide: A natural source of stratospheric NO, J. Atmos. Sci., 28, 1095, 1971.

McElroy, M.B., Sources and sinks for nitrous oxide, pp. 345-364, in: Nicolet, M. and Aikin, A.C. (eds), Proceedings of the Nato Advanced Study Institute on Atmospheric Ozone: Its Variation and Human Influences, U.S. Department of Transportation, Washington, D.C. 20591, 1980.

McKeen, S.A., S.C. Liu and C.S. Kiang, On the chemistry of stratospheric SO_2 from volcanic eruptions, in press, J. Geophys. Res., 1984.

Meira, L.G., Jr., Rocket measurements of upper atmospheric nitric oxide and their consequences to the lower ionosphere, J. Geophys. Res., 76, 202, 1971.

Menzies, R.T., Remote measurement of ClO in the stratosphere, Geophys. Res. Lett., 6, 151, 1979.

Migeotte, M.V., Methane in the earth's atmosphere, Astrophys. J., 400, 1948.

Miller, C., D.L. Filkin, A.J. Owens, J.M. Steed and J.P. Jesson, A two-dimensional model of stratosphere chemistry and transport, J. Geophys. Res., 86, 12039, 1981.

Millier, F., B.A. Emery and R.G. Roble, OSO-8 lower mesospheric ozone number density profiles, pp. 572-579, in: London J. (ed.), Proceedings of the Quadrennial International Ozone Symposium, Boulder, Colorado, 1980, Volume I, National Center for Atmospheric Research, Boulder, Colorado, 1981.

Molina, J.S. and F.S. Rowland, Stratospheric sink for chlorofluoromethanes: Chlorine atom-catalyzed destruction of ozone, Nature, 249, 810, 1974a.

Molina, J.S. and F.S. Rowland, Predicted present stratospheric abundances of chlorine species from photodissociation of carbon tetrachloride, Geophys. Res. Letters, 1, 309, 1974b.

Mount, G.H., D.W. Rusch, J.F. Noxon, J.M. Zawodny and C.A. Barth, Measurements of stratospheric NO_2 from the solar mesosphere explorer satellite 1. An overview of the results, J. Geophys. Res., 89, 1327, 1984.

Mroz, E.J., A.L. Lazrus and J. Bonelli, Direct measurements of stratospheric fluoride, Geophys. Res. Letters, 4, 149, 1977.

Mumma, M.J., J.D. Rogers, T. Kostiuk, D. Deming, J.J. Hillman and D. Zipory, Is there any chlorine monoxide in the stratosphere?, Science, 221,

268, 1983.

Murcray, D.G., D.D. Barker, J.N. Brooks, A.Goldman and W.J. Williams, Seasonal and latitudinal variations of the stratospheric concentration of HNO_3, Geophys. Res. Letters, 2, 223, 1975.

Murcray, D.G., A. Goldman, W.J. Williams, F.H. Murcray, F.S. Bonono, G.M. Bradford, G.R. Cole, P.L. Hanst, and M.J. Molina, Upper limit for stratospheric $ClONO_2$ from balloon-borne infrared measurements, Geophys. Res. Letters, 4, 227, 1977.

Naudet, J.P., D. Hugenin, P. Rigaud and D. Cariolle, Stratospheric observations of NO_3 and its experimental and theoretical distribution between 20 and 40 km, Planet. Space Sci., 29, 707, 1981.

Newell, R. E., The circulation of the upper atmosphere, Scient. Am., 210, 62, 1964.

Nicolet, M., Contribution a l'etude de la structure de l'ionosphere, Mem. Inst. Meteorol. Belg., 19, 83, 1945.

Nicolet, M., Ozone and hydrogen reactions, Ann. Geophys., 26, 531, 1970a.

Nicolet, M., The origin of nitric oxide in the terrestrial atmosphere, Planet. Space Sci., 18, 1111, 1970b.

Nicolet, M., Aeronomic reactions of hydrogen and ozone, pp. 1-51, in: Fiocco, G. (ed.), *Mesospheric Models and Related Experiments,* D. Reidel Publishing Company, (Dordrecht, Holland), 1971.

Nicolet, M. and E. Vergison, L'oxyde azoteux dans la stratosphere, Aeronomica Acta, A-91, 26 pp., 1971.

Nicolet, M. and W. Peetermans, The production of nitric oxide in the stratosphere by oxidation of nitrous oxide, Annls. Geophys., 28, 751, 1972.

Nicolet, M., Stratospheric ozone: An introduction to its study, Rev. Geophys. Space Phys., 13, 593, 1975.

Nicolet, M., On the production of nitric oxide by cosmic rays in the mesosphere and stratosphere, Planet. Space Sci., 23, 637, 1975.

Nicolet, M., The chemical equations of stratospheric and mesospheric ozone; pp. 823-864, in: Nicolet, M. and Aikin, A.C. (eds), Proceedings of the Nato Advanced Study Institute on Atmospheric Ozone: Its Variation and Human Influences, U.S. Department of Transportation, Washington, D.C. 20591, 1980.

Noxon, J.F., R.B. Norton and W.R. Henderson, Observations of atmosphere NO_3, Geophys. Res. Lett., 5, 675, 1978.

Noxon, J.F., Stratospheric NO_2, 2. Global behavior, J. Geophys. Res., 84, 5067, 1979.

Noxon, J.F., E.C. Whipple, Jr. and R.S. Hyde, Stratospheric NO_2, 1. Observational method and behavior at midlatitude, J. Geophys. Res., 84, 5047, 1979.

Noxon, J.F., Correction, J. Geophys. Res., 85, 4560, 1980.

Noxon, J.F., A global study of $O_2(^1\Delta_g)$ airglow day and twilight, Planet. Space Sci., 30, 545, 1982.

Paetzold, H.K., F. Piscalar and H. Zschorner, Secular variation of the stratospheric ozone layer over middle Europe during the solar cycles from 1951 to 1972, Nature, 240 (101), 106, 1972.

Paneth, F.A., The chemical composition of the atmosphere, Quart. J. Roy. Meteorol. Soc., 63, 433, 1937.

Patel, C.K.L., E.G. Burkhardt and C.A. Lambert, Spectroscopic measurements of stratospheric nitric oxide and water vapor, Science, 184, 1173, 1974.

Penndorf, R., In Report No. FAA-EE-78-29, NTIS, Springfield, VA, 1978.

Perner, D., D.H. Ehhalt, H.W. Patz, V. Platt, E.P. Roth and A. Volz, OH-Radicals in the lower troposphere, Geophys. Res. Lett., 3, 466, 1976.

Pyle, J. A., A.M. Zavody, J.E. Harries, and P.H. Moffat, Derivation of OH concentration from satellite infrared measurements of NO_2 and HNO_3, Nature, 305, 690, 1983.

Radford, H.E., M.M. Litvak, C.A. Gottlieb, S.K. Rosenthal and A.E. Lilley, Mesospheric water vapor measured from ground-based microwave observations, J. Geophys. Res., 82, 472, 1977.

Ramanathan, K.R., Bi-annual variation of atmospheric ozone over the tropics, Quart. J. Roy. Met. Soc., 89, 540, 1963.

Raper, O.F., C.B. Farmer, R.A. Toth and B.D. Robbins, The vertical distribution of HCl in the stratosphere, Geophys. Res. Lett., 4, 531, 1977.

Ridley, B.A. and H.I. Schiff, Stratospheric odd-nitrogen: Nitric oxide measurements at 32 N in autumn, J. Geophys. Res., 86, 3167, 1981.

Ridley, B.A., M. McFarland, J.T. Bruin, H.I. Schiff, and J.C. McConnell, Sunrise measurements of stratospheric nitric oxide, An. J. Phys., 55, 212, 1977.

Rogers, J.W., A.T. Stair, Jr., T.C. Degges, C.L. Wyatt and D.J. Baker, Rocketborne measurement of mesospheric H_2O in the auroral zone, Geophys. Res. Lett., 4, 366, 1977.

Roscoe, H.K., J.R. Drummond and R.F. Jarnot, Infrared measurements of stratospheric composition, III. The daytime changes of NO and NO_2, Proc. Roy. Soc. London, A375, 507, 1981.

Roscoe, H.K., A tentative observation of stratospheric N_2O_5, Geophys. Res. Lett., 9, 901, 1982.

Rosen, J.M., D.J. Hofmann and J. Laby, Stratospheric aerosol measurements, II: The worldwide distribution, J. Atmos. Sci., 32, 1457, 1975.

Rowland, F.S. and M.J. Molina, Chlorofluoromethanes in the environment, Rev. Geophys. Space Phys., 13, 1-35, 1975.

Rudolph, J., D.H. Ehhalt and A. Tonissen, Vertical profiles of ethane and propane in the stratosphere, J. Geophys. Res., 86, 7767, 1981.

Russell, J.M., S. Solomon, L.L. Gordley, E. E. Remsberg and L.B. Callis, The variability of stratospheric and mesospheric NO_2 in the polar winter night observed by LIMS, in press, J. Geophys. Res., 1984.

Sandalls, F.J. and S.A. Penkett, Measurements of carbonyl sulfide and carbon disulfide in the atmosphere, Atmos. Environ., 11, 197, 1977.

Schmeltekopf, A.L., P.D. Goldan, W.R. Henderson, W.J. Harrop, T.L. Thompson, F.C. Fehsenfeld, H.I. Schiff, P.J. Crutzen, I.S.A. Isaken and E.E. Ferguson, Measurements of stratospheric $CFCl_3$, CF_2Cl_2 and N_{2O}, Geophys. Res. Lett., 2, 393-396, 1975.

Schmeltekopf, A.L., Data as given by N. Sundararaman, Summary of upper atmospheric data, NASA Report FAA-EQ-77-2, 1976.

Schmeltekopf, A.L., D.L. Albritton, P.J. Crutzen, P.D. Goldan, W.J. Harrop, W.R. Henderson, J.R. McAfee, M. McFarland, H.I. Schiff, T.L. Thompson, D.L. Hofmann and N.T. Kjome, Stratospheric nitrous oxide altitude profiles at various latitudes, J. Atmos. Sci., 34, 729-736, 1977.

Schmidt, U., Molecular hydrogen in the atmosphere, Tellus, 26, 78, 1974.

Schmidt, U., G. Kulessa, and E. P. Roth, The Atmospheric H_2 cycle, pp. 307-322, in: Nicolet, M. and Aikin, A.C. (eds), Proceedings of the Nato Advanced Study Institute on Atmospheric Ozone: Its Variation and Human Influences, U.S. Department of Transportation, Washington, D.C. 20591, 1980.

Scholz, T.G., D.H. Ehhalt, L.E. Heidt and E.A. Martell, Water vapor molecular hydrogen, methane, and tritium concentrations near the stratopause, J. Geophys. Res., 75, 3049, 1970.

Schutz, K., C. Junge, R. Beck and B. Albrecht, Studies of atmospheric N_2O, J. Geophys. Res., 75, 2230, 1970.

Seiler, W. and P. Warneck, Decrease of the carbon monoxide mixing ratio at the tropopause, J. Geophys. Res., 77, 3204, 1972.

Seiler, W., The cycle of the atmospheric CO, Tellus, 26, 117, 1974.

Seiler, W. and P.J. Crutzen, Estimates of gross and net fluxes of carbon between the biosphere and the atmosphere from biomass burning, Climatic Change, 2, 207, 1980.

Shah, G.M., Quasi-biennial oscillation in ozone, J. Atmos. Sci., 24, 396, 1967.

Solomon, S., P.J. Crutzen and R.G. Roble, Photochemical coupling between the thermosphere and the lower atmosphere, I. Odd nitrogen from 50 to 120 km, J. Geophys. Res., 87, 7206, 1982a.

Solomon, S., G.C. Reid, R.G. Roble and P.J. Crutzen, Photochemical coupling between the thermosphere and the lower atmosphere, II. D-region ion chemistry and the winter anomaly, J. Geophys. Res., 87, 7721, 1982b.

Solomon, S., E.E. Ferguson, D.W. Fahey and P.J. Crutzen, On the chemistry of H_2O, H_2 and meteoritic ios in the mesosphere and lower thermosphere, Planet. Space Sci., 1117, 1982c.

Solomon, S. and R.R. Garcia, On the distribution of nitrogen dioxide in the high latitude stratosphere, J. Geophys. Res., 88, 5229, 1983a.

Solomon, S. and R.R. Garcia, Simulation of NO_x partitioning along isobaric parcel trajectories, J. Geophys. Res., 88, 5497, 1983b.

Stolarski, R.S. and R.J. Cicerone, Stratospheric chlorine: A possible sink for ozone, Canad. J. Chem., 52, 1610, 1974.

Swider, W., Daytime nitric oxide at the base of the thermosphere, J. Geophys. Res.,, 83, 4407, 1978.

Syed, M.Q. and A.W. Harrison, Seasonal trend of stratospheric NO_2 over Calgary, Can. J. Phys., 59, 1278, 1981.

Sze, N.D. and M.K.W. Ko, Photochemistry of COS, CS_2, CH_3SCH_3 and H_2S: Implications for the atmospheric sulfur cycle, Atmos. Environ., 14, 1223, 1980.

Thomas, L., The composition of the mesosphere and lower thermosphere, Phil. Trans. R. Soc. London A, 296, 243, 1980.

Thomas, R.J., C.A. Barth, G.J. Rottman, D.W. Rusch, G.H. Mount, G.M. Lawrence, R.W. Sanders, G.E. Thomas, and L.E. Clemens, Ozone density in the mesosphere (50-90 km) measured by the SME near infrared spectrometer, Geophys. Res. Lett., 10, 245, 1983.

Tisone, G.C., Measurements of NO densities during sunrise at Kauai, J. Geophys. Res., 78, 746, 1973.

Tohmatsu, T. and N. Iwagami, Measurement of nitric oxide distribution in the upper atmosphere, Space. Res., 15, 241, 1975.

Tohmatsu, T. and N. Iwagami, Measurement of nitric oxide abundance in equatorial upper atmosphere, J. Geomag. Geo-Elect., 28, 343-358, 1976.

Tung, K.K., On the two-dimensional transport of stratospheric trace gases in isentropic coordinates, J. Atmos. Sci., 39, 2330, 1982.

Turco, R.P. and R.C. Whitten, A comparison of several computational techniques for solving some common aeronomic problems, J. Geophys. Res., 79, 3179, 1974.

Turco, R.P., P. Hamill, O.B. Toon, R.C. Whitten and C.S. Kiang, A one-dimensional model describing aerosol formation and evolution in the stratosphere, I. Physical processes and mathematical analogs, J. Atmos. Sci., 36, 399, 1979.

Turco, R.P., R.C. Whitten, O.B. Toon, J.B. Pollack and P. Hamill, OCS, stratospheric aerosols and climate, Nature, 283, 283, 1980.

Turco, R.P., R.C. Whitten and O.B. Toon, Stratospheric aerosols: Observations and theory, Rev. Geophys. Space Sci., 20, 233 1982.

Van Hemelrijck, E., Atomic oxygen determination from a nitric oxide point release in the equatorial lower atmosphere, J. Atmos. Terr. Phys., 43, 345, 1981.

Vaughn, G., Diurnal variations of mesospheric ozone, Nature, 296, 133, 1982.

Volz, A., D.H. Ehhalt and H. Cosatto, The vertical distribution of CFM and related species in the stratosphere, Pageoph, 116, 545-553, 1978.

Wang, C.C., L.I. Davis, Jr., C.H. Wu, S. Japan, H. Niki and B. Weinstock, Hydroxyl radical concentrations measured in ambient air, Science, 189, 797, 1975.

Waters, J.W., J.J. Gustincic, and P.N. Swanson, *Atmospheric Water Vapor* (A. Deepak, ed., Academic Press, N.Y.), 1980.

Waters, J.W., J.C. Hardy, R.F. Jarnot and H.M. Pickett, Chlorine monoxide radical, ozone and hydrogen peroxide: Stratospheric measurements by microwave limb sounding, Science, 214, 61, 1981.

Weeks, L.H. and L.G. Smith, A rocket measurement of ozone near sunrise, Planet. Space Sci., 16, 1189, 1968.

Whitten, R.C. O.B. Toon and R.P. Turco, The stratospheric sulfate aerosol layer: Processes, models observations and simulations, Pageoph., 118, 86, 1980.

Whitten, R.C. (ed), *The Stratospheric Sulfate Aerosol Layer, Springer Verlag (New York), 1982.*

Willett, H.C., The relationship of total atmospheric ozone to the sunspot cycle, J. Geophys. Res., 67, 661, 1962.

Willett, H.C. and J. Prohaska, Further evidence of sunspot-ozone relationships, J. Atmos. Sci., 22, 493-497, 1965.

Williams, W.J., J.J. Kostus, A. Goldman and D.G. Murcray, Measurements of the stratospheric mixing ratio of HCl using an infrared absorption technique, Geophys. Res. Lett., 3, 383, 1976.

Witt, G., J.E. Dye, and N. Wilhelm, Rocket-borne measurements of scattered sunlight in the mesosphere, J. Atmos. Terr. Phys., 38, 223, 1976.

Wofsy, S.C., J.C. McConnell and M.B. McElroy, Atmospheric CH_4, CO and CO_2, J. Geophys. Res., 77, 4477, 1972.

Wofsy, S.C, M.B. McElroy, and Y. L. Yung, The chemistry of atmospheric bromine, Geophys. Res. Lett., 2, 215, l975.

World Meteorological Organization (WMC), The stratosphere 1981: Theory and measurements, Report No. 11, WMO Global ozone research and monitoring project, Geneva, Switzerland, 1982.

Yung, Y.L., J.P. Pinto, R.T. Watson and S.P. Sander, Atmospheric bromine and ozone perturbations in the lower stratosphere, J. Atmos. Sci., 37, 2, 339, 1980.

Zafonte, L., N.E. Hester, E.R. Stephens, O.C. Taylor, Background and vertical atmospheric measurements of fluorocarbon 11 and fluorocarbon 12 over southern California, Atm. Environm., 9, 1007, 1975.

Zander, R., Recent observations of HF and HCl in the upper stratosphere, Geophys. Res. Lett., 8, 413, 1981.

Zander, R, H. Leclercq, and L.D. Kaplan, Concentration of carbon monoxide in the upper stratosphere, Geophys. Res. Lett., 8, 365, 1981.

Zbinden, P.A., M.A. Hidalgo, P. Eberhardt and J. Geiss, Mass spectrometer measurements of the positive ion composition in the D and E regions of the ionosphere, Planet. Space Sci., 23, 1621, 1975.

Zimmerman, P.R., R.B. Chatfield, J. Fishman, P.J. Crutzen and P.L. Hanst, Estimates on the production of CO and H_2 from the oxidation of hydrocarbon emissions from vegetation, Geophys. Res. Lett., 5, 679, 1978.

Chapter 6

The Ions

6.1 Introduction

Atmospheric atoms and molecules can be ionized either by short wavelength solar radiation (UV and x-rays), or by precipitating energetic particles:

$$X + h\nu \rightarrow X^+ + e \tag{6.1}$$

$$X + e^* \rightarrow X^+ + 2e \tag{6.2}$$

where e^* represents an energetic electron. These processes are the starting points for a series of reactions which determine the structure of the ionosphere. Several types of general reactions of importance in ion chemistry should be outlined in order to study the ionosphere. For example, different ions may be produced from the initial (also called *primary*) particles through charge exchange reactions with neutral molecules:

$$X^+ + Y \rightarrow Y^+ + X \tag{6.3}$$

Negative ions can be formed by electron attachment on neutral species:

$$e + Z + M \rightarrow Z^- + M \tag{6.4}$$

Electrons can be released from negative ions, either by photodetachment or by collisional detachment:

$$Z^- + h\nu \rightarrow Z + e \tag{6.5}$$

$$Z^- + M \rightarrow Z + M + e \tag{6.6}$$

Positively and negatively charged particles can recombine:

$$X^+ + e \rightarrow \text{neutral products} \tag{6.7}$$

$$X^+ + Y^- \rightarrow \text{neutral products} \tag{6.8}$$

We will examine the detailed chemistry of each of these processes below.

At the end of the 19th century it was suggested that an electrically conducting atmospheric layer could explain the observed diurnal variation in the terrestrial magnetic field. In 1901, Marconi established the first radiowave transmission between Europe and North America. Later, Kennelly and Heaviside independently suggested that this communication was possible only because of the reflection of radio signals by a conducting layer near 80 km altitude. Radiowave methods later led to the first quantitative studies of this layer, through analysis of emitted signals reflected to the surface. Systematic sounding of this type showed that the base of the ionosphere was located at about 50 km. Present understanding is also due to numerous rocket and balloon observations as a function of altitude. Such studies have revealed, for example, that electrons are probably present in very small abundances below the stratopause, and that positive and negative ions dominate the ion composition of the stratosphere.

The ionosphere is a weakly ionized fluid of net neutral charge. Understanding the formation of the ions requires knowledge of the spectral distribution of solar radiation at short wavelengths, the structure of solar and galactic cosmic rays, as well as the chemical composition of the atmosphere and its physical characteristics such as pressure, temperature, and transport. The variations of solar activity must also be considered.

It is customary to divide the ionosphere into a number of characteristic layers based on the mean vertical profile of electron density.

The D-region is located between 60 and 85 km. In this layer, ionization results mostly from photoionization of NO by Lyman α radiation (121.6 nm). High energy cosmic rays also contribute to the ionization of O_2 and N_2 below 70 km. Radio waves are readily absorbed in the D-region, as will be discussed in §6.6.

The E-region extends from about 85 to 130 km. It is produced by ionization of molecular and atomic oxygen, and molecular nitrogen, by x-rays and Lyman β radiation. Although the predominant ions in this layer are O_2^+ and NO^+, relatively large concentrations of Fe^+, Ca^+, Si^+, and Mg^+ are also observed. These ions are produced by meteor ablation between 85 and 130 km.

The F-region begins above 130 km and is sometimes subdivided into two layers, F_1 and F_2. It is primarily produced by ionization of atomic oxygen and molecular nitrogen by extreme ultraviolet radiation (9-91 nm). The atomic oxygen ion, O^+, dominates. The electron density attains its maximum of about 10^6 cm^{-3} in this layer. The F region plays an important role in the transmission of certain radio waves, which can be reflected or refracted if their frequency is less than 5-10 MHz, or transmitted if their frequency is above this limit (television waves).

In the D, E, and F_1 layers, the electron concentration is greatest at maximum solar elevation (local noon). At night, the electrons disappear almost entirely in the D region, and their density is reduced by a factor of 100 in the

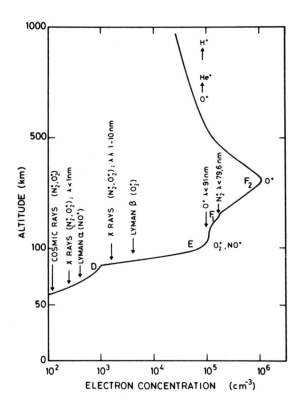

Fig. 6.1. Definition of the ionospheric layers based on the vertical distribution of electron density. From Banks and Kockarts (1973). (Copyright by Academic Press).

E and F_1 regions. In these layers, the lifetime of the ions is short compared to the transport time scale, and the charged particle concentrations are therefore determined by a photochemical equilibrium between production (by ionization processes) and loss (by recombination of positive ions with electrons or with negative ions). In the F_2 region, on the other hand, the electron density is no longer due to a simple equilibrium between ionization and recombination and the effects of transport by molecular diffusion must be considered, particularly above 300 km. The structure of this part of the ionosphere is complex, and varies considerably with geomagnetic latitude.

Above the F layer, the electron density rapidly decreases. Helium He^+ and then hydrogen H^+ ions dominate, and the effect of the magnetic field on these charged particles becomes more important. As a result, the particles no longer occur in horizontal layers; rather they tend to be aligned by the earth's magnetic field.

Figure 6.1 shows the shape of the mean vertical distribution of electrons below 1000 km and indicates the dominant ions in each layer. In this chapter, we will discuss only the ions in the middle atmosphere, i. e., those observed in the D and lower E regions, and in the stratosphere.

To study the behavior of the ions and electrons in a simple fashion, we first write the following equations of continuity, assuming only a single ion of each type:

$$\frac{dn^+}{dt} = nI - \alpha_D n^+ n_e - \alpha_i n^+ n^- \tag{6.9}$$

$$\frac{dn^-}{dt} = a_e n^2 n_e - n^- (d_p + f_d n + \alpha_i n^+) \tag{6.10}$$

$$\frac{dn_e}{dt} = nI - \alpha_D n^+ n_e - a_e n^2 n_e + n^- (d_p + f_d n) \tag{6.11}$$

where

n = concentration of the neutral species which ionize (e. g., NO, O, etc.)

n^+ = concentration of the positive ion

n^- = concentration of the negative ion

n_e = electron concentration

α_D = electron-ion recombination rate

α_i = ion-ion recombination rate

a_e = electron-neutral attachment rate

d_p = negative ion photodetachment rate

f_d = negative ion collisional detachment rate

I = ionization frequency (formation of an electron and a positive ion).

Further, electrical neutrality is assumed:

$$n^+ = n^- + n_e = (1 + \lambda) n_e \tag{6.12}$$

where

$$\lambda = n^- / n_e \tag{6.13}$$

is the ratio of negative ions to electrons. Expressions (6.9) and (6.10) can now be transformed:

$$\frac{dn_e}{dt} = \frac{nI}{1 + \lambda} - (\alpha_D + \lambda \alpha_i) n_e^2 - \frac{n_e}{1 + \lambda} \frac{d\lambda}{dt} \tag{6.14}$$

$$\frac{dn^-}{dt} = \frac{a_e n^2 n_e}{\lambda} - n_e [d_p + f_d n + (1 + \lambda) \alpha_i n_e] - \frac{n_e}{\lambda} \frac{d\lambda}{dt} \tag{6.15}$$

eliminating dn_e/dt, we obtain the ratio, λ:

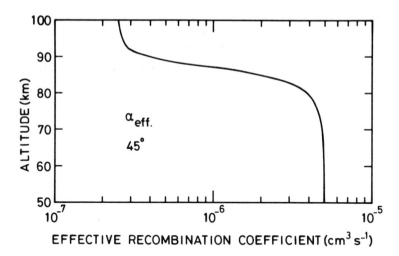

Fig. 6.2a. Vertical distribution of the effective recombination coefficient obtained with a mathematical model. From Brasseur (1982).

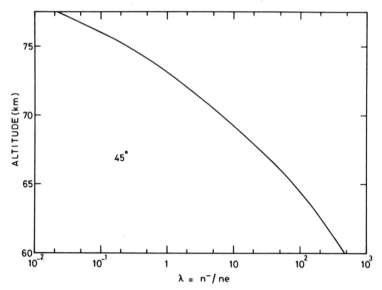

Fig. 6.2b. Ratio of negative ions to electrons as a function of altitude as calculated with a mathematical model for low solar activity. From Brasseur (1982).

$$\frac{1}{\lambda(1 + \lambda)}\frac{d\lambda}{dt} = \left[\frac{a_e n^2}{\lambda} + (d_p + f_d n)\right] - \left[\frac{nI}{(1 + \lambda)n_e} + (\alpha_i - \alpha_D)n_e\right] \quad (6.16)$$

which is a Riccati differential equation. Analysis of approximate magnitude of the respective terms in equation (6.16) shows that is is possible to neglect the contribution of the ionization and recombination terms, so that the equation reduces approximately to

$$\frac{d\lambda}{dt} = (1 + \lambda)a_d n^2 - (1 + \lambda)\lambda(d_p + f_d n) \tag{6.17}$$

Since the time constant for negative ion formation by attachment (a_e) is at most about an hour at the top of the middle atmosphere (and is much less at lower altitudes), photochemical steady state can be assumed:

$$\lambda = \frac{a_e n^2}{d_p + f_d n} \tag{6.18}$$

This shows that the ratio of the negative ions to electrons depends principally on the rate of attachment of electrons on neutral particles, and on the rate of detachment by solar radiation and collisions.

The continuity equation for electrons in the D region can be written:

$$\frac{dn_e}{dt} = \frac{nI}{1 + \lambda} - (\alpha_D + \lambda\alpha_i)n_e^2 \tag{6.19a}$$

but, in the E region, where $\lambda << 1$,

$$\frac{dn_e}{dt} = nI - \alpha_D n_e^2 \tag{6.19b}$$

A more realistic analysis considers formation of several ions (j), and involves more complex expressions of the type

$$\frac{dn_e}{dt} = \sum_j \frac{n_j I_j}{1 + \lambda} - \frac{n_e}{1 + \lambda}\sum_j (\alpha_{D,j} + \lambda\alpha_{i,j})n_j^+ \tag{6.20}$$

with

$$\sum_j n_j^+ = (1 + \lambda)n_e \tag{6.21}$$

An effective recombination rate can then be defined:

$$\alpha_{eff} = \frac{\sum_j (\alpha_{D,j} + \lambda\alpha_{ij})n_j^+}{\sum_j n_j^+} \tag{6.22}$$

such that the equation for electrons becomes

$$\frac{dn_e}{dt} = \sum_j \frac{n_j I_j}{1 + \lambda} - \alpha_{eff} n_e^2 \tag{6.23}$$

Since the lifetime of electrons in this part of the atmosphere is quite short, we can assume photochemical equilibrium, and the electron concentration is given by

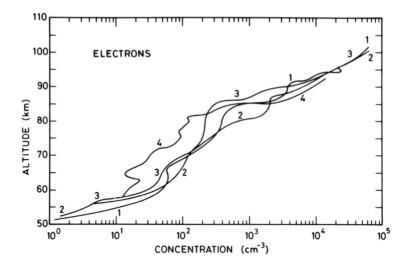

Fig. 6.3. Vertical distributions of electron density observed during daytime at Wallops Island, Virginia during different seasons, for a zenith angle of 60 degrees. Curve 1: April, 1964; Curve 2: June, 1965; Curve 3: September, 1965; Curve 4: December, 1965. From Mechtly and Smith (1968). (Copyrighted by Pergamon Press).

$$n_e = \left(\frac{\sum\limits_{j} n_j I_j}{(1 + \lambda)\alpha_{eff}} \right)^{1/2} \tag{6.24}$$

Figure 6.2 presents the shape of the vertical distributions of λ and α_{eff}, and Figure 6.3 shows some observed electron density profiles in the D and lower E regions.

6.2 Formation of ions in the middle atmosphere

6.2.1. Effect of solar radiation

Examination of the photoionization thresholds and absorption cross sections for atmospheric constituents shows that the solar radiation capable of producing ionization is primarily absorbed by O_2, N_2, and O at altitudes above 100 km. In the region below 100 km, soft x-rays (1-10 nm) produce some ionization in the E-region, and hard x-rays (< 1 nm) contribute to D-region ionization, but a large fraction of the ionizing solar radiation is provided by the Lyman α line at 121.57 nm, the Lyman β line at 102.6 nm, the C III line at 97.7 nm and a few other lines in the solar emission continuum in the far ultraviolet. Table 6.1 gives some typical values of the incident fluxes of these lines for different levels of solar activity.

Table 6.2 presents the photoionization thresholds for various atmospheric molecules. The photoionization limit of molecular oxygen is 102.8 nm, implying that oxygen is ionized by the Lyman β and C III lines and by solar x-rays. On the other hand, molecular nitrogen can only be ionized by wavelengths less than 79.58 nm, and is therefore unaffected by the principal ultraviolet lines which are available below 100 km.

Table 6.1. Flux of ionizing solar radiation $\left(cm^{-2} s^{-1}\right)$ for different levels of solar activity.

Wavelength interval (nm)	Very quiet sun	Quiet sun	Moderate solar activity	High solar activity
102.6 (L$-\beta$)	3.5(9)	5.0(9)	8.0(9)	1.2(10)
97.7 (C III)	4.4(9)	5.0(9)	6.0(9)	1.0(10)
91-79.6	7.5(9)	1.0(10)	1.3(10)	1.5(10)
79.6-73.2	1.0(9)	1.3(9)	1.5(9)	2.0(9)
73.2-66.5	5.0(8)	6.0(8)	8.0(8)	1.0(9)
66.5-37.5	4.0(9)	6.0(9)	8.0(9)	1.2(10)
37.5-27.5	7.5(9)	1.0(10)	1.5(10)	2.0(10)
27.5-15.0	7.5(9)	1.0(10)	1.5 (10)	2.0(10)
15.0-8.0	5.0(8)	7.5(8)	1.0(9)	2.0(9)
8.0-6.0	2.5(7)	5.0(7)	1.0(8)	1.5(8)
6.0-4.1	2.5(7)	5.0(7)	1.0(8)	1.5(8)
4.1-3.1	7.5(6)	1.5(7)	3.0(7)	4.5(7)
0.8-0.5	2.9(2)	2.9(3)	2.9(4)	2.9(5)
0.33-0.5	2.0(1)	2.0(2)	2.0(3)	2.0(4)
0.15-0.33	1.0(0)	1.0(1)	1.0(2)	1.0(3)

Table 6.2 Ionization potentials of atoms and molecules (wavelength in nm).

Na	241.2	NO	134.0	H_2O	98.5	O	91.0
Al	207.1	CH_3	126.0	O_3	96.9	CO_2	89.9
Ca	202.8	NH_3	122.1	N_2O	96.1	CO	88.5
Mg	162.2	CH	111.7	CH_4	95.4	N	85.2
Si	152.1	O_2	102.8	OH	94.0	N_2	79.6
C	110.0	SO_2	100.8	H	91.1	Ar	78.7

Given the ionization (σ) and absorption (K) cross sections of N_2, O_2, and O for different spectral intervals, the photoionization coefficient can be calculated using the following expression:

$$I(X) = \sum \sigma_i(X) q_{\infty,i} \, e^{-\tau} \quad (X = N_2, O_2 \text{ or } O) \tag{6.25}$$

where

$$\tau_i = K_{i,O_2}(O_2) + K_{i,N_2}(N_2) + K_{i,O}(O)$$

and where the other symbols have been defined in Chapter 4.

Stewart (1970) suggested analytic expressions which can be used to obtain the photoionization coefficients of N_2, O_2, and O to within 15% accuracy. Defining the effective column density as:

$$N^*(z,\chi) = N(N_2,z,\chi) + N(O_2,z,\chi) + 0.8 \times N(O,z,\chi) \tag{6.26}$$

where $N(N_2,z,\chi)$ represents the total column abundance of N_2 (molec cm^{-2}) along the solar ray path, and $N(O_2,z,\chi)$ and $N(O,z,\chi)$ are defined similarly. Then

$$I_{X^+}^{-1} \approx A_o + A_1(N^*(z,\chi) \times 10^{-17}) + A_2(N^*(z,\chi) \times 10^{-17}))^\alpha \tag{6.27}$$

Using this expression, the photoionization frequencies of N_2^+, O_2^+, and O^+ can be estimated for mean solar activity by substituting the appropriate constants given below:

Constituent	A_o	A_1	A_2	α
$I_{N_2^+}$	2.2(6)	2.2(6)	6.4(6)	2.0
$I_{O_2^+}$	1.1(6)	1.0(6)	1.3(6)	2.5
I_{O^+}	2.3(6)	3.2(6)	1.0(6)	2.4

Solar ultraviolet radiation represents an important source of ions in the E and upper D-regions because of its effects on O_2. However, for typical conditions, most of the ionization in the D-region is due to the effect of the solar Lyman α ray on nitric oxide (Nicolet, 1945). The ionization potential of the NO molecule is only 9.25 eV, which corresponds to a wavelength of 134 nm. In the spectral region of the Lyman α line an atmospheric window exists due to the low absorption cross section of O_2 in this interval, and thus the ionizing radiation can penetrate relatively far into the mesosphere. The rate of production of the NO^+ ion is given by

$$P(NO^+) = I_{NO}(NO) \tag{6.28}$$

and can be estimated by

$$I_{NO} = (6 \pm 2) \times 10^{-7} \exp[-1 \times 10^{-20} N(O_2, z, \chi)] \ (s^{-1}) \qquad (6.29)$$

This expression assumes that the incident flux (q_∞) of Lyman α is $3 \pm 1 \times 10^{11} cm^{-2} s^{-1}$, $\sigma_{NO} = 2.0 \times 10^{-18} cm^{-2}$, and $\sigma_{O_2} = 1.0 \times 10^{-20} cm^{-2}$. $N(O_2, z, \chi)$ represents the integrated oxygen abundance along the line of sight, as above. It should be noted that even at night, when the direct incident flux of Lyman α is zero, there is still a considerable flux of scattered Lyman α reflected by the hydrogen geocorona in the uppermost levels of the atmosphere. Its intensity is about 100 to 1000 times weaker than the direct daytime flux, and thus contributes significantly to D-region ionization only during the night. A similar effect occurs for Lyman β, and is important in the E-region.

The morphology of the D-region is therefore closely related to the distribution of nitric oxide, which is strongly tied to dynamics (Brasseur and Nicolet, 1973; Solomon et al., 1982a,b; Brasseur and De Baets, 1984). Nitric oxide is produced in large amounts in the stratosphere and thermosphere, while in the mesosphere its distribution is heavily dependent on the competition between transport from these neighboring regions and destruction by photochemical processes (see Chapter 5).

The absorption of solar radiation in the spectral region from 102.7 [photoionization limit of $O_2(^1\Sigma_g^-)$] to 111.8 nm [photoionization limit of $O_2(^1\Delta_g)$] represents an additional source of D-region ionization as suggested by Hunten and McElroy (1968). In this region, however, the incident radiation is absorbed both by O_2 and CO_2, so that the rate of formation of the O_2^+ ion

$$P(O_2^+) = I_{O_2(^1\Delta_g)}(O_2(^1\Delta_g)) \qquad (6.30)$$

is much smaller than the rate of ionization of NO. The rate of production has been determined by Paulsen et al. (1972) using the solar fluxes reported by Hall and Hinteregger (1970). The following approximate expression can be applied from 70 to 90 km.

$$I_{O_2(^1\Delta_g)} = 0.549 \times 10^{-9} \exp[-2.406 \times 10^{-20} N(O_2, z, \chi)]$$

$$+ 2.6 \times 10^{-9} \exp[-8.508 \times 10^{-20} N(O_2, z, \chi)] \ (s^{-1}) \qquad (6.31)$$

Hard x-rays ($\lambda < 1$ nm) also penetrate into the D-region. The production of ionization by x-rays is somewhat complex, since it involves the absorption of an x-ray photon by a neutral particle, leading to the production of photoelectrons which are sufficiently energetic to ionize other particles. It thus becomes necessary to distinguish between primary and secondary ionization. The mean number of secondary ion pairs produced depends on the energy of the incident photon. For example, this number has been estimated at about 45 for $\lambda = 0.6$ nm, 75 for $\lambda = 0.4$ nm , and 165 for $\lambda = 0.2$ nm (Nicolet and Aikin, 1960).

For low levels of solar activity, this source of ions is small compared to that produced by the Lyman α line. On the other hand, its contribution becomes more important during disturbed conditions: from 0.1 to 0.8 nm, for example, the solar irradiance is about 1000 times more intense for active conditions, and may increase by an additional factor of 100 during solar flares.

The ionization rate due to x-rays in the D-regions is thus quite variable. During solar flares, the x-ray contribution is reflected in observed changes in ion densities. However, the typical ion and electron densities observed in the D-region generally exhibit rather low variability, which tends to confirm the stability of the dominant ion sources.

Soft x-rays are only important in the upper part of the middle atmosphere ($z > 85$ km). Their flux also varies somewhat with solar activity. Table 6.1 provides some numerical estimates of these fluxes and their variations.

6.2.2. The effect of energetic particles

• *General*

Energetic charged particles can penetrate into the earth's atmosphere and ionize atmospheric species. They propagate along helical trajectories following the earth's geomagnetic field. In polar regions (magnetic latitude $\Lambda > 75°$) the magnetic field lines are open and cosmic particles (e. g., galactic) can easily

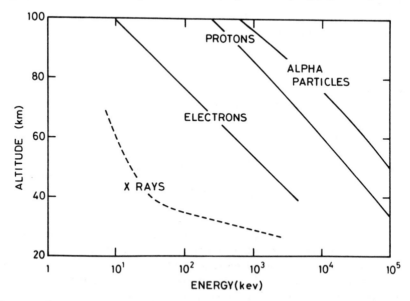

Fig. 6.4. Depth of atmospheric penetration for different charged particles and x-rays as a function of energy. Adapted from Potemra (1974) and Thorne (1977a).

enter the atmosphere. The depth of penetration of charged particles depends on their mass and energy. Figure 6.4 depicts the approximate depth of atmospheric penetration for different types of charged particles as a function of their energy. This diagram also shows the altitude of bremsstrahlung x-ray penetration produced by precipitation of high energy particles. In the auroral zone (70° < Λ < 75°), low energy particles (mostly electrons from 1-10 keV) precipitate into the atmosphere from the magnetospheric plasmasheet. These particles provide the optical displays known as the aurora, but do not penetrate much below 100 km. In the subauroral zone, (Λ ⩽ 70°) particles can be accelerated in the radiation belt, attaining energies as high as a few MeV, and can thus penetrate into the upper stratosphere. Figure 6.5 shows the configuration of the earth's geomagnetic field and the zones where charged particle precipitation occurs. It should be noted that high energy particles produce substantial fluxes of secondary electrons (energies from 10 to 100 eV) which are responsible for a considerable fraction of the energy transfer from the primary particle to the atmosphere. Knowledge of both the primary and secondary electron fluxes as a function of energy (also called the particle spectrum) and their cross sections for interaction with the principal atmospheric species is thus an important part of evaluating the impact of particle precipitation in aeronomic processes.

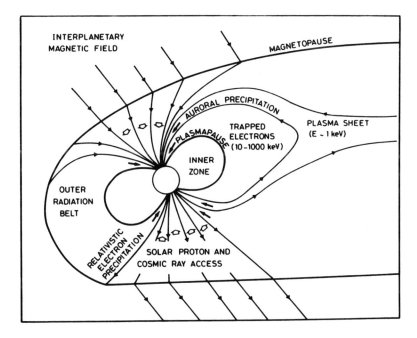

Fig. 6.5. Schematic representation of the structure of the magnetosphere. From Thorne (1980). (Copyrighted by Birkhauser-Verlag).

• *Cosmic rays*

    The ionization of atmospheric constituents by galactic cosmic rays provides the dominant source of ions in the lower mesosphere, stratosphere, and troposphere. The rate of ion pair production increases exponentially with penetration into the atmosphere, i. e., in proportion to the atmospheric density (see Fig. 6.6) to reach a maximum near 10 to 15 km (Fig. 6.7). The production is larger at high latitude than in the tropics, and is modulated by solar activity (see eqns. 6.32 to 6.35, below). Galactic cosmic rays are produced outside the solar system, and consist mostly of protons (about 83 %) and α particles (about 12%). They tend to follow the magnetic field lines as they approach the earth, and to penetrate near the magnetic poles.

    This is the primary reason for the observed meridional gradient in cosmic ray fluxes, particularly for less energetic particles which are more strongly

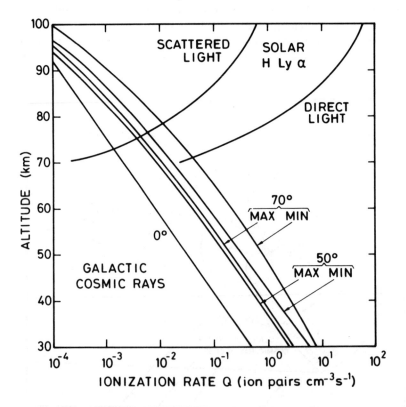

Fig. 6.6. Rate of ionization provided by cosmic rays at different geomagnetic latitudes (0, 50, and 70°) and for minimum and maximum levels of solar activity. These values are compared to the ion pair production producted by direct and diffuse Lyman α. From Rosenberg and Lanzerotti (1979).

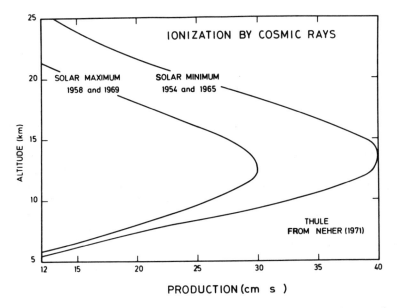

Fig. 6.7. Rate of ionization by cosmic rays in the lower stratosphere and troposphere. (From Brasseur and Nicolet, 1973, copyrighted by Pergamon Press).

influenced by the magnetic field. During periods of high solar activity, the galactic cosmic rays tend to be removed from the solar system by the intense solar wind. The amplitude of cosmic ray precipitation thus decreases, as does the associated ion pair production. The reverse is true at solar minimum.

Heaps (1978a) has derived a convenient parameterization of the rate of ion pair production by cosmic rays as a function of latitude, altitude, and solar activity level. For latitudes less than 53°, the following expressions can be used: First, define

$$X_1 = 1.74 \times 10^{-18} \tag{6.32a}$$

$$X_2 = 1.93 \times 10^{-17} \text{ (solar max)}, \quad 2.84 \times 10^{-17} \text{ (solar min)} \tag{6.32b}$$

$$X_3 = 0.6 + 0.8 \,|\cos\phi\,| \tag{6.32c}$$

then, for total number density (M) greater than 3×10^{17} molec cm^{-3},

$$Q_{CR} = \left(X_1 + X_2(|\sin\phi\,|^4)\right) 3 \times 10^{17^{(1-X_3)}} (M)^{X_3} \tag{6.33a}$$

while for (M) less than 3×10^{17},

$$Q_{CR} = (X_1 + X_2 |\sin\phi\,|^4)(M) \tag{6.33b}$$

where ϕ is the latitude. For latitudes greater than 53°

$$X_1 = 1.44 \times 10^{-17}; \quad X_2 = 4.92 \times 10^{-18} \qquad (6.34)$$

$$Q_{CR} = X_1(M) \text{ (solar max)}; \quad Q_{CR} = (X_1 + X_2)(M) \text{ (solar min)} \qquad (6.35)$$

- *Magnetospheric electrons and x-rays
  produced by bremsstrahlung*

Electron precipitation from the earth's radiation belt is sporadically observed to produce ionization at high latitudes. As we have already mentioned, the electron penetration depth depends directly on their energy and knowledge of their corresponding spectrum is thus necessary. Significant fluxes at energies exceeding 100 keV are often observed, and their precipitation is generally associated with geomagnetic storms in the subauroral radiation belt. The induced ionization events are generally referred to as relativistic electron precipitation (REP). Their lifetime is short (a few hours) and their frequency is not well known. Thorne (1977b) estimated that this phenomenon probably occurs between 1 and 10 percent of the time and may even be more intense in dark

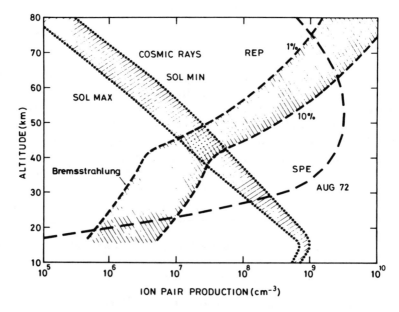

Fig. 6.8. Rate of ion pair production by magnetospheric electrons (REP) in the subauroral zone. These values correspond to an annual average, and are compared to the effect of cosmic rays and the total ion pair production associated with the solar proton event (SPE) of August, 1972. From Thorne (1977a). (Copyright by Reidel Publishing Company).

regions. An estimate of the associated time averaged mean ionization rate Q is provided by Figure 6.8. The effect of associated bremsstrahlung (x-rays) is also indicated. In any case, in the subauroral zone, this mechanism of ion pair production is of considerable importance. It should be noted, however, that the belt only extends over about 7 percent of the globe.

The effect of this process on ionospheric structure and trace neutral species distributions is not yet well established.

• *Solar proton events*

During large solar flares, heavy particles (mostly protons of energies from 10-300 MeV) can be emitted from the surface of the sun in substantial amounts. These particles can produce intense ionization in the earth's D-region, primarily at high latitudes (polar cap). Although this phenomenon is not frequent, it often lasts for several days. These events can change the ionization rate from its normal value of about $10 \text{ cm}^{-3} \text{ s}^{-1}$ to 10^4 or $10^5 \text{cm}^{-3} \text{ s}^{-1}$. As we shall see below, these events can also alter the composition of the neutral atmosphere. Figure 6.9 presents the vertical distribution of the ion pair production rates associated with some of the solar proton events of recent years. The events of August, 1972 and July, 1982 were particularly intense.

• *Ionization rate calculation*

The calculation of the rate of ion pair production due to corpuscular radiation requires knowledge of the particle energy spectrum and the energy degradation rate as they pass through the atmosphere. If $j(z,E)dE$ represents the

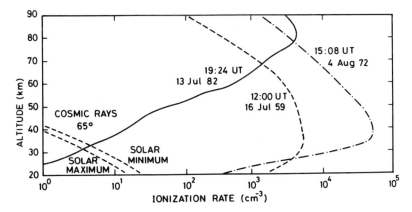

Fig. 6.9. Ionization rates associated with some solar proton events, compared to that due to cosmic rays. Adapted from Solomon et al. (1983).

flux of ionizing particles at altitude z of energies from E to E + dE (the differential flux j(E) is expressed in $cm^{-2}s^{-1}eV^{-1}sr^{-1}$) and if dE/ds is the energy loss per particle in an inelastic collision, then the ionization rate at altitude z is given by (Dubach and Barker, 1971):

$$Q\,(z) = \frac{\rho(z)}{W} \int_E \int_\Omega \frac{dE}{ds}\, j(z,E)\, dE\, d\Omega \qquad (6.36)$$

where W is the mean energy required per ion pair formation (W $\approx$ 35 eV), $\rho(z)$ is the air density (g cm^{-3}), and Ω is the solid angle over which equation (6.36) is integrated.

The ionization rates of the major species, molecular nitrogen and oxygen, can be determined from the rate of total ion pair production, Q, if it is assumed that the fraction of ionization is proportional to the mass of the target particle (Rusch et al., 1981):

$$Q_{N_2} = Q\frac{.88(N_2)}{.88(N_2) + (O_2)} = .77Q \qquad (6.37a)$$

$$Q_{O_2} = Q\frac{(O_2)}{.88(N_2) + (O_2)} \qquad (6.37b)$$

The ion chemistry is initiated by the impact of energetic secondary electrons, e^*, on the nitrogen molecule:

$$e^* + N_2 \rightarrow N_2^+ + 2e \qquad (6.38a)$$

$$e^* + N_2 \rightarrow N + N^+ + 2e \qquad (6.38b)$$

$$e^* + N_2 \rightarrow N + N + e \qquad (6.38c)$$

and on the oxygen molecule:

$$e^* + O_2 \rightarrow O_2^+ + 2e \qquad (6.39a)$$

$$e^* + O_2 \rightarrow O + O^+ + 2e \qquad (6.39b)$$

The ratio of the peak cross section for simple ionization versus that for dissociative ionization is 0.76:0.24 for N_2 (eqns. 6.38a,b), and 0.67:0.33 for O_2 (eqns. 6.39a,b) (Rapp et al., 1965). Assuming that the rate of production is proportional to these cross section ratios, we obtain the following approximate expressions (Rusch et al, 1981):

$$P_{N_2^+} = 0.76 \times 0.77Q = 0.585Q \qquad (6.40a)$$

$$P_{N^+} = 0.24 \times 0.77Q = 0.185Q \qquad (6.40b)$$

$$P_{O_2^+} = 0.67 \times 0.23\,Q = 0.154\,Q \qquad \text{(6.40c)}$$

$$P_{O^+} = 0.33 \times 0.23\,Q = 0.076\,Q \qquad \text{(6.40d)}$$

It should also be noted that the reactions discussed above can also lead to the formation of neutral oxygen and nitrogen atoms, thus initiating processes which can affect the neutral chemistry. These will be discussed in §6.5.

6.2.3 Comparison of different ionization processes

Summarizing some of the statements made previously, we note that the ionization above about 90 km is due primarily to the effect of solar ultraviolet radiation (Lyman β and extreme UV) and x-rays on the major species, particularly molecular oxygen. From 60 to 90 km, ionization is produced mainly through the effect of solar Lyman α radiation on nitric oxide, with photoionization of $O_2(^1\Delta_g)$ and hard x-rays ($\lambda < 1$ nm) playing a secondary role under normal conditions. The effect of cosmic rays dominates below about 60 km. Figures 6.10a and 6.10b compare the contributions of each of these terms as a function of altitude during the day and the night, respectively, for quiet solar conditions.

In the absence of solar radiation, the ionization is due primarily to the precipitation of high energy particles (e. g. see Vampola and Gorney, 1983), diffuse Lyman α radiation and cosmic rays. The effect of x-rays emitted by certain stars (Sco XR1, GX333-2.5) plays only a minor role.

As we have already mentioned, sudden perturbations on the sun (increases in x-ray production, precipitation of energetic protons or electrons) can greatly

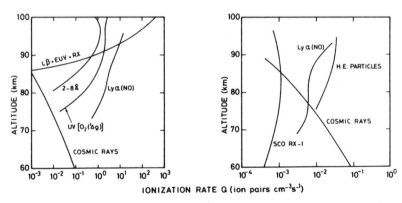

Fig. 6.10a and 6.10b. Ionization rates produced by various sources in the region from 60 to 100 km, during the day and at night. From Thomas, (1974). (Copyright by the American Geophysical Union).

modify the ionization rate and the morphology of the ionosphere. Table 6.3 presents order of magnitude estimates of the energy penetrating the middle atmosphere, during both quiet and disturbed periods.

6.3 Positive ion chemistry

6.3.1 Positive ions in the E-region

E-region ions are produced primarily by photoionization of nitrogen and molecular oxygen by solar ultraviolet radiation and x-rays. Small amounts of NO^+ and O^+ are also produced by direct ionization.

The primary O_2^+ ion is present in large abundances because its destruction rate by dissociative recombination

$$(\alpha_{O_2^+}); \quad O_2^+ + e \rightarrow O + O \tag{6.41}$$

and by charge exchange with NO

$$(\gamma_5); \quad O_2^+ + NO \rightarrow NO^+ + O_2 \tag{6.42}$$

Table 6.3. Sources of ionization in the middle atmosphere.
(From Rosenberg and Lanzerotti, 1979).

Permanent sources	Flux (ergs cm^{-2} s^{-1})
Galactic cosmic rays	1(-3) to 1(-2)
Cosmic x-rays: 0.1 to 1 nm	4(-9)
Solar x-rays: weak activity, $\lambda <$ 1nm	1(-3) to 1(-1)
Solar x-rays: weak activity, λ 1-10nm	1(-1) to 1
Solar H Lyman α: direct	6
Solar H Lyman α: scattered by the geocorona	6(-3) to 6(-2)
Magnetospheric electrons: auroral zones	1(-1) to 1
Magnetospheric electrons: mid-latitudes	1(-4) to 1(-3)

Sporadic sources	
Solar cosmic rays (SPE)	1(-3) to 1 (50 on 8/2/72)
Solar x-rays: solar flares < 1 nm	< 3
Solar x-rays: solar flares 1-10 nm	< 35
Cosmic x-rays: SCO X-1 0.1-1 nm	4(-7)
Magnetospheric electrons: auroral zones	1 to 1(3)
Magnetospheric electrons: mid-latitudes	1(-3) to 1 (-2)

is relatively slow. In contrast, the N_2^+ ion is very short lived because of a rapid charge exchange reaction with oxygen:

$$(\gamma_9); \quad N_2^+ + O_2 \rightarrow N_2 + O_2^+ \tag{6.43}$$

which provides an additional source of O_2^+. N_2^+ also reacts with atomic oxygen, which is present in relatively large amounts in the E-region:

$$(\gamma_3); \quad N_2^+ + O \rightarrow NO^+ + N \tag{6.44}$$

to produce NO^+. The rates of reactions (6.43) and (6.44) are comparable, implying that the relative importance of each process is dependent on the abundances of O_2 and O. Below about 95-100 km, reaction (6.43) dominates. The dissociative recombination of NO^+ must also be considered:

$$(\alpha_{NO^+}); \quad NO^+ + e \rightarrow N(^4S, 25\%; {}^2D, 75\%) + O \tag{6.45}$$

The equations for the ionic constituents of the E-region can now be written. Neglecting minor reactions and eliminating the density of N_2^+, we can write

$$(O_2^+) = \frac{I_{N_2^+}(N_2) + I_{O_2^+}(O_2)}{\alpha_{O_2^+}(n_e) + \gamma_5(NO)} \tag{6.46}$$

and

$$(NO^+) = \frac{I_{N_2^+}(N_2)\dfrac{\gamma_3(O)}{\gamma_3(O) + \gamma_9(O_2)} + \gamma_5(NO)(O_2^+)}{\alpha_{NO^+}(n_e)} \tag{6.47}$$

The equation of electrical neutrality must also be considered in ionospheric models. In the E-region, we may write:

$$(n_e) \approx (O_2^+) + (NO^+) \tag{6.48}$$

Figure 6.11 presents a vertical distribution of O_2^+ and NO^+ as observed by Keneshea et al. (1970). Figure 6.12 presents a schematic diagram of E-region ion chemistry. For completeness, it should be noted that the O^+ produced by photoionization is destroyed rapidly by reaction with O_2, N_2, and CO_2, and thus plays a negligible role in the middle atmosphere. Its maximum density reaches only about 10 percent of that of O_2^+. All of these polyatomic ions are produced directly by ionization or charge transfer with neutral species, and are referred to as *molecular* ions.

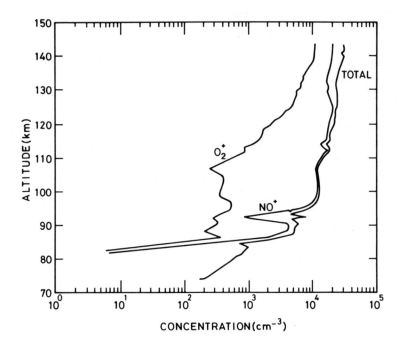

Fig. 6.11. Observations of the vertical distributions of O_2^+ and NO^+. From Keneshea et al. (1970). (Copyright by the American Geophysical Union).

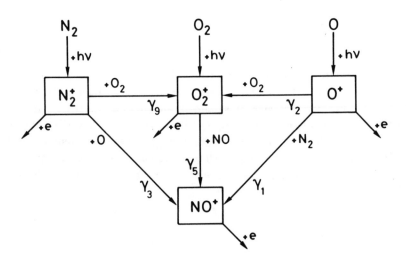

Fig. 6.12. Schematic diagram of E-region ion chemistry.

• *Metal ions*

Gas phase metal ions have been observed at mid-latitudes by several investigators. Meteor ablation appears to be the source of these ions, producing large amounts of Mg^+, Fe^+, etc. These species are present in a permanent layer from 85 to 100 km.

In addition, at altitudes near 100 - 105 km, thin layers (1-3 km thick) of greatly enhanced metal ion densities are found to extend over horizonatal areas of a few hundred km. This phenomenon is called a sporadic E layer, and arises primarily from the effects of wind shears in the presence of a magnetic field. The electron density in these layers is considerably larger than in the nearby regions, leading to fast recombination of O_2^+ and NO^+, reducing their densities relative to those of the metal ions, which have extremely low recombination coefficients.

Part of the ionization of the metallic species may occur during ablation, but most of the ions are probably produced by photoionization and charge transfer from O_2^+ and NO^+. Many metal ions participate in a class of general reactions which largely determine the ratios between ions and neutral metals, and metal oxide ions to metal ions. These reactions are shown in Table 6.4, and the metals for which each reaction is believed to occur are indicated.

In most cases, these reactions have not been studied in the laboratory, and can only be estimated based on thermochemical information regarding their exothermicity (see Chapter 2 and Murad, 1978). Notable exceptions are provided by magnesium and silicon, which have been intensively examined in laboratory studies (Ferguson et al., 1981a,b). Assuming only the reaction chemistry shown in Table 6.4, the continuity equation for X^+ can be written:

$$\frac{d(X^+)}{dt} = m_4(XO^+)(O) + m_7(X)(NO^+) + m_8(X)(O_2^+) + m_9(X)$$

$$- m_2(X^+)(O_2)(M) - m_6(X^+)(O_2) \tag{6.49}$$

Assuming steady state and retaining only the most important terms, we find

$$\frac{(X^+)}{(X)} \approx \frac{m_7(NO^+) + m_8(O_2^+) + m_9}{m_1(O_3) + m_2(O_2)(M) + m_6(O_2)} \tag{6.50}$$

While for XO^+,

$$\frac{d(XO^+)}{dt} = m_1(X^+)(O_3) + m_3(XO_2^+)(O) + m_5(X)(O) + m_6(X^+)(O_2)$$

$$+ m_{11}(XO)(NO^+) + m_{12}(XO)(O_2^+) - m_{13}(XO^+)(e) - m_4(XO^+)(O) \tag{6.51}$$

and for XO_2^+,

Table 6.4. General metal ion reactions

Reaction†	Rate constant	Metals
$X^+ + O_3 \rightarrow XO^+ + O_2$	m_1	Al, Fe, Mg, Si, Ti, Sc
$X^+ + O_2 + M \rightarrow XO_2^+ + M$	m_2	Al, Fe, Mg, Na, Si, Ti, Sc
$XO_2^+ + O \rightarrow XO^+ + O_2$	m_3	Al, Fe, Mg, Na, Si, Ti, Sc
$XO^+ + O \rightarrow X^+ + O_2$	m_4	Al, Fe, Mg, Na, Si
$X + O \rightarrow XO^+ + e$	m_5	Ti, Sc
$X^+ + O_2 \rightarrow XO^+O$	m_6	Ti, Sc
$NO^+ + X \rightarrow X^+ + NO$	m_7	Al, Fe, Mg, Na, Si, Ti, Sc
$O_2^+ + X \rightarrow X^+ + O_2$	m_8	Al, Fe, Mg, Na, Si, Ti, Sc
$X + h\nu \rightarrow X^+ + e$	m_9	Al, Fe, Mg, Na, Si, Ti, Sc
$XO + h\nu \rightarrow XO^+ + e$	m_{10}	Al, Fe, Mg, Na, Si, Ti, Sc
$NO^+ + XO \rightarrow XO^+ + NO$	m_{11}	Fe, Mg, Na, Ti, Sc
$O_2^+ + XO \rightarrow XO^+ + O_2$	m_{12}	Al, Fe, Mg, Na, Si, Ti, Sc
$XO^+ + e \rightarrow X + O$	m_{13}	Al, Fe, Mg, Na, Si
$XO^+ + e \rightarrow XO + h\nu$	m_{14}	Sc, Ti

† X represents any of the indicated metals

$$\frac{d(XO_2^+)}{dt} = m_2(X^+)(O_2)(M) - m_3(XO_2^+)(O) \tag{6.52}$$

so that

$$\frac{(XO_2^+)}{(X^+)} = \frac{m_2(O_2)(M)}{m_3(O)} \tag{6.53}$$

and

$$\frac{(XO^+)}{(X^+)} = \frac{m_1(O_3) + m_2(O_2)(M) + m_6(O_2)}{m_4(O) + m_{13}(e)}$$

$$+ \frac{m_5(M)(O) + m_{11}(MO)(NO^+) + m_{12}(MO)(O_2^+)}{m_4(O)(X^+) + m_{13}(e)(X^+)} \tag{6.54}$$

Si^+ reacts rapidly with H_2O, so that this scheme should be appropriately modified for that metal (see, e. g., Solomon et al., 1982). Additional association reactions are probably also important for the alkali metals (e.g., Sze et al., 1982; Jegou et al., 1984).

Using these ratios along with observations of Si^+, Fe^+, and Mg^+, it can be shown that the inferred relative abundances of total Mg, Si, and Fe, are not far removed from those expected based on their cosmic abundances (Goldberg and Aikin, 1973). On the other hand, Murad (1978) has indicated that observations of the ratios of AlO^+/Al, FeO^+/Fe, and NaO^+/Na, for example, are far outside of expectations based on present chemistry. These discrepancies may be due to incomplete knowledge of the ion chemistry, or perhaps to incorrect assignments of the observed ion masses. Higher resolution mass spectrometers may help resolve this question, but at present many aspects of metal ion chemistry are poorly understood.

6.3.2 Positive ions in the D-region

In 1945 Nicolet suggested that D-region ionization must be due to photoionization of nitric oxide by Lyman α (see §6.2). The first theoretical study of the chemistry of this layer by Nicolet and Aikin (1960) led to the expectation that the D-region should be composed primarily of O_2^+ and NO^+. However, the first rocket-borne mass spectrometric measurements by Narcisi and Bailey (1965) revealed that in addition to O_2^+ and NO^+, other species of mass 19 and 37, corresponding to the H_3O^+ and $H_5O_2^+$ clusters, were present in larger abundances than the primary ions below about 82 km. The presence of relatively

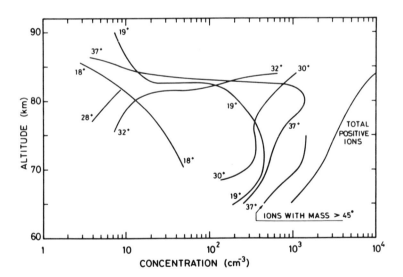

Fig. 6.13. First mass spectrometric observations of positive ions from 65 to 90 km (Narcisi and Bailey, 1965, copyrighted by the American Geophysical Union).

large amounts of heavy ions, mass numbers greater than 45, was also established by these measurements. The altitude profiles of D-region ions observed by this experiment are shown in Figure 6.13. Later studies showed that below about 80 km, the D-region composition is dominated by heavy $H^+(H_2O)_n$ cluster ions. The altitude at which the cross over from cluster to molecular ions occurs varies from about 70 to 90 km according to season and latitude. The hydration order of the ions (n) is dependent on geophysical conditions (particularly temperature) as well as the atmospheric water vapor content. The most abundant ions generally display hydration orders from 2 to 4, but at cold temperatures near the mesopause hydration orders of 8 or 9 are not uncommon. Bjorn and Arnold (1981) observed ions as large as $H^+(H_2O)_{20}$ near a very cold summer mesopause at high latitude. These authors also suggest that such ions may be important in the formation of noctilucent clouds.

The first chain of reactions which must be considered involves O_2^+ as the primary ion. This species can react with molecular oxygen to form O_4^+:

$$O_2^+ + O_2 + M \rightarrow O_4^+ + M \qquad (6.55)$$

which can react with atomic oxygen to reform O_2^+ and O_3:

$$O_4^+ + O \rightarrow O_2^+ + O_3 \qquad (6.56)$$

O_4^+ can also react with water vapor to form the first hydrated species:

$$O_4^+ + H_2O \rightarrow O_2^+ \cdot H_2O + O_2 \qquad (6.57)$$

The $O_2^+ \cdot H_2O$ ion is very short lived because of the fast hydration reactions

$$O_2^+ \cdot H_2O + H_2O \rightarrow H_3O^+ \cdot OH + O_2 \qquad (6.58)$$

$$H_3O^+ \cdot OH + H_2O \rightarrow H_3O^+ \cdot H_2O + OH \qquad (6.59)$$

In each case, a proton hydrate of the form $H_3O^+ \cdot (H_2O)_n$ and an OH particle are produced. Successive hydration can occur through the following equilibria:

$$H_3O^+ \cdot (H_2O)_n + H_2O + M \Leftrightarrow H_3O^+ \cdot (H_2O)_{n+1} + M \qquad (6.60)$$

It should be noted that reaction (6.56) can interrupt the hydration chain. Thus, the rapid decrease in hydrated ions which is observed above 80 km must be due at least in part to the rapid increase in atomic oxygen which occurs in the lower thermosphere. The reaction scheme just described does not completely explain the observed D-region ion distributions, since the primary ion at these altitudes is predominantly NO^+ rather than O_2^+.

Another chain of reactions was found to resolve this discrepancy (Ferguson, 1971; Reid, 1977), explaining how proton hydrate formation occurred starting from NO^+ as the primary ion. Although the direct hydration processes

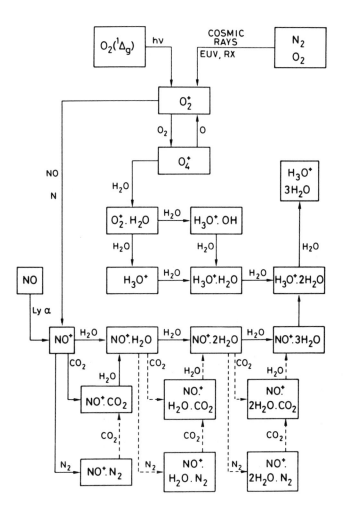

Fig. 6.14. Schematic diagram of D-region positive ion chemistry. From Ferguson (1979). (Copyrighted by Plenum).

$$NO^+ + H_2O + M \Leftrightarrow NO^+ \cdot H_2O + M \tag{6.61}$$

$$NO^+ \cdot (H_2O)_n + H_2O + M \Leftrightarrow NO^+ \cdot (H_2O)_{n+1} + M \tag{6.62}$$

were found to be too slow to explain the observations, it was shown that the equivalent reaction can occur via two indirect sequences involving CO_2 and N_2:

$$NO^+ + CO_2 + M \Leftrightarrow NO^+ \cdot CO_2 + M \tag{6.63}$$

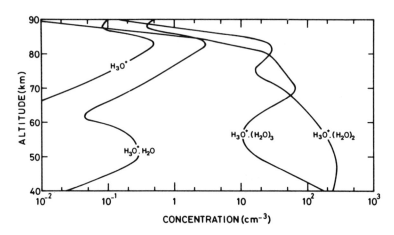

Fig. 6.15a. Vertical distributions of some of the cluster ions from 40 to 90 km. From Brasseur (1982).

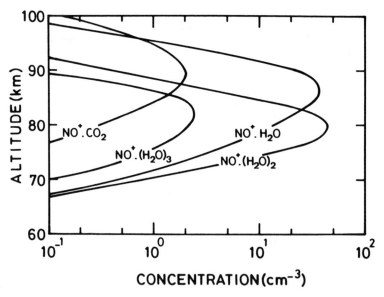

Fig. 6.15b. Vertical distribution of ions of the type $NO^+ \cdot X$ from 60 to 100 km. From Brasseur (1982).

$$NO^+ \cdot CO_2 + H_2O \rightarrow NO^+ \cdot H_2O + CO_2 \qquad (6.64)$$

and

$$NO^+ + N_2 + M \Leftrightarrow NO^+ \cdot N_2 + M \qquad (6.65)$$

$$NO^+ \cdot N_2 + CO_2 \rightarrow NO^+ \cdot CO_2 + N_2 \qquad (6.66)$$

followed by

$$NO^+ \cdot CO_2 + H_2O \rightarrow NO^+ \cdot H_2O + CO_2$$

Similar processes occur for higher hydration (e.g., involving $NO^+ \cdot H_2O \cdot CO_2$, see Fig. 6.14) until NO^+ is converted to $NO^+ \cdot (H_2O)_3$. This species reacts rapidly with water vapor to produce $H_3O^+ \cdot (H_2O)_2$:

$$NO^+ \cdot (H_2O)_3 + H_2O \rightarrow H_3O^+ \cdot (H_2O)_2 + HNO_2 \qquad (6.67)$$

Higher order clusters can then be formed via reaction (6.60). Most of the equilibria are strongly temperature sensitive, implying that the composition of the D-region should be quite variable with season and latitude, and that sporadic changes associated with local temperature modulations should be observed.

Figure 6.14 presents a schematic diagram of the aeronomy of positive ions in the D-region as it is presently understood. Recombination of these ions with negative ions and electrons must also be considered. The recombination coefficient of positive ions with electrons should be between about 10^{-7} and 10^{-5} cm^3 s^{-1}; the coefficient is faster for clusters than for molecular ions such as NO^+ and O_2^+ (Leu et al., 1973). Figures 6.15a and b present model distributions of the principal ions in the D-region for low solar activity (Brasseur and DeBaets, 1984).

6.3.3. Positive ions in the stratosphere

In the stratosphere, ionization is produced by cosmic rays. Many of the reaction processes are similar to those occurring in the D-region. The rate of formation of NO^+, however, becomes negligible; the precursor ions are O_2^+ and N_2^+ (which immediately forms O_2^+ by charge exchange with O_2). The reaction chain leading to water vapor clusters is modified by the presence of certain stratospheric neutral species. The reaction of O_4^+ with O_3 must be mentioned:

$$O_4^+ + O_3 \rightarrow O_5^+ + O_2 \qquad (6.68)$$

and this process competes with reaction with H_2O, reaction (6.57). The reaction with atomic oxygen is negligible in the stratosphere due to its reduced abundances there. O_5^+ is almost as abundant as O_4^+ near 25 to 30 km, where ozone densities are greatest. O_5^+ reacts rapidly with H_2O:

$$O_5^+ + H_2O \rightarrow O_2^+ \cdot H_2O + O_3 \qquad (6.69)$$

and the ozone molecule destroyed by reaction (6.68) is thus reproduced. The

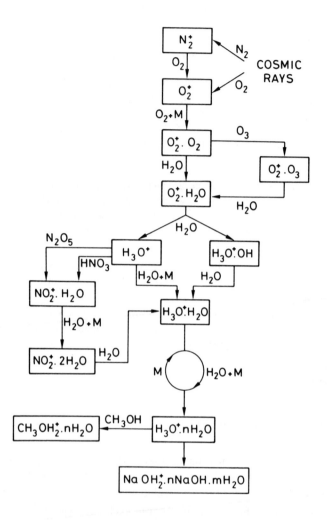

Fig. 6.16. Schematic of the aeronomic reactions of positive ions in the strato-sphere. From Ferguson (1979). (Copyright by Plenum).

formation chain leading to water cluster ions can then continue as in the D-region (see Fig. 6.16).

The presence of stratospheric species such as HNO_3 and N_2O_5 can alter the chain through reactions of the form:

$$H_3O^+ + N_2O_5 \rightarrow NO_2^+ \cdot H_2O + HNO_3 \tag{6.70}$$

$$H_3O^+ + HNO_3 \rightarrow NO_2^+ \cdot H_2O + H_2O \tag{6.71}$$

followed by

$$NO_2^+\cdot H_2O + H_2O + M \rightarrow NO_2^+\cdot(H_2O)_2 + M \qquad (6.72)$$

$$NO_2^+\cdot(H_2O)_2 + H_2O \rightarrow H_3O^+\cdot H_2O + HNO_3 \qquad (6.73)$$

These processes accelerate the conversion of the first hydrated water cluster ion to the second. It should be noted that reaction (6.70) converts N_2O_5 to HNO_3. Boehringer et al. (1983) have shown that this process is probably not fast enough to be important in stratospheric neutral chemistry, however.

The water cluster ions are destroyed by recombination with negative ions in the stratosphere, as are all positive ions. Other processes can also occur, such as the reaction with ammonia:

$$H_3O^+\cdot(H_2O)_n + NH_3 \rightarrow NH_4^+\cdot(H_2O)_m + (n-m+1)H_2O \qquad (6.74)$$

(where m=0,1,2,...n). This process is quite rapid, particularly for n=1,2, and 3 (Fehsenfeld and Ferguson, 1973). However, the stratospheric ammonia abundance is small due to its large solubility and removal in clouds. Reaction (6.74) may be important in the troposphere. The reaction with methanol:

$$H_3O^+\cdot(H_2O)_n + CH_3OH \rightarrow CH_3OH_2^+\cdot(H_2O)_m + (n-m+1)H_2O \qquad (6.75)$$

could play a role, but the vertical distribution of methanol is not well known. Ions of the form $CH_3OH_2^+\cdot(H_2O)_m$ have not been detected by in-situ measurements, implying that the stratospheric abundance of methanol must be small.

The first detection of stratospheric positive ions by rocket borne mass spectrometer (Arnold et al., 1977) showed that above 45 km the most abundant ions were proton hydrates $(H_3O^+\cdot(H_2O)_n)$ while below that altitude non-proton hydrates of masses 29± 2, 42± 2, 60± 2, and 80± 2 were dominant. In a balloon experiment designed to avoid the possible problem of ion fragmentation due to shock waves in front of rocket borne mass spectrometers, Arijs et al. (1978) found that in addition to proton hydrates and the non-proton hydrates mentioned above, ions with a mass number of 96 ± 2 were also observed.

Arnold et al. (1977) proposed that the non-proton hydrates were formed by reactions of proton hydrates with formaldehyde, which possesses a large proton affinity. Laboratory studies by Fehsenfeld et al. (1978) showed that the reaction of formaldehyde with higher order proton hydrates is endothermic, so that this interpretation was rejected. Ferguson (1978) suggested that proton hydrates might react rapidly with NaOH. This hypothesis was motivated in part by the observation of a sodium layer near 90 km due to meteor ablation, and supported by model calculations by Liu and Reid (1979). If this sodium could reach the stratosphere, it should be present there as NaOH, a species with a large proton affinity. Reactions of the type

$$H_3O^+\cdot(H_2O)_n + NaOH \rightarrow NaOH_2^+\cdot(H_2O)_m + (n-m+1)H_2O \qquad (6.76)$$

should therefore proceed rapidly. Potassium hydrates of the form KOH_2^+ would be expected to form in a similar fashion. Ions of the type $MgOH^+\cdot(H_2O)_m$ can

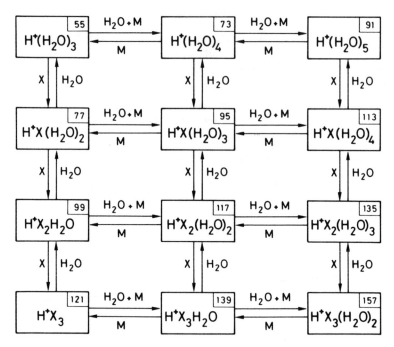

Fig. 6.17. Schematic diagram of stratospheric positive ion reactions proposed by Arnold (1980).

also be formed via

$$H_3O^+ \cdot (H_2O)_n + MgO \rightarrow MgOH^+ \cdot (H_2O)_m + (n-m+1)H_2O \qquad (6.77)$$

$MgOH^+$ will charge exchange with NaOH, however:

$$MgOH^+ + NaOH \rightarrow NaOH_2^+ + MgO \qquad (6.78)$$

so that although magnesium is about 10 times more abundant than sodium its role in ion composition is probably limited.

Other observations by Arnold et al. (1978) and Arijs et al. (1978; 1980) confirmed the presence of clusters of the form $H^+X_n(H_2O)_m$, where X is the unidentified molecule. The likely reaction scheme is presented in Figure 6.17. Arijs et al. (1980) unambiguously showed that the mass of X was equal to 41, implying that the identity of X could not be NaOH (mass 40). MgOH (mass 41) is also impossible because the isotopic ratio which would be expected to result from the ^{25}Mg and ^{26}Mg isotopes is not observed. Indeed, the failure to observe evidence of metal ion clusters in the stratosphere implies that these species are not present there in very large amounts, at least in the gaseous phase.

Theoretical and experimental observations, as well as laboratory studies by Boehringer and Arnold (1981) have indicated that the identity of X is almost certainly CH_3CN (mass 41). This species has been detected in the troposphere (where it is probably produced at least in part by forest fires) by Becker and Ionescu (1983), but the complete budget of its sources and sinks is at present poorly understood. Recent studies of the fractional ion abundances of proton hydrates and non-proton hydrates have been used to infer the profile of CH_3CN from about 15 to 45 km (Arnold et al., 1981; Henschen and Arnold, 1981; Arijs et al., 1982a; Arijs et al., 1983a,b). These profiles have been fitted to model calculations of the neutral chemistry and transport of CH_3CN (Brasseur et al., 1983).

Comprehensive reviews of the chemistry and observations of stratospheric positive ions are provided by Arnold (1980) and Arijs (1983).

6.4. Negative ion chemistry

6.4.1 Negative ions in the D-region

The first studies of the ionosphere using radiowave techniques demonstrated that electrons are nearly absent below 65 or 70 km during the day and 75 or 80 km at night. However, electrical neutrality requires a balance between charged particles. The observation of positive ions implies that equal quantities of a negatively charged particle must be present.

Johnson et al. (1958) performed the first observations of negative ions, and suggested that the most abundant negative ion in the D-region is NO_2^- at mass 46. Since that time, only a very few measurements have been performed. Narcisi et al. (1971) and Arnold et al. (1971) observed the D-region at night. The presence of a transition zone above which the concentration of negative ions rapidly decreases was established by observation but the altitude of this layer seemed to vary considerably with time (Arnold and Krankowsky, 1977; Arnold, 1980).

This transition zone between electrons and negative ions appears to be located near 70 km during the day, and 5 to 10 km higher at night. But the few available observations are not in agreement as to the composition of the D-region negative ions. Narcisi et al. suggested that the dominant ions were hydrated clusters, probably of the type $NO_3^- \cdot (H_2O)_n$, but Arnold et al. found that the most abundant species were CO_3^-, HCO_3^- and Cl^-. The chloride ion could be a result of contamination from the rocket, but the possibility of a natural source for this ion can not be ruled out. Figure 6.18 shows some vertical profiles suggested by Arnold et al. (1971). Further observations are badly needed.

The chemistry of negative ions has been intensively studied in the laboratory since the 1960's by Ferguson and coworkers. A schematic diagram of the

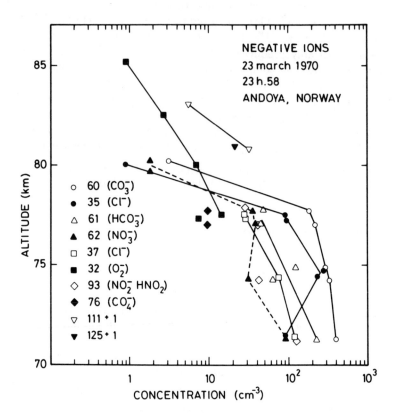

Fig. 6.18. Negative ions from 70 to 85 km. (Arnold et al., 1971). (Copyrighted by Pergamon Press).

suggested chemistry resulting from their experiments is presented in Figure 6.19. The chain of reactions is initiated by electron attachment on an oxygen molecule:

$$e + O_2 + M \rightarrow O_2^- + M \tag{6.79}$$

The rate of this reaction varies with the atmospheric density (M), leading to a nearly complete disappearance of free electrons below a certain altitude. The attached electron can be liberated, however, by several reactions:

$$O_2^- + O \rightarrow O_3 + e \tag{6.80}$$

$$O_2^- + O_2(^1\Delta_g) \rightarrow 2O_2 + e \tag{6.81}$$

These reactions lead to a balance between negative ions and electrons. The increasing densities of atomic oxygen in the upper mesosphere result in increased electron abundances there through reaction (6.80). The chain leading

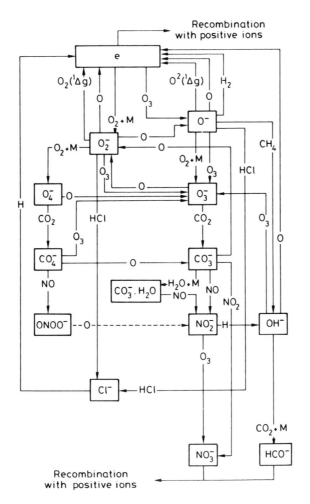

Fig. 6.19. Schematic diagram of the negative ion chemistry of the D-region. From Ferguson (1979). (Copyrighted by Plenum).

to stable negative ions occurs through the following sequence:

$$O_2^- + O_3 \rightarrow O_3^- + O_2 \qquad (6.82)$$

$$O_3^- + CO_2 \rightarrow CO_3^- + O_2 \qquad (6.83)$$

The CO_3^- ion can also be formed by a second sequence:

$$O_2^- + O_2 + M \rightarrow O_4^- + M \qquad (6.84)$$

$$O_4^- + CO_2 \rightarrow CO_4^- + O_2 \tag{6.85}$$

$$CO_4^- + O \rightarrow CO_3^- + O_2 \tag{6.86}$$

This last reaction competes with the following processes:

$$CO_4^- + O_3 \rightarrow O_3^- + CO_2 + O_2 \tag{6.87}$$

$$O_3^- + CO_2 \rightarrow CO_3^- + O_2 \tag{6.88}$$

The CO_3^- ion is rapidly destroyed by reaction with atomic oxygen

$$CO_3^- + O \rightarrow O_2^- + CO_2 \tag{6.89}$$

which interrupts the formation of negative ions. However, the sequence above can also be followed by:

$$CO_3^- + NO \rightarrow NO_2^- + CO_2 \tag{6.90}$$

$$NO_2^- + O_3 \rightarrow NO_3^- + O_2 \tag{6.91}$$

which leads to the formation of NO_3^-. This species is particularly stable because of its very high electron affinity. Reaction (6.90) is relatively slow and controls the rate of formation of the terminal NO_3^- ion. Reaction (6.91) is fast, and thus the NO_2^- ion can only be present in small quantities.

The OH^- ion is produced by the reaction

$$NO_2^- + H \rightarrow OH^- + NO \tag{6.92}$$

and leads to the eventual formation of HCO_3^-, another terminal ion, through

$$OH^- + CO_2 + M \rightarrow HCO_3^- + M \tag{6.93}$$

if it is not destroyed by these reactions:

$$OH^- + O \rightarrow HO_2 + e \tag{6.94}$$

$$OH^- + H \rightarrow H_2O + e \tag{6.95}$$

Figure 6.19 shows that Cl^- can also be formed as follows:

$$O_2^- + HCl \rightarrow Cl^- + HO_2 \tag{6.96}$$

$$O^- + HCl \rightarrow Cl^- + OH \tag{6.97}$$

$$NO_2^- + HCl \rightarrow Cl^- + HNO_2 \tag{6.98}$$

Cl^- is destroyed by the chemical detachment process:

$$Cl^- + H \rightarrow HCl + e \tag{6.99}$$

Figure 6.19 does not indicate any formation paths for ions of the type

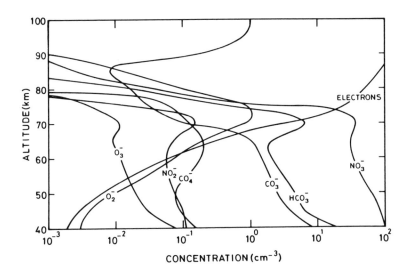

Fig. 6.20. Vertical distributions of the principal negative ions from 40 to 100 km as calculated by a mathematical model. From Brasseur (1982).

$X^- \cdot (H_2O)_n$. The formation of such ions by reaction with water vapor should be possible but would not greatly change the schematic picture just described (Fehsenfeld and Ferguson, 1974). For example, $CO_3^- \cdot H_2O$ would react with NO to accelerate the conversion of CO_3^- to NO_3^-. It should be noted that, as in the case of NO^+, the formation of clusters ions with O_2, N_2, CO_2 and H_2O could play an important part in producing ions of the form $X^- \cdot (H_2O)_n$, but these reactions have not yet been studied in the laboratory.

The eventual destruction of negative ions, particularly the terminal ions, occurs by recombination with positive ions. The rate of this reaction has not been measured for all of the species which are actually present in the atmosphere, but a mean rate for reactions of this type, i. e.

$$(\alpha); \quad X^- + Y^+ \rightarrow \text{neutrals}$$

of 6×10^{-8} cm^3 s^{-1} can be adopted (Smith et al., 1976). Note, however, that this reaction is also believed to exhibit a pressure dependence which may be important in the lower stratosphere (Smith et al., 1981). Figure 6.20 shows an example of the vertical distributions of negative ions calculated in a one dimensional model.

6.4.2 Negative ions in the stratosphere

In order to describe the behavior of negative ions in the stratosphere, it is useful to adopt the D-region scheme just described and consider the changes which occur as a result of differences in pressure, temperature, and composition. The chain of reactions leading to NO_3^- from O_2^- can still occur, and is even more efficient because the low densities of atomic oxygen there decrease the frequency of the interrupting reactions. Further, additional reactions facilitate the formation of NO_3^-:

$$CO_3^- + NO_2 \rightarrow NO_3^- + CO_2 \tag{6.100}$$

This reaction is 20 times faster than 6.90 and becomes dominant in the stratosphere, where the densities of NO and NO_2 are comparable. This mechanism also avoids the chain breaking reactions presented in § 6.4.1. Further, the rapid reactions of O^-, O_2^-, CO_3^-, NO_2^- and Cl^- with HNO_3, and the reaction of the latter three species with N_2O_5, lead to additional production of NO_3^-. The formation of hydrated ions, for example by reaction with O_3^-, can reduce the formation rate of NO_3^-, however. Figure 6.21 presents a schematic diagram of these processes.

The possibility of destruction of NO_3^- must also be considered. The hydration of this ion must proceed as in the D-region. Kinetic studies (Fehsenfeld et al. 1975) showed that HNO_3 can substitute for H_2O in a cluster ion. It is

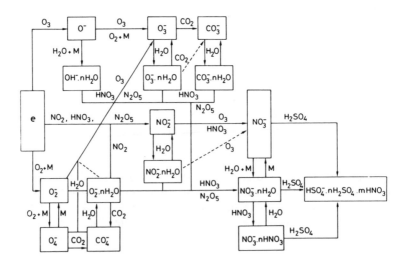

Fig. 6.21. Schematic diagram of the negative ion chemistry in the stratosphere. From Ferguson (1979). (Copyright by Plenum).

thus reasonable that ions of the form $NO_3^- \cdot (H_2O)_n (HNO_3)_m$ should be formed. $NO_3^- \cdot (HNO_3)_n$ reacts rapidly, however, with H_2SO_4 (Viggiano et al., 1980), at least for n=0,1, or 2. Thus HSO_4^- or a cluster of the form $HSO_4^- \cdot (H_2SO_4)_m \cdot (HNO_3)_n \cdot (H_2O)_p$ should be the terminal ion in the stratosphere. Figure 6.22 presents a schematic diagram of the corresponding reactions proposed by Arijs et al. (1981). The first observations by Arnold and Henschen (1978), and later by Arijs et al. (1981) confirmed the presence of some of these cluster ions; the most abundant species observed at 35 km were $HSO_4^- \cdot (H_2SO_4)_3$, $HSO_4^- \cdot (HNO_3)$, $NO_3^- \cdot (HNO_3)_2$, and $HSO_4^- \cdot (H_2SO_4)_2$. Studies of

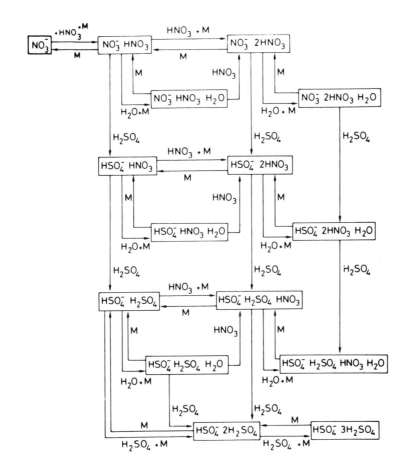

Fig. 6.22. Schematic diagram of the negative ion chemistry of the stratosphere suggested by Arijs et al. (1981) based on in-situ observations. (Copyrighted by the American Geophysical Union).

the fractional abundance of these ions can be used to infer the vertical profile of neutral H_2SO_4, as discussed briefly in Chapter 5.

Other polar molecules are also likely to be important. Investigations of minor mass peaks in negative ion spectra (McCrumb and Arnold, 1981; Arijs et al., 1982) have revealed the presence of ions which may contain other ligands in addition to HNO_3 and H_2SO_4, such as HOCl and HCl.

A model of the positive and negative ion distributions in the stratosphere was presented by Brasseur and Chatel (1983). Theoretical studies of both positive and negative ions are hampered by the lack of laboratory data regarding some of the relevant rate constants.

In conclusion, the chemistry of negative ions in both the stratosphere and mesosphere remains at present in a primitive state. Systematic observations as a function of altitude are needed, as well as laboratory kinetic studies, before a deeper understanding can be acquired.

6.5. Effect of ionic processes on neutral constituents

Reactions between ions and neutral species, can, in principle, affect the budgets of neutral species as mentioned in Chapter 5.

In the stratosphere, the present understanding of the ion composition and chemistry is not adequate to completely describe the possible interactions. The ion production by cosmic rays appears, however, to be too small to modify the neutral composition, except near the tropopause, where the ionization and decomposition of N_2 leads to a small production of nitrogen oxides (Warneck, 1972; Nicolet and Peetermans, 1972; Brasseur and Nicolet, 1973).

During intense solar proton events, the number of ion pairs produced can increase greatly even at relatively low altitudes, perturbing the neutral atmosphere and leading to a considerable increase in the rate of production of nitrogen and hydrogen oxides (see, e. g. Crutzen et al., 1975; Jackman et al., 1980; Swider and Keneshea, 1973). This will be discussed in more detail in Chapter 7.

In principle, ions can provide a loss mechanism for species formed at ground level which are long lived in the lower atmosphere. For example, ion reactions with chlorofluorocarbons which would lead to a lifetime less than a hundred years could be important. However, the most abundant ions, $H_3O^+ \cdot (H_2O)_n$ and $NH_4^+ \cdot (H_2O)_n$ do not react with CF_2Cl_2 and $CFCl_3$.

Only the reaction

$$H_3O^+ + CFCl_3 \rightarrow CCl_3^+ + HF + H_2O \qquad (6.101)$$

proceeds rapidly, but H_3O^+ is not present in large enough amounts to be important.

In the thermosphere, the effect of the ions in producing neutral nitrogen oxides is very important. In addition to photolysis,

$$J_{N_2}; \quad N_2 + h\nu(80 < \lambda < 125\text{nm}) \to 2N(^2D \text{ or } ^4S)[F_o] \quad (6.102)$$

(where F_o represents the fraction of nitrogen atoms produced in the ^2D state) ionic reactions can lead to the production of nitrogen atoms. The reaction of N_2^+ with atomic oxygen is one such source

$$(\gamma_3); \quad N_2^+ + O \to NO^+ + N(^2D \text{ or } ^4S)[F_1] \quad (6.103)$$

as well as recombination:

$$(\alpha_{N_2^+}); \quad N_2^+ + e \to 2N(^2D \text{ or } ^4S)[F_2] \quad (6.104)$$

The recombination of NO^+

$$(\alpha_{NO^+}); \quad NO^+ + e \to N(^2D;^4S) + O[F_3] \quad (6.105)$$

and the following reactions must also be considered:

$$(\gamma_1); \quad O^+ + N_2 \to NO^+ + N(^4S) \quad (6.106)$$

$$(\gamma_{12}); \quad N^+ + O_2 \to O_2^+ + N(^2D;^4S)[F_4] \quad (6.107)$$

$$(\gamma_{12}^+); \quad N^+ + O_2 \to NO^+ + O \quad (6.108)$$

Fast secondary electrons produced by cosmic and solar particles can dissociate or dissociatively ionize the nitrogen molecule:

$$N_2 + e^* \to 2N(^2D;^4S) + e[F_5] \quad (6.109)$$

$$N_2 + e^* \to N(^2D;^4S) + N^+ + e[F_6] \quad (6.110)$$

The fractional production of the nitrogen atoms in each electronic state is not yet known with accuracy. This parameter represents an extremely important factor in determining the density of nitric oxide in the E and D regions, as discussed below (see also, Norton and Barth, 1971; Rees and Roble, 1978; Solomon, 1983).

Almost all of the atoms which are produced in the ^2D state react with O_2 in the middle atmosphere, producing nitric oxide:

$$(b_7^*); \quad N(^2D) + O_2 \to NO + O \quad (6.111)$$

Deactivation by atomic oxygen and electrons play only a minor role below 100 km:

$$N(^2D) + O \to N(^4S) + O \quad (6.112)$$

$$N(^2D) + e \to N(^4S) + e \quad (6.113)$$

On the other hand, the reaction of the ground state atom with molecular oxygen is relatively slow:

$$(b_7); \quad N(^4S) + O_2 \rightarrow NO + O \tag{6.114}$$

so that almost all of these nitrogen atoms are destroyed in the lower thermosphere by the rapid reaction with NO:

$$(b_6); N(^4S) + NO \rightarrow N_2 + O \tag{6.115}$$

Therefore, when dissociation of molecular nitrogen yields one $N(^4S)$ and one $N(^2D)$ atom in the lower thermosphere, the net odd nitrogen production is extremely small: almost every $N(^2D)$ atom produces one NO molecule, but almost every $N(^4S)$ atom immediately destroys one at these altitudes. Net production is provided only by the very small fraction of $N(^4S)$ atoms which react with oxygen rather than with NO. Even very small deviations from a 50%-50% branching ratio can have large effects on lower thermospheric NO densities, making this parameter an extremely important one in NO_x chemistry. The only laboratory study concerning the state of the product atoms produced in these reactions is that of Kley et al. (1977), who found that $F_3 = 0.76$. Laboratory studies by Zipf and McLaughlin (1978) suggest that $F_5 = 0.5$. To model the

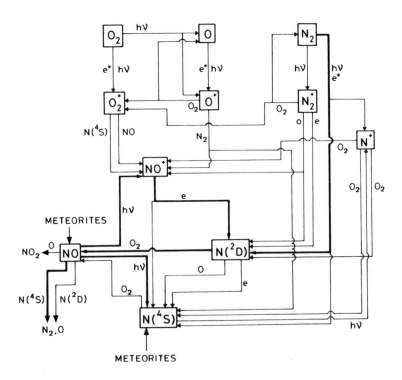

Fig. 6.23. Chemistry of nitrogen compounds in the thermosphere and the interactions between neutrals and ions. From Massie (1979).

effects of these processes, indirect estimates based on airglow studies are often adopted. Rusch et al. (1981) adopted, for example, $F_1 = 1$, $F_2 = 0.5$, and $F_4 = 0$. Massie (1979) assumed $F_1 = 1$, $F_2 = 1$, $F_4 = 1$, and $F_5 = 0.75$. Figure 6.23 shows the chemistry of nitrogen compounds in the thermosphere. We can now evaluate the rates of production P_{NO}, $P_{N(^4S)}$ and $P_{N(^2D)}$ which were referred to in equations (5.161) and (5.162).

$$P_{N(^2D)} = 2F_o J_{N_2}(N_2) + F_1 \gamma_3 (N_2^+)(O) + 2\alpha_{N_2^+}(N_2^+)n_e$$

$$+ F_3 \alpha_{NO^+}(NO^+)n_e + 2P_D F_5 + P_I F_6 \qquad (6.116)$$

$$P_{N(^4S)} = (1-F_o)2J_{N_2}(N_2) + (1-F_1)\gamma_3(N_2^+)(O) + 2\alpha_{N_2^+}(1-F_2)(N_2^+)n_e$$

$$+ (1-F_3)\alpha_{NO^+}(NO^+)n_e + \gamma_{12}(N^+)(O_2) + \gamma_1(O^+)(N_2)$$

$$+ 2P_D(1-F_5) + P_I(1-F_6) \qquad (6.117)$$

$$P_{NO} = 2b_{39}(N_2O)(O^1D) + P_M(NO) + \gamma_1(O^+)(N_2) \qquad (6.118)$$

where P_D is the rate of production of nitrogen atoms due to dissociation of N_2 by energetic electrons (6.109), P_I is the production due to dissociative ionization (6.110), and P_M is the rate of odd nitrogen production associated with meteors. The rate of production of the nitrogen atom by dissociative ionization is equal to the rate of production of N^+ by the same process (see eqn. 6.40b):

$$P_I = P_{N^+} = 0.185Q \qquad (6.119)$$

where Q is the rate of ion pair production. Note that the N^+ and NO^+ ions formed by these reactions will ultimately yield reactive nitrogen, either by reaction (6.107) or by reaction (6.108) followed by (6.104). The relative values of the cross section for dissociation (Winters, 1966; Niehaus, 1967) and dissociative ionization (Rapp et al., 1965) imply a rate for P_D of about $0.85Q$ (Rusch et al., 1981).

The O^+ ion can also produce nitrogen atoms by reaction (6.106) with nitrogen and dissociative recombination of NO^+. This contribution is only about $0.03\ Q$ in the middle atmosphere, however. Each ion pair produced during particle precipitation in the atmosphere (cosmic rays, magnetospheric electrons, and solar proton events) results in a total gross odd nitrogen production of about $1.25\ Q$ (see, e.g., Jackman et al., 1979; Rusch et al., 1981). The net production depends on the fractional production of N^4S and N^2D atoms.

Below the mesopause, where water cluster ions are present, the ions can contribute to the formation of hydrogen free radicals (see Fig. 6.24). Referring to the reactions in §6.3.2, we note that the $O_2^+ \cdot H_2O$ ion can lead to the

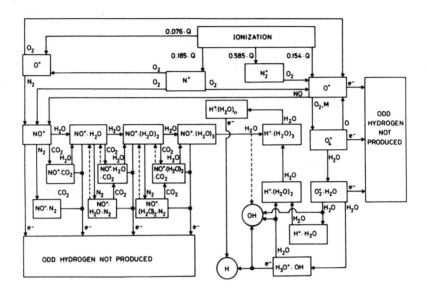

Fig. 6.24. Schematic diagram of the ion chemistry leading to odd hydrogen production. From Solomon et al. (1981). (Copyright by Pergamon Press).

following chain, producing one H and one OH radical:

$$O_2^+ \cdot H_2O + H_2O \rightarrow H_3O^+ \cdot OH + O_2$$

$$H_3O^+ \cdot OH + e \rightarrow H + OH + H_2O$$

$$\boxed{\text{Net:} \ H_2O \rightarrow H + OH}$$

or

$$O_2^+ \cdot H_2O + H_2O \rightarrow H_3O^+ \cdot OH + O_2$$

$$H_3O^+ \cdot OH + H_2O \rightarrow H_3O^+ \cdot H_2O + OH$$

$$H_3O^+ \cdot H_2O + nH_2O \rightarrow H_3O^+ \cdot (H_2O)_{n+1}$$

$$H_3O^+ \cdot (H_2O)_{n+1} + e \rightarrow H + (n+2)H_2O$$

$$\boxed{\text{Net:} \ H_2O \rightarrow H + OH}$$

Other secondary chains must be considered. The processes and rates of formation of hydrogen radical are described by Heaps (1978b) and Solomon et al.

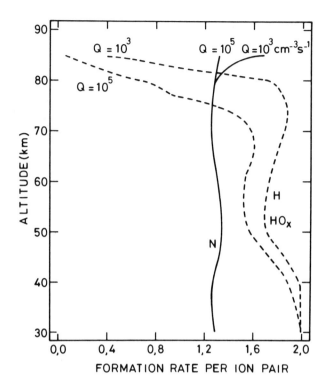

Fig. 6.25. Rate of production of nitrogen atoms and odd hydrogen radicals per ion pair. From Heaps (1978b).

(1981). Below about 75 km, nearly two odd hydrogen particles are formed per O_2^+ ionization; above that altitude, odd hydrogen production depends on the ionization rate and the atomic oxygen and water vapor densities. Figure 6.24 presents a schematic diagram of the positive ion chemistry responsible for odd hydrogen production in the D-region. Note that odd hydrogen production is much more efficient if the primary ion is O_2^+ than if the primary ion is NO^+ (i. e., during particle precipitation as opposed to quiet times). Figure 6.25 presents a calculated vertical distribution of the rates of production of HO_x and N per ion pair for two values of Q.

6.6 Radio waves in the lower ionosphere

The study of the ionosphere has largely been performed by observation of the properties of radio wave propagation through this medium. In this work we will not attempt to present an exhaustive study of the interaction of electromagnetic waves and a weakly ionized plasma, nor will we review the many techniques used for such studies. The object of this section is to indicate the basic principles which underlie studies of this sort and to present an elementary description of magneto-ionic theory (Appleton, 1927; Hartree, 1931), and finally to show how this approach yields information about the properties of the D-region, particularly in winter.

The conditions of propagation of an electromagnetic wave in the ionospheric medium are determined by the Maxwell and Lorentz equations. These equations determine the properties of the electric (E) and magnetic (H) fields, as well as the displacement (D) and the induction (B) as a function of the electric charge and current densities (J) (see, for example, Budden, 1961; Davies, 1965):

$$\vec{\nabla} \cdot \vec{D} = \rho \tag{6.120}$$

$$\vec{\nabla} \cdot \vec{B} = 0 \tag{6.121}$$

$$\vec{\nabla} \times \vec{H} = J + \frac{\partial \vec{D}}{\partial t} \tag{6.122}$$

$$\vec{\nabla} \times \vec{E} = -\frac{\partial \vec{B}}{\partial t} \tag{6.123}$$

with

$$\vec{D} = \epsilon_o \vec{E} + \vec{P} \tag{6.124}$$

$$\vec{B} = \mu_o \vec{H} \tag{6.125}$$

where $\vec{P}$ is the polarization vector, ϵ_o is the permittivity of free space, ρ is the charge density, and μ_o is the magnetic permeability. The Lorentz equation describes the flow of charge as a function of the electric and induction fields. For electrons, one may write

$$m_e \frac{d\vec{V}}{dt} = -e(\vec{E} + \vec{V} \times \vec{B}) - m_e \nu_c \vec{V} \tag{6.126}$$

where m_e is the mass of the electron, $\vec{V}$ is its speed, -e is the electrical charge and ν_c is the collision frequency between electrons and neutrals.

If one assumes that a plane transverse wave

$$E = E_o \exp i(\omega_r t - kx) \tag{6.127}$$

of amplitude E_o and frequency ω_r propagates as a function of time t in the x direction with a propagation factor $k = 2\pi/\lambda$, λ being the actual wavelength in the medium, then it can be shown (after some mathematical manipulation), that the index of refraction of the ionized medium is given by the Appleton formula:

$$n^2 = 1 - \cfrac{X}{1 - iZ - \cfrac{Y_T}{2(1-X-iZ)} \pm \left[\cfrac{Y_T^4}{4(1-X-iZ)^2} + Y_L^2 \right]^{1/2}} \qquad (6.128)$$

where $X = n_e e^2 / \epsilon_0 m_e \omega_r^2$, (a dimensionless parameter proportional to the electron concentration n_e), $Y_L = eB_L / m_e \omega_r$, and $Y_T = eB_T / m_e \omega_r$ represent the effects of the longitudinal (L) and transverse (T) magnetic field components and $Z = \nu_c / \omega_r$ is a dimensionless parameter describing the effect of electron-neutral collisions. This formula shows that the refractive index is a complex number when the collision frequency cannot be neglected compared to the frequency of the wave. The real part of the refractive index describes the refracting and dispersive properties of the medium and the imaginary part its absorption properties. The Appleton formula also shows that if Y_L or Y_T are different from zero, two values can be assigned to n^2, indicating that the presence of a magnetic field introduces anisotropy into the medium, so that two propagation modes exist for each frequency (ordinary and extraordinary wave); this is a phenomenon similar to the optical birefringence of certain crystals.

The most simple form of the Appleton equation is obtained by neglecting the effects of the magnetic field and collisions. This equation is adequate to describe the propagation of high frequency waves (above 30 MHz) in the ionosphere, and can be used to obtain an approximate understanding of the propagation characteristics of lower frequency waves. In this case, one obtains

$$n^2 = 1 - \frac{K n_e}{f^2} \qquad (6.129)$$

where K is a constant which is equal to 80.5 if the wave frequency f is expressed in Hertz. Coupling this expression with Snell's law (which describes the propagation of the electromagnetic radiation), the level of reflection of the wave can be derived. Further, by tuning the wave frequency of the transmitter, information on the vertical profile of electron density can be deduced.

In the D-region, collisions cannot be neglected except for very high frequency propagation. Neglecting the effect of the magnetic field, the refractive index then becomes:

$$n^2 = 1 - \frac{X}{1 - iZ} = 1 - \frac{X}{1 + Z^2} - \frac{iXZ}{1 + Z^2} \qquad (6.130)$$

It follows that, to this order of approximation, a wave traveling through an ionized medium with negligible refraction is progressively attenuated with an absorption coefficient given by

$$K = \frac{e^2}{2\epsilon_0 m_e c} \frac{n_e \nu_c}{\omega_r^2 + \nu_c^2} \qquad (6.131)$$

which is often expressed in units of nepers per meter (a neper corresponds to 8.69 dB). The Appleton-Hartree theory used to obtain this expression employs an invariant collision frequency for all electrons. However, as pointed out by Phelps and Pack (1959), the cold plasma approximation is not valid in the D-region and the collision frequency should vary with the velocity of the electrons. Sen and Wyller (1960) generalized the Appleton-Hartree theory by assuming a realistic electron energy distribution. In practical applications, however, the above expressions can still be used, provided that ν_c is replaced by an effective collision frequency ν_{eff}, such that the ratio ν_{eff}/ν_c, varies with the radio wave frequency.

Study of the amplitude of waves in the MHz range, reflected in the E and F regions, gives a measure of D-region absorption. Techniques involving vertical reflection of pulses (called the A1 method) or oblique reflection of continuous waves (A3 method) provide useful indications of some properties of the D-region, particularly the change in electron concentration with time at a give point for a layer where $\nu_c \approx \omega_r$. Such systematic observations of absorption have established the nature of variations which are related to ionization

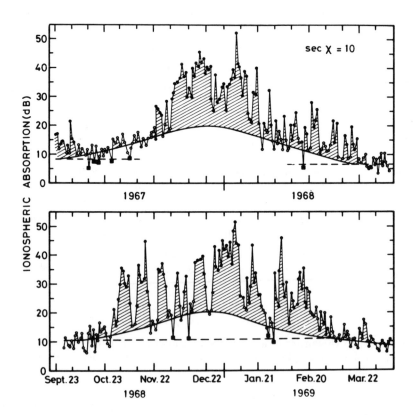

Fig. 6.26. Time series of D-region absorption during summer and winter at mid-latitude. From Schwentek (1971). (Copyright by Pergamon Press).

processes. For example, a variation with zenith angle is found, of the form:

$$L = C \cos^a \chi \qquad (6.132)$$

where a is typically between 0.6 and 1.0 at mid-latitudes, and C is a constant which varies with season, geomagnetic activity, etc. A variation with solar activity has also been noted (see, e. g., Appleton and Piggott, 1954). It has been found that while the absorption varies systematically with the mean zenith angle in summer, abnormally high absorption is observed in winter. In the absence of changes in the atmospheric composition, one would expect to obtain lower absorption in winter than in summer because of the larger wintertime solar zenith angle (this is sometimes referred to as "smooth winter anomaly"). In addition, unexpected intense absorption events lasting a few days are often observed in winter (see Fig. 6.26). This effect was observed by Appleton as early as 1937, and is often referred to as "sporadic winter anomaly". It has been the subject of considerable research in recent years. The phenomenon is clearly related to an increase in electron concentration (Mechtly and Shirke, 1968) and NO^+ ion density (Zbinden et al., 1975). Under these conditions, the level at which the water cluster ion densities equal that of the molecular ions (NO^+ and O_2^+) is typically 5 km below normal.

The observations obtained during a measurement campaign at El Arenosillo (37N) (European winter anomaly campaign 1975/6, see Offermann, 1979) as well as model studies (e. g. Koshelev, 1979; Solomon et al., 1982b) have indicated that the origin of the winter anomaly probably lies in sporadic injection of large amounts of nitric oxide from the thermosphere to the mesosphere in association with changes in atmospheric dynamics. D-region models which include the effects of both chemistry and dynamics (Solomon et al., 1982a,b; Brasseur and De Baets, 1984) show that ionization is more intense in winter than in summer, even under normal conditions, as a result of the greater NO densities in the winter season. The observed changes are consistent with the observations of a "smooth winter anomaly" underlying the sporadic enhancements. Measurements of NO densities during sporadic winter anomaly events (Beran and Bangert, 1979; Beynon et al., 1976; Arnold and Krankowsky, 1979; Aikin et al., 1977) indicate that the NO concentration is greatly elevated during these conditions. The sporadic winter anomaly may be related to atmospheric heating, which decreases the positive ion-electron recombination coefficient because the concentrations of the cluster ions decrease while those of the molecular ions increase (Danilov and Taubenheim, 1983). It has also been noted that the NO enhancement and winter anomalous absorption may be related to transport processes occurring in planetary waves (Labitzke et al., 1979; Kawahira, 1982).

References

Aikin, A.C., R.A. Goldberg, W. Jones and J.A. Kane, Observations of the mid-latitude lower ionosphere in Winter, J. Geophys. Res., 82, 1869, 1977.

Appleton, E.V., URSI Proc. Washington, 1927.

Appleton, E. and W.R. Piggott, Ionospheric absorption measurements during a sunspot cycle, J. Atm. Terr. Phys., 8, 141, 1954.

Arijs, E., J. Ingels and D. Nevejans, Mass spectrometric measurement of positive ion composition in the stratosphere, Nature, 271, 642, 1978.

Arijs, E., D. Nevejans and J. Ingels, Unambiguous mass determination of major stratospheric positive ions, Nature, 288, 684, 1980.

Arijs, E., D. Nevejans, P. Frederick and J. Ingels, Negative ion composition measurements in the stratosphere, Geophys. Res. Lett., 8, 1, 121-124, 1981.

Arijs, E., D. Nevejans and J. Ingels, Stratospheric positive ion composition measurements, ion abundances and related trace gas detection, J. Atmos. Terr. Phys., 44, 43, 1982a.

Arijs, E., D. Nevejans, P. Frederick and J. Ingels, Stratospheric negative ion composition measurements, ion abundances and related trace gas detection, J. Atm. Terr. Phys., 44, 681, 1982b.

Arijs, E., Positive and negative ions in the stratosphere, Annales Geophysicae, 1, 149, 1983.

Arijs, E., D. Nevejans and J. Ingels, Positive ion composition measurements and acetonitrile in the upper stratosphere, Nature, 303, 314, 1983.

Arijs, E., D. Nevejans, J. Ingels and P. Frederick, Positive ion composition measurements between 33 and 20km altitude, Annales Geophysicae, 2, 161, 1983.

Arnold, F., J. Kissel, H. Wieder and J. Zhringer, Negative ions in the lower ionosphere : a mass spectrometric measurement, J. Atmos. Terr. Phys., 33, 1669, 1971.

Arnold, F. and D. Krankowsky, Ion composition and electron - and ion - loss processes in the earth's atmosphere, pp. 93-127, in : Grandal, B. and Holtet, J.A., (eds.), Dynamical and chemical coupling between the neutral and ionized atmosphere, D. Reidel Publishing Company, Dordrecht - Holland, 1977.

Arnold, F., D. Krankowsky and K.H. Marien, First mass spectrometric measurements of positive ions in the stratosphere, Nature, 267, 30, 1977.

Arnold, F., H. Boehringer and G. Henschen, Composition measurements of stratospheric positive ions, Geophys. Res. Lett., 5, 653, 1978.

Arnold, F., H. Boehringer and G. Henschen, Composition measurements of stratospheric positive ions, Geophys. Res. Lett., 5, 653, 1978.

Arnold, F. and G. Henschen, First mass analysis of stratospheric negative ions, Nature, 275, 521, 1978.

Arnold, F. and D. Krankowsky, Mid-latitude lower ionosphere structure and composition measurements during winter, J. Atmos. Terr. Phys., 41, 1127, 1979.

Arnold, F., The middle atmosphere ionized component, Proceedings of the ESA-symposium on rocket - and balloon - programmes held at Bourne-mouth (1980).

Arnold, F., G. Henschen and E.E. Ferguson, Mass spectrometric measurements of fractional ion abundances in the stratosphere. Positive ions, Planet. Space Sci., 29, 185, 1981.

Banks, P., and G. Kockarts, *Aeronomy*, (Academic Press, New York), 1973.

Becker, K.H. and A. Ionescu, Acetonitrile in the lower troposphere, Geophys. Res. Lett., 9, 1349, 1982.

Beran, D. and W. Bangert, Trace constituents in the mesosphere and lower thermosphere during winter anomaly events, J. Atmos. Terr. Phys., 41, 1091, 1979.

Beynon, W. J. G., E. R. Williams, F. Arnold, D. Krankowsky, W. C. Bain and P. H. G. Dickinson, D-region rocket measurements in winter anomaly conditions, Nature, 261, 118, 1976.

Bjorn, L. G., and F. Arnold, Mass spectrometric detection of precondensation nuclei at the arctic summer mesopause, Geophys. Res. Lett., 8, 1167, 1981.

Boehringer, H. and F. Arnold, Acetonitrile in the stratosphere, implicatiions from laboratory studies, Nature, 290, 321, 1981.

Boehringer, H., D. W. Fahey, F. C. Fehsenfeld, and E. E. Ferguson, The role of ion-molecule reactions in the conversion of N_2O_5 to HNO_3 in the strato-sphere, Planet. Space Sci., 31, 185, 1983.

Brasseur, G. and M. Nicolet, Chemospheric processes of nitric oxide in the mesosphere and stratosphere, Planet. Space Sci., 21, 939, 1973.

Brasseur, G., *Physique et chimie de l'atmosphere moyenne*, (Masson, Paris), 1982.

Brasseur, G., E. Arijs, A. De Rudder, D. Nevejans and J. Ingels, Acetonitrile in the atmosphere, Geophys. Res. Lett., 10, 725 1983.

Brasseur, G. and A. Chatel, Modelling of stratospheric ions : a first attempt, Annales Geophysicae, 1, 173, 1983.

Brasseur, G. and P. De Baets, Minor constituents in the mesosphere and lower thermosphere, to be submitted to J. Geophys. Res., 1984.

Budden, K.G., *Radio waves in the ionosphere*, (Cambridge Univ. Press), 1961.

Crutzen, P. J., I. S. A. Isaksen and G. C. Reid, Solar proton events: stratos-pheric sources of nitric oxide, Science, 189, 457, 1975.

Danilov, A.D. and J. Taubenheim, NO and temperature control of the D-region, Space Science Reviews, 34, 413, 1983.

Davies, K., Ionospheric radio propagation, National Bureau of Standard, Mono-graph 80, 1965 and Dover Publications, Inc., New York, N.Y., 1966.

Dubach, J. and W.A. Barker, Charged particle induced ionization rates in planetary atmospheres, J. Atmos. Terr. Phys., 33, 1287, 1971.

Fehsenfeld, F.C. and E.E. Ferguson, Laboratory studies of negative ion reactions with atmospheric trace constituent, J. Chem. Phys., 61, 3181, 1974.

Fehsenfeld, F.C., C.J. Howard, and A.L. Schmeltekopf, Gas phase ion chemistry of HNO_3, J. Chem. Phys., 63, 2835, 1975.

Fehsenfeld, D., F.C.Dotan, D. Albritton, C. Howard and E. Ferguson, Stratospheric positive ion chemistry of formaldehyde and methanol, J. Geophys. Res., 83, 1333, 1978.

Fehsenfeld, F.C. and E.E. Ferguson, Thermal energy positive ion reactions in a wet atmosphere containing ammonia, J. Chem. Phys., 59, 6272, 1983.

Ferguson, E. E., D-region ion chemistry, Rev. Geophys. Space Phys., 9, 997, 1971.

Ferguson, E., Sodium hydroxide in the stratosphere, Geophys. Res. Lett., 5, 1035, 1978.

Ferguson, E.E., Ion-molecule reactions in the atmosphere, pp. 377- 403, in : Ausloos, P. (ed.) *Kinetics of ion - molecule reactions,* Plenum Press, (New York), 1979.

Ferguson, E.E., D.W. Fahey, F.C. Fehsenfeld and D.L. Albritton, Silicon ion chemistry in the ionosphere, Planet. Space Sci., 29, 307, 1981a.

Ferguson, E.E., B.R. Rowe, D.W. Fahey and F.C. Fehsenfeld, Magnesium ion chemistry in the stratosphere, Planet. Space Sci., 29, 479, 1981b.

Goldberg, R.A. and A.C. Aikin, Comet Encke : Meteor metallic ion identification by mass spectrometer, Science, 180, 294, 1973.

Hall, L.A. and H.E. Hinteregger, Solar radiation in the extreme ultraviolet and its variation with solar rotation, J. Geophys. Res., 75, 6959, 1970.

Hartree, D.R., The propagation of electro-magnetic waves in a refracting medium in a magnetic field, Proc. Cambridge Phil. Soc., 27, 143, 1931.

Heaps, M. G., A parameterization of cosmic ray ionization, Planet. Space Sci., 26, 513, 1978a.

Heaps, M.G., U.S. Army Atmospheric Sci. Lab. Report ASL-TR-0012, 1978b.

Henschen, G. and F. Arnold, Extended positive ion composition measurements in the stratosphere. Implication for neutral trace gases, Geophys. Res. Lett., 8, 999, 1981.

Hunten, D.M. and M.B. McElroy, Metastable $O_2(^1\Delta_g)$ as a major source of ions in the D-region, J. Geophys. Res., 73, 2421, 1968.

Jackman, C. H., H. S. Porter and J. E. Frederick, Upper limits on production rate of NO per ion pair, Nature, 280, 170, 1979.

Jackman, C. H., J. E. Frederick and R. S. Stolarski, Production of odd nitrogen in the stratosphere and mesosphere: an intercomparison of source strengths, J. Geophys. Res., 85, 7495, 1980.

Jegou, J. P., C. Granier, M. L. Chanin and G. Megie, General theory of the alkali metals present in the Earth's upper atmosphere, in press, Annales

Geophysicae, 1984.

Johnson, C.Y., E.B. Meadows and J.C. Holmes, Ion composition of the arctic ionosphere, J. Geophys. Res., 63, 443, 1958.

Kawahira, K., An observational study of the D-region winter anomaly and sudden stratospheric warmings, J. Atm. Terr. Phys., 44, 947, 1982.

Keneshea, T.J., R.S. Narcisi, and W. Swider, Diurnal model of the E region, J. Geophys., 75, 845, 1970.

Kley, D., G.M. Lawrence and E.J. Stone, The Yield of $N(^2D)$ atoms in the dissociative recombination of NO^+, J. Chem. Phys., 66, 4157, 1977.

Koshelev, V.V., Variations of transport conditions and winter anomaly in the D-ionospheric region, J. Atmos. Terr. Phys., 41, 431, 1979.

Labitzke, K., K. Paetzoldt and H. Schwentek, Planetary waves in the strato- and mesosphere during the Western European Winter anomaly campaign 1975/76 and their relation to ionospheric absorption, J. Atmos. Terr. Phys., 41, 1149, 1979.

Leu, M.T., M.A. Biondi and R. Johnsen, Measurements of the recombination of elections with $H_3O^+(H_2O)_n$ series ions, Phys. Rev., A7, 292, 1973.

Liu, S. C., and G. C. Reid, Sodium and other minor constituents of meteoritic origin in the atmosphere, Geophys. Res. Lett., 6, 283, 1979.

Massie, S.T., PhD thesis, University of Colorado, Boulder, CO, 1979.

Mc Crumb, J.L. and F. Arnold, High sensitivity detection of negative ions in the stratosphere, Nature, 294, 136, 1981.

Mechtly, E.A. and J.S. Shirke, Rocket electron concentration measurements on winter days of normal and anomalous absorption, J. Geophys. Res., 73, 6243, 1968.

Mechtly, E.A. and L.G. Smith, Seasonal variation of the lower ionosphere at Wallops Island during the IQSY, J. Atmos. Terr. Phys., 30, 1555, 1968.

Murad, E., Problems in the chemistry of metallic species in the D region, J. Geophys. Res., 5525, 1978.

Narcisi, R.S. and A.D. Bailey, Mass spectrometric measurements of positive ions at altitudes from 64 to 112 kilometers, J. Geophys. Res., 70, 3687, 1965.

Narcisi, R.S., A.D. Bailey, L. Della Lucca, C. Sherman and D.M. Thomas, Mass spectrometric measurements of negative ions in the D- and lower E-regions, J. Atmos. Terr. Phys., 33, 1147, 1971.

Nicolet, M., Contribution a l'etude de la structure de l'ionosphere, Mm. Inst. Mtor. Belge, n 19, 83, 1945.

Nicolet, M. and A.C. Aikin, The formation of the D region of the ionosphere, J. Geophys. Res., 65, 5, 1960.

Nicolet, M. and W. Peetermans, The production of nitric oxide in the stratosphere of oxidation of nitrous oxide, Ann. Geophys., 28, 751, 1972.

Niehaus, A., Excitation and dissociation of molecules by electron bombardment. Measurement of the formation probability for neutral fragments as a function of electron energy (in German), Z. Naturf., 22a, 690, 1967.

Norton, R. B. and C. A. Barth, Theory of nitric oxide in the Earth's atmosphere, J. Geophys. Res., 75, 3903, 1970.

Offermann, D., An integrated GBR campaign for the study of the D region winter anomaly in western Europe 1975/76, J. Atm. Terr. Phys., 41, 1047, 1979.

Park, C. and G.P. Menees, Odd nitrogen production by meteoroids, J. Geophys. Res., 83, 4029, 1978.

Paulsen, D.E., R.E. Huffman and J.C. Larrabee, Improved photoionization of $O_2(^1\Delta_g)$ in the D region, Radio Science, 7, 51, 1972.

Phelps, A.V. and J.L. Pack, Electron collision frequencies in nitrogen and in the lower ionosphere, Phys. Rev. Lett., 3, 340, 1959.

Potemra, T.A., Ionizing radiation affecting the lower ionosphere, pp. 21-37, in : Holtet, J.A. (ed.), ELF-VLF radio wave propagation, D. Reidel Publishing Company, Dordrecht - Holland, 1974.

Rapp, D., P. Englander-Golden and D.D. Briglia, Cross sections for dissociative ionization of molecules by electron impact, J. Chem. Phys., 42, 4081, 1965.

Reagan, J.B., R.C. Gunton, J.E. Evans, R.W. Nightingale, R.G. Johnson, W.L. Imhof and R.E. Meyerrott, Effects of the August 1972 solar particle events on stratospheric ozone in : Lockheed Report D 630455, 1978.

Rees, M. H., and R. G. Roble, Morphology of N and NO in auroral substorms, Planet. Space Sci., 27, 453, 1978.

Reid, G. C., The production of water cluster positive ions in the quiet daytime D-region, J. Geophys. Res., 25, 275, 1977.

Rosenberg, T.J. and L.J. Lanzerotti, Direct energy inputs to the middle atmosphere, pp. 43-70, in : Maynard, N.C. (ed.), Middle atmosphere electrodynamics, NASA CP-2090, 1979.

Rusch, D.W., J.-C. Gerard, S. Solomon, P.J. Crutzen, and G. C. Reid, The effect of particle precipitation events on the neutral and ion chemistry of the middle atmosphere. - I. Odd nitrogen, Planet. Sp. Sci., 29, 767, 1981.

Schwentek, H., Ionospheric absorption between 53 N and 53 S observed on board ship, J. Atm. Terr. Phys., 38, 89, 1976.

Sen, H.K. and A.A. Wyller, On the generalizations of the Appleton- Hartree magnetoionic formulas, J. Geophys. Res., 65, 3931, 1960.

Smith, D., N.G. Adams and M.J. Church, Mutual neutralization rates of ionospherically important ions, Planet. Space Sci., 24, 697, 1976.

Smith, D., N.G. Adams and E. Alge, Ion-ion mutual neutralization and ion-neutral switching reactions of some stratospheric ions, Planet. Space Sci., 4, 449, 1981.

Solomon, S., D.W. Rusch, J.-C. Gerard, G.C. Reid, and P.J. Crutzen, The effect of particle precipitation events on the neutral and ion chemistry of the middle atmosphere : II. Odd nitrogen, Planet. Sp. Sci., 29, 885-893, 1981.

Solomon, S., E.E. Ferguson, D.W. Fahey and P.J. Crutzen, On the chemistry of H_2O, H_2 and meteoritic ions in the mesosphere and lower thermosphere, Planet. Space Sci., 30, 1117, 1982.

Solomon, S., P.J. Crutzen and R.G. Roble, Photochemical coupling between the thermosphere and the lower atmosphere. I. Odd nitrogen from 50 to 120 km, J. Geophys. Res., 87, 7206, 1982a.

Solomon, S., G.C. Reid, R.G. Roble and P.J. Crutzen, Photochemical coupling between the thermosphere and the lower atmosphere. II. D-region ion chemistry and the winter anomaly, J. Geophys. Res., 87, 7221, 1982b.

Solomon, S., G. C. Reid, D. W. Rusch, and R. J. Thomas, Mesospheric ozone depletion during the solar proton event of July 13, 1982, Geophys. Res. Lett., 10, 257, 1983.

Stewart, A.I., Photoionization coefficients and photoelectron impact excitation efficiences in the daytime ionosphere, J. Geophys. Res., 75, 31, 1970.

Swider, W. and T. J. Keneshea, Decrease of ozone and atomic oxygen in the lower mesosphere during a PCA event, Planet. Space Sci., 21, 1969, 1973.

Sze, N. D., M. K. W. Ko, W. Swider, and E. Murad, Atmospheric sodium chemistry I. The altitude region 70-100 km, Geophys. Res. Lett., 9, 1187, 1982.

Taubenheim, J., Meteorological control of the D region, Space Science Reviews, 34, 397, 1983.

Thomas, L., Recent developments and outstanding problems in the theory of the D region, Radio Sci., 9, 121, 1974.

Thorne, R.M., Influence of relativistic electron precipitation on the lower ionosphere and stratosphere, pp. 161-168, in : Grandal, B. and Holtet, J.A. (eds.), *Dynamical and chemical coupling between the neutral and ionized atmosphere,* D. Reidel Publishing Company, (Dordrecht-Holland), 1977a.

Thorne, R.M., Energetic radiation belt electron precipitation : a natural depletion mechanism for stratospheric ozone, Science, 195, 187, 1977b.

Thorne, R.M., The importance of energetic particle precipitation on the chemical composition of the middle atmosphere, Pageoph., 118, 128, 1980.

Vampola, A. L., and D. J. Gorney, Electron energy deposition in the middle atmosphere, J. Geophys. Res., 88, 6267, 1983.

Viggiano, A.A., R.A. Perry, D.L. Albritton, E.E. Ferguson and F.C. Fehsenfeld, The role of H_2SO_4 in stratospheric negative ion chemistry, J. Geophys. Res., 85, 4551, 1980.

Warneck, P., Cosmic radiation as a source of odd nitrogen in the stratosphere, J. Geophys. Res., 77, 33, 6589, 1972.

Winters, H.F., Ionic absorption and dissociation cross section for nitrogen (Σ), J. Chem. Phys., 44, 1472, 1966.

Zbinden, P.A., M.A. Hidalgo, P. Eberhardt and J. Geiss, Mass spectrometer measurements of the positive ion composition in the D and E regions of the ionosphere, Planet. Space Sci., 23, 1621, 1975.

Chapter 7

Possible Perturbations
and Atmospheric Responses

7.1 Introduction

In the preceding chapters, we have described some of the fundamental processes which are responsible for the present state and the evolution of the middle atmosphere. These processes are not independent, however, because the atmosphere is a highly interactive system characterized by non-linear coupling. Any perturbation to the atmospheric system will have chemical, thermal, and dynamic feedbacks, and the relative amplitude of each must be evaluated. In particular, modification of the middle atmosphere can have important influences on climate.

The complete study of atmospheric interactions and the assessment of perturbations requires the simultaneous solution of the equations of energy, mass, and momentum, as well as the radiative transfer and chemical kinetic equations in three dimensions, including all appropriate spatial scales and specifying realistic initial and boundary conditions. This approach is not currently very practical, even with the most powerful computers. It is therefore useful to introduce simplifications and parameterizations. For example, three-dimensional models of the general circulation can be coupled with a simplified radiative transfer code which can provide a realistic spatial and temporal distribution of temperature without requiring excessive computer time and storage (e. g., Fels et al., 1980). In some of these sophisticated three-dimensional models, simplified chemical schemes are presently used to calculate the distribution of optically active species such as ozone (Mahlman et al., 1980). These models have progressively developed along with computer technology, providing more and more detailed descriptions of the various phenomena of importance in the atmosphere, thus leading to more complete understanding of atmospheric responses to perturbations.

On the other hand, it is also possible to construct very simple models based only on energy conservation (Budyko, 1969; Sellers, 1969). This approach is often used to study the evolution of climate by defining global

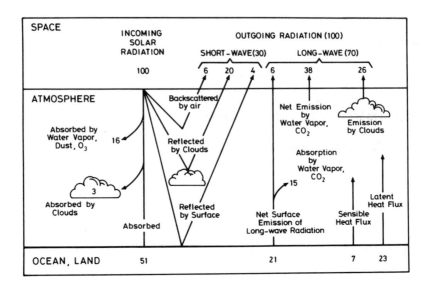

Fig. 7.1. Radiation budget in the Earth-atmosphere system. From National Academy of Sciences (1975).

parameters without regard for the details of atmospheric processes. By averaging the fundamental equations over spatial and temporal scales, the remaining variables may be only, for example, the global mean atmospheric temperature or relative humidity. The models then treat only an energy balance equation (e. g., Schneider and Dickinson, 1974; North et al., 1981) which relates the variation in temperature $\overline{T}$ of the earth-atmosphere system to the difference between absorbed solar radiation and the energy emitted by the system (Fraedrich, 1979):

$$C\frac{d\overline{T}}{dt} = Q_0[1 - A(\overline{T})] - \sigma\, g(T)\overline{T}^4 \qquad (7.1)$$

where C is the heat capacity of the system, Q_0 is the amount of radiative energy reaching the earth, averaged over the year and over the entire surface of the world (this parameter is about a quarter of the solar constant because of the Earth's sphericity, i. e., about 340 Wm^{-2}). $A(\overline{T})$ is the planetary albedo, which is about 0.33; its dependence on temperature describes the (poorly known) relationship between $\overline{T}$ and the characteristic cloud and snow cover. σ is the Stefan-Boltzmann constant and $g(T)$ is the "grayness factor". (This expression is a somewhat more complete version of the simple energy balance used in eqn. 1.1 to estimate the temperature of the primitive Earth.) Such an equation can be used to roughly describe the equilibrium climatic state, its stability, and response to perturbations.

The analysis of the equilibrium energy balance demonstrates the relative importance of several parameters (see Fig. 7.1). The short wave radiation

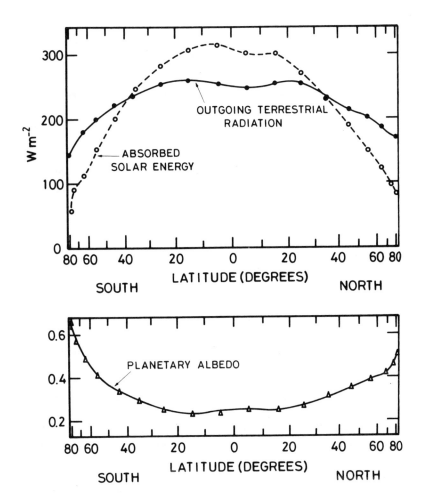

Fig. 7.2. (upper) Solar energy absorbed by the Earth-atmosphere system and outgoing terrestrial radiative energy as a function of latitude. (lower) Planetary albedo measured by satellite as a function of latitude. From Von der Haar and Suomi (1971). (Copyright by the American Geophysical Union.)

from the sun (100%) is absorbed by the atmosphere (about 19%) and is reflected back to space (about 20% by the clouds, 4% by the surface, and 6% by atmospheric molecules for a total of 30%). The rest of the incoming energy (51%) reaches the surface, where it heats the earth-atmosphere system, causing it to emit long wave infrared radiation. The infrared radiation emitted by the surface is about 21%, that emitted by atmospheric gases is 38%, and 26% by the clouds, leading to total outgoing radiation of 70%. In other words, when the earth-atmosphere system receives 100 units of solar energy, 30 units are reflected to space as short wave radiation and 70 units are emitted as long wave radiation. The 21% emitted as infrared radiation by the surface clearly does not equal the 51% absorbed by it. The difference (30%) is emitted as sensible (7%) and latent (23%) heat, which is converted to radiative energy in the atmosphere, and contributes to the outgoing radiation originating from the trace gases and the clouds.

Although the earth-atmosphere climatic system is in global energy balance, local radiative imbalances do occur, leading to atmospheric transport processes. Figure 7.2, for example, shows that the energy absorbed by the system ($0.67Q_o$ on the average) varies with the solar zenith angle, and is greater in tropical than in polar regions. On the other hand, the energy emitted by the surface ($0.15Q_o$) and by the atmosphere ($0.52Q_o$) depends on the temperature, and its latitude variation is not very great. To establish an equilibrium, a net transport of energy from the equator to the pole must occur via the atmospheric and oceanic circulation systems. At $30°$ latitude, the zonally averaged flux is about $4.0x10^{19}$ kilocalories per year (London and Sasamori, 1971). Some climate models (Budyko, 1969; Sellers, 1969) include these differential latitude effects and parameterize the energy flux as a function of the meridional temperature gradient.

Energy balance models have been used to study glaciation cycles (Ghil, 1981; Nicolis, 1982) and to examine climate stability to various perturbations (Budyko, 1969; Ghil, 1976; Nicolis and Nicolis, 1979; Benzi et al., 1982).

There are also a series of intermediate models. Radiative models (see Chapter 4) with convective adjustment (Manabe and Wetherald, 1967) can be used to determine the global mean vertical temperature profile as a function of the atmospheric water vapor, ozone, or carbon dioxide content. They can also include, in a very simplified fashion, the effects of clouds or aerosols. In these models, an exact equilibrium between the energy received by the earth-atmosphere system and the emitted energy is established. The global mean surface temperature is an important climatic parameter which can be determined by such models, but the convective adjustment technique arbitrarily specifies the vertical temperature gradient in the troposphere, where dynamic instability leads to convection. Such models are often used to estimate the effect of a change in the amount of CO_2 or ozone on the climate.

Two-dimensional dynamical and thermal models of the atmosphere can be used to determine the zonally averaged atmospheric structure, given a distribution of the sources of heat (calculated or observed). The first of these was a classical Eulerian two-dimensional model presented by Eliassen (1952), who used a parameterization of the eddy fluxes of momentum and heat (see eqns. 3.43 and 3.45). These parameters are often obtained by a statistical analysis of meteorological observations, and therefore are generally assumed to be independent of the perturbation. Self consistent two and three-dimensional models provide a means of examining the important feedbacks between radiation and dynamics without resort to such assumptions (e. g., Fels et al., 1980).

The study of atmospheric perturbations is aimed at examining the effects of modifications in the energy balance of the earth-atmosphere system. These perturbations can be external to the system: for example, a variation in the solar flux due to solar activity. They may also be produced by man as a result of agricultural and industrial activities, or they may be related to natural phenomena such as volcanoes. In this case the change in the chemical structure of the atmosphere can have an influence on the absorption and emission of radiation and can therefore modify the energy balance which we have just briefly described.

Another very important reason for concern regarding chemical perturbations in the middle atmosphere is the possible reduction in the ozone content with a related increase in the solar radiation reaching the Earth's surface. The enhancement in the incident irradiance occurs between about 280 and 300 nm, a spectral range commonly called UV-B. Since organic macromolecules such as DNA strongly absorb in this spectral region, a reduction in the ozone content could have important biological consequences.

The photobiological effects of an increase in UV-B radiation are numerous and largely detrimental. They influence many varieties of flora and fauna, including man. The exposure of human skin to UV-B, for example, can lead to several diseases, and facilitates the development of skin cancer. The link between UV-B exposure and skin cancer is not yet well understood, but is supported by a good deal of evidence, such as a larger incidence of skin cancer at lower latitudes.

The absorption cross section of macromolecules is highly dependent on the energy of the incident photon. The efficiency of photons in provoking biological responses such as DNA damage, photosynthesis inhibition, erythema (sunburn), etc. are characterized by an "action spectrum" as a function of wavelength. Most of these action spectra exhibit high values below 300 nm and decrease sharply above this wavelength, precisely where the solar irradiance at ground level begins to increase rapidly. The biological dose results from the spectral integration of the solar irradiance weighted by the action spectrum. The radiation amplification factor (RAF) is the ratio of percentage increase in biologically active dosage to the percentage decrease in total ozone abundance. This parameter is, for example, about 2.3 for DNA damage and 1.7 for human erythema according to Rundel (1983), which means that for each 1% reduction in the total ozone column about a 2% increase in these forms of biological damage will result.

The protection of the ozone layer against possible anthropogenic perturbations is therefore of vital importance to the climatic and biological equilibria on the planet surface.

7.2 The importance of coupling in the study of perturbations

The geographical and temporal distributions of minor constituents depend not only on the chemical reactions which contribute to their formation and destruction, but also on the temperature, which determines the rates of

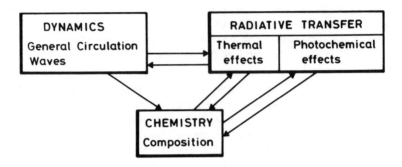

Fig. 7.3. Coupling between chemistry, radiation, and dynamics in the middle atmosphere

many chemical processes, and on atmospheric dynamics, which can transport chemical species. Similarly, the stratospheric and mesospheric temperature is determined by the densities of optically active chemical constituents, whose distribution can depend on chemical and photochemical processes. Finally, the atmospheric circulation is related to the non-uniform absorption of solar energy; the direction and strength of the mean winds is related to thermal gradients. Figure 7.3 schematically depicts the possible interactions. These are described by the fundamental equations which were discussed in Chapters 3, 4, and 5.

The study of atmospheric responses to any external perturbation (variation of the solar flux, energetic particle precipitation, volcanic and anthropogenic emissions, etc.) must involve these coupling mechanisms. The ozone-temperature interaction, for example, provides an important example of such coupling. It is well known that the primary source of heat in the stratosphere and mesosphere is the absorption of ultraviolet radiation by ozone. If the concentration of this gas changes as a result of some perturbation, (and ignoring the feedback mechanisms, for the purpose of illustration) then the temperature will change in response, according to the relation described by Dickinson (1973):

$$\Delta T = \frac{P_H}{\alpha} \frac{\Delta(O_3)}{(O_3)} \tag{7.2}$$

where P_H is the total gross heating rate, α is the relaxation rate (see expression 4.89) and $\Delta(O_3)/(O_3)$ is the relative ozone change. At 50 km, for example, $P \approx 10 K/day$ and $\alpha \approx 0.2\ day^{-1}$, so that for an ozone reduction of 20%, a temperature decrease of about 10K will occur. This estimate is only approximate, since it assumes a linear variation of the temperature with changes in the heating rate (which is only valid for small temperature variations) and because it neglects the fact that the cooling rate is also related to the ozone density through its emission in the 9.6μm band (modifying the value of α).

Similarly, a change in temperature will modify the temperature dependent chemical kinetic rates and the ozone density. If a pure oxygen atmosphere is assumed (see Chapter 5), this interaction is described by equation (5.32), such that, to a first approximation,

$$\frac{\Delta(O_3)}{(O_3)} = \frac{1}{2} \frac{\Delta(k_2/k_3)}{(k_2/k_3)} \tag{7.3}$$

i. e., if $k_2 = 1.1 \times 10^{-34} \exp(510/T) \text{cm}^6 \text{ s}^{-1}$, and $k_3 = 1.9 \times 10^{-11} \exp(-2300/T) \text{cm}^3 \text{ s}^{-1}$ (Hampson and Garvin, 1978), then

$$\frac{\Delta(O_3)}{(O_3)} = \frac{-1405}{T^2} \Delta T \tag{7.4}$$

or

$$\frac{d \ln(O_3)}{dT^{-1}} = -1405K \tag{7.5}$$

At the stratopause (T = 270 K), a decrease in temperature of 10 K results in a 19% increase in ozone.

In the real atmosphere, however, both effects occur simultaneously, and there is a strong feedback between the two (see Luther et al., 1977). If an

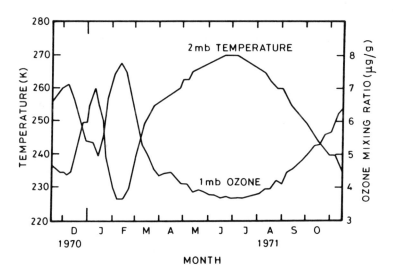

Fig. 7.4. Zonally averaged temperature at the 2 mb level (from the SCR experiment) and ozone mixing ratio at 1 mb (from the BUV experiment), from Nov. 1970 to Nov. 1971. From Krueger et al. (1980). (Copyright by the Royal Society).

ozone perturbation occurs (for example, as a result of a chemical change such as a solar proton event, see §7.4) then the associated change in temperature, assuming a pure oxygen atmosphere, will be given by:

$$\Delta T = \frac{P_H/\alpha}{1 + \dfrac{P_H}{\alpha}(1405/T^2)} \frac{\Delta(O_3)}{(O_3)} \tag{7.6}$$

instead of expression (7.2). Let us assume that a perturbation is applied which would yield a 10% ozone decrease in the absence of coupling. As the ozone concentration begins to decrease, however, the temperature decreases. This causes the rate of ozone decrease to lessen, so that the equilibrium temperature change is only about 2.5K (instead of 5K), and the equilibrium ozone change only about 5% (instead of 10%). Similarly, for a temperature increase of 10K, induced, for example, by dynamical forcing, the ozone at the stratopause will be depleted by 9.5% instead of 19% as determined when the feedback mechanism is ignored. In the middle atmosphere, the driving mechanism over short time scales (days or weeks) is usually the dynamic variability. For example, the presence of planetary waves introduces local temperature perturbations and an associated anticorrelation between temperature and ozone concentration in those regions where ozone is photochemically controlled.

This behavior has been routinely observed in satellite studies (see, e. g., Barnett et al., 1975; Krueger et al., 1980; Ghazi and Barnett, 1980). For example, Figure 7.4 (Krueger et al., 1979) shows the zonally averaged 2 mb temperatures observed by the SCR experiment (Barnett, 1974) and the ozone mixing ratios obtained by the BUV experiment onboard Nimbus 4 (Heath, 1980). The strong anticorrelation is clearly seen in spite of the different pressure levels used.

This qualitative analysis illustrates that the ozone-temperature relationship in the upper stratosphere provides a stabilizing feedback: a reduction in ozone by a chemical perturbation, for example, will be partly compensated by a coupling with the temperature.

Equation (7.5) was derived for the pure oxygen atmosphere. A more realistic analysis must also include the effects of hydrogen, nitrogen and chlorine compounds. Haigh and Pyle (1982) suggest that the following expression can be used:

$$\frac{d\ln(O_3)}{dT^{-1}} = \frac{a + bf_1 + cf_2 + df_3}{2 + f_1 + f_2 + f_3} \tag{7.7}$$

where f_1, f_2, and f_3 are the ratios of the rate of destruction of odd oxygen due to HO_x, NO_x, and ClO_x, respectively, compared to the the rate of destruction by the pure oxygen reactions. These factors depend on the chemical state of the atmosphere, and, in particular, on the amplitude of the perturbation. The other constants depend only on the chemical kinetics and are approximately given by $a = 2810K$, $b = 510K$, $c = 1200K$ (Haigh and Pyle, 1982). The value of d is small (less than 275K) but is difficult to establish because of complex chemical coupling processes. This relationship shows, however, that consideration of all of the chemical families will reduce the sensitivity of ozone to the temperature, but it will increase the sensitivity of the upper stratospheric temperature to ozone depletion. Further, the sensitivity of ozone to temperature will decrease most strongly if the amount of atmospheric chlorine increases; therefore, the increase of chlorine associated with human activities will reduce the stabilizing feedback between ozone and temperature. Perturbation calculations must include these non-linear effects for each individual case.

Ozone is not the only atmospheric chemical species which depends sensitively on the thermal structure. For example, the chemistry of odd nitrogen is related to the temperature, particularly in the mesosphere through the strong temperature sensitivity of the reaction $N(^4S) + O_2 \rightarrow NO + O$. Thus aeronomic studies of perturbations should include both the temperature and minor constituent distributions.

The nature of all of the interactions between chemistry, temperature, and dynamics is not yet completely understood. One-dimensional (and many two-dimensional) models are based on empirical transport coefficients whose variations with respect to atmospheric perturbations are not known. The modification of the circulation associated with changes in ozone density can be evaluated with three dimensional or self-consistent two-dimensional models (e. g., Schoeberl and Strobel, 1978), and these studies have suggested that the interactions are weak. More detailed studies including climatic parameters such as the cloud feedback, ocean interaction and albedo changes are still needed to definitively resolve this question, however.

7.3 The effect of changes in the solar irradiance

Changes in the solar irradiance can modify the physical structure and chemical composition of the middle atmosphere. In principle, perturbations can also occur at low altitudes, resulting in meteorological and climatological effects. However,

"despite a massive literature on the subject, there is at present little or no convincing evidence of statistically significant or practically useful correlations between sunspot cycles and the weather or climate" (Pittock, 1978).

Part of the reason for the lack of an obvious sun-weather connection is probably the fact that the temporal variation of the total amount of solar energy incident at the top of the Earth's atmosphere (the solar constant) varies little with solar activity. For example, the solar constant measured between 1975 (minimum of the 11 year solar cycle) and 1982 varied by only ± 2 Wm^{-2} about its mean value of 1367 Wm^{-2}, or only $\pm 0.15\%$ (Brusa and Frohlich, 1982). The measurement of such a weak variation is limited by the accuracy and precision of the measuring instruments, but it is clear that the global total energy input to the atmosphere has changed little over solar cycle 21.

Even if the variation in the solar constant is small, the possibility of solar influences on tropospheric dynamics cannot be ruled out. For example, Gage and Reid (1982) have found that a clear correlation exists between the height of the tropical tropopause and the solar activity level. Based on a study of the normal seasonal variation of tropopause heights (due to the seasonal change in the distance from the Earth to the sun and its influence on solar radiation received, see eqns. 4.2-4.3), they conclude that a change of only about 0.5% in solar radiation incident on the tropical ocean surface may be responsible for observed variations in tropical tropopause heights over the 11 year solar cycle. It is possible that the tropopause height may be associated with other climatic effects. Further, the tropical tropopause plays an important role in the transport of trace species from the troposphere to the stratosphere (see Chapter 3), particularly water vapor, so that the middle atmosphere could also be influenced by these processes (Mastenbrook and Oltmans, 1983).

On the other hand, much larger variations are observed in the ultraviolet and extreme ultraviolet portions of the solar spectrum. These wavelengths contribute little to the solar constant, but control the stratospheric and mesospheric temperature ($\lambda < 310$nm), and the photochemistry of the middle atmosphere. In this section, we will examine the likely response of the middle atmosphere to estimates of the variation of the solar flux for the relevant wavelengths. A detailed analysis of the temporal variability of the solar spectrum as a function of wavelength was presented in Chapter 4.

The following different types of solar activity variations can influence the solar emission behavior:

- 1). Short term variations (seconds to hours) related to solar flares and disruptive prominences.

- 2). Intermediate term variations (days to weeks) due to active regions and the solar rotation (27 days).

- 3). Long term variations over the 11 year sunspot cycle or 22 year magnetic field cycle.

In the ultraviolet region from 120 to 210 nm, variations of the solar flux with the 27 day rotation period of the sun can be observed (Rottman, 1983). Numerous measurements obtained since 1970 show that the irradiance in the same wavelength region also varies over the 11 year solar cycle. In both cases, the observed variability is largest at short wavelengths and decreases with increasing wavelength. Above 210 nm, the solar variability is smaller than the accuracy of the measuring instruments.

In the middle atmosphere, the variability of the emission of electromagnetic radiation from the sun can have two direct consequences. First, it can produce a change in the atmospheric heating rate by molecular oxygen and ozone. The resulting temperature variation is probably small for altitudes below about 90 km, where the major heating is provided by ozone absorption in the Hartley band (near 250 nm), since the amplitude of the solar flux variation at these wavelengths is quite small. The change in temperature will also depend on the ozone-temperature feedback discussed earlier. Model calculations (e. g., Brasseur and Simon, 1981; Vupputuri, 1982; Garcia et al., 1984) indicate a maximum stratospheric temperature change of about 2 to 3K near the stratopause, in reasonable agreement with a statistical analysis by Quiroz (1979).

The second effect which must be considered is the modification of the rates of photodissociation and photoionization of atmospheric constituents, particularly in the upper atmosphere, where the short wavelength radiation exhibiting the largest solar activity variability can be found. It must also be noted that for species which absorb radiation above 200 nm, a secondary modification of the photodissociation rate can occur through changes in the total ozone column which controls the penetration of this radiation.

The variability of the Lyman α line and shorter wavelengths of extreme ultraviolet radiation is probably about a factor of two over the course of the 11 year solar cycle. The Lyman α line penetrates deeply into the upper mesosphere, and is largely responsible for the photodissociation of water vapor and methane at those altitudes, producing odd hydrogen radicals which can catalytically destroy odd oxygen. Lyman α also controls the rate of photoionization of NO in the D-region. Extreme ultraviolet radiation also produces nitric oxide in the thermosphere, which can be transported down into the mesosphere and

SOLAR CYCLE AND POTENTIAL RESPONSES OF THE MIDDLE ATMOSPHERE

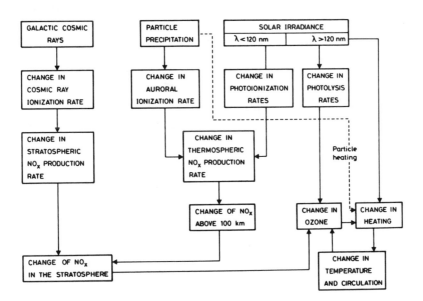

Fig. 7.5. Potential responses of the middle atmosphere to variabilities in solar activity over the 11 year solar cycle. Adapted from Garcia et al. (1984). (Copyright by Pergamon Press).

probably also into the stratosphere (see Chapter 5). Thus a change in the chemistry of the thermosphere and upper mesosphere in connection with the 11 year cycle is likely (see, e. g., Allen et al., 1984; Garcia et al., l984; Brasseur and DeBaets, l982). The possible propagation of these effects into the lower atmosphere is at present uncertain, but will be examined in more detail below.

The radiation in the Schumann-Runge bands can penetrate to the stratopause, and its variability can therefore influence the entire mesosphere. The rate of photodissociation of molecular oxygen in the $\delta(0-0)$ and $\delta(1-0)$ bands can vary by about 15-20 percent over the 11 year solar cycle. The variability of the Herzberg continuum, which penetrates into the stratosphere, is considerably weaker. Since an important part of the photodissociation in this spectral region comes from the atmospheric window near 200 nm, we adopt an upper limit for the variation of the photodissociation rate in the Herzberg continuum of 15%. This number applies both to the photodissociation of molecular oxygen and to the source gases, N_2O, $CFCl_3$, etc., thus controlling the variability of the production of odd oxygen, odd nitrogen, and odd chlorine. As we have already mentioned, a feedback exists between changes in ozone and their effect on the depth of penetration of ultraviolet radiation into the stratosphere. The concentration of ozone depends on the ratio between two photolysis rates, J_{O_2} and J_{O_3} (c. f., eqn. 5.32). Both of these are likely to increase along with solar activity, thus tending to cancel one another. However, J_{O_3} depends primarily on the solar flux at wavelengths near 250 nm, where the variability is relatively small,

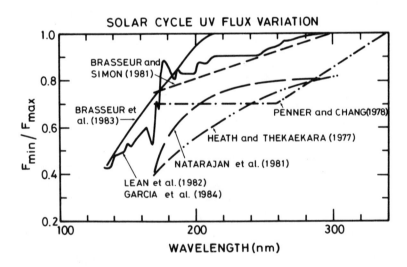

Fig. 7.6. Examples of solar irradiance variability based on observations or adopted in models to assess the effect of the 11 year cycle of solar activity. Adapted from Garcia et al. (l984).

while J_{O_2} is determined by wavelengths less than 242 nm, whose variation with solar activity is larger. The net effect should be a slight increase in upper stratospheric ozone, but it should be emphasized that temperature and dynamical feedbacks must also be considered. Quantitative estimates of these effects require a detailed calculation allowing for the complete spectra of these molecules, as well as the solar variability as a function of wavelength.

High solar activity levels are also accompanied by increased injection of energetic particles, such as aurorae and solar proton events. These events increase atmospheric ionization, and produce Joule heating and chemical perturbations (see §6.5). These processes can influence the dynamics and chemistry of the thermosphere, and possibly also those of lower levels through long range transport. The potential coupled responses of the middle atmosphere to changes in solar activity over the 11 year solar cycle are schematically indicated in Figure 7.5.

The first suggestions of a connection between the middle atmosphere and solar activity were based on observations of ozone (e. g., Willett, 1962; Paetzold, 1969; Angell and Korshover, 1976), temperature (Zerefos and Crutzen, 1975;

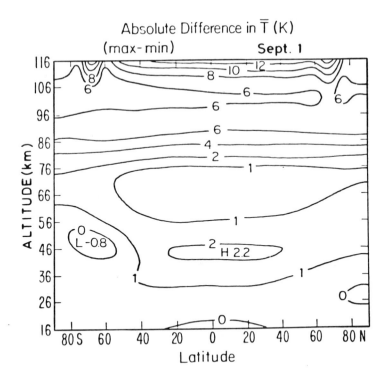

Fig. 7.7. Calculated variation in temperature associated with the 11 year cycle of solar activity. From Garcia et al. (1984). (Copyright by Pergamon Press).

Angell and Korshover, 1978a; Quiroz, 1979) and atmospheric circulation (Hines, 1974; Ebel and Batz, 1977; Nastrom and Belmont, 1978). The validity of some of these apparent correlations has since been questioned (London and Haurwitz, 1963; London and Reber, 1979). Dutsch (1979) has concluded that no valid statistical correlation between total ozone and solar activity has yet been obtained.

The potential response of the atmosphere to solar activity perturbations has been studied with one-dimensional (e. g. Penner and Chang, 1980; Natarajan et al., 1981; DeBaets et al., 1982, Allen et al., 1984), and two-dimensional models (Brasseur and Simon, 1981; Vupputuri, 1982; Brasseur et al., 1983; Garcia et al., 1984). It is difficult to compare these studies because each adopts different assumed solar flux variations (Fig. 7.6). We present below some of the results from the study by Garcia et al., to provide examples of quantitative estimates of the magnitudes of some of the perturbing influences outlined above. We note that the change in solar flux for wavelengths longer than 210 nm adopted by these authors is probably an upper limit. Therefore, many of the calculated stratospheric changes should be regarded as upper limits.

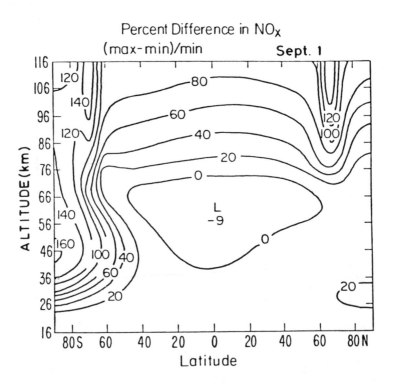

Fig. 7.8. Relative variation in NO_x associated with the 11 year cycle of solar activity as calculated in the study of Garcia et al. (1984). (Copyright by Pergamon Press).

Figure 7.7 shows the calculated temperature response to the 11 year cycle as determined by Garcia et al. for September 21. A calculated temperature increase from solar minimum to solar maximum is obtained in the lower thermosphere, where the effects of heating by molecular oxygen (and energetic particles in the conjugate auroral zones) is most intense. Much smaller changes are found in the mesosphere and stratosphere, where ozone heating dominates. Thus the largest thermal response is predicted to occur in the thermosphere, with much smaller variations at lower altitudes. It must, however, be noted that fairly small temperature and ozone variations can perhaps influence planetary wave propagation into the middle atmosphere, with associated changes in stratospheric dynamics (Geller and Alpert, 1980).

Similarly, the largest photochemical responses to changing solar activity are also predicted to occur at thermospheric heights. In particular, due to increases in particle precipitation and extreme ultraviolet fluxes, thermospheric nitric oxide should be expected to exhibit large increases from solar minimum to solar maximum, with possible propagation to lower altitudes as a result of transport, especially in the winter season. These effects are shown in Figure 7.8.

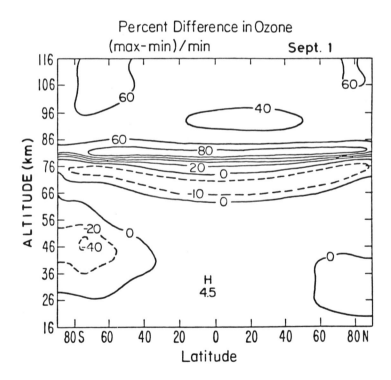

Fig. 7.9. Ozone density variation over the course of the 11 year cycle of solar activity as calculated by Garcia et al. (1984). (Copyright by Pergamon Press).

Calculated changes in ozone density are shown in Figure 7.9. In general, ozone increases are calculated for solar maximum as a result of increased molecular oxygen photodissociation, particularly at higher altitudes, where the solar variability is most intense. The maximum amplitude of the calculated stratospheric variation is at most about 5% in tropical latitudes; it is related to variations in solar radiation at wavelengths longer than 200 nm, as discussed above. The calculated ozone reduction obtained in the upper stratosphere in high latitude spring reflects the effects of increased odd nitrogen produced in the thermosphere and transported downward (see Fig. 7.8). The decreases in mesospheric ozone are due to increases in the catalytic destruction of odd oxygen by odd hydrogen, due mostly to the increased photodissociation of water vapor by Lyman α.

The associated variation in the ozone column abundance as obtained from this study is no more than a few percent, maximizing in the tropics. Although these changes are small, they may have significant consequences. Ramanathan et al. (1976), and Ramanathan and Dickinson (1979) have shown that a decrease of the order of 10 percent in the ozone column could reduce the surface temperature by a few tenths of a degree K. These studies indicate the possible impact of the middle atmosphere on climate, and underscore the need to study the entire atmosphere as a coupled system.

7.4 Particle precipitation

As has already been mentioned in Chapter 6, protons, electrons and alpha particles ejected from the surface of the sun can precipitate into the Earth's atmosphere and produce ionization at high latitudes. In particular, solar disturbances can be associated with the production of large quantities of charged particles which result in aurorae, relativistic electron precipitation (REP's) and solar proton events (referred to as SPE or polar cap absorption, PCA events). Section 6.5 described the ion chemistry which leads to the possibility of interactions between charged particles and neutral constituents. These interactions occur because of the production of NO_x during particle precipitation events throughout the middle atmosphere, and of HO_x production for altitudes below about 85 km (see, e. g. Rusch et al., 1981; Solomon et al., 1981). Particularly during solar proton events, the incoming particles may be of sufficiently high energy to produce appreciable amounts of ionization (and NO_x and HO_x) in the stratosphere and mesosphere. Since the NO_x free radicals play an important role in the destruction of stratospheric ozone, while the HO_x radicals dominate ozone destruction in the mesosphere (see Chapter 5), their production in particle precipitation events may be expected to influence the atmospheric ozone density if that production is large enough to perturb the naturally occuring abundances of these species. A perturbation may be manifested in two ways: either by a sudden increase in radical concentrations, and a corresponding decrease in ozone abundance, during and immediately after such an event, or by a gradual accumulation of the radical densities as a result of repeated events,

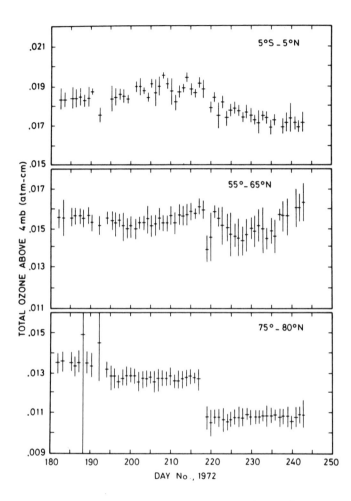

Fig. 7.10. Ozone density changes during the August, 1972 solar proton event observed by the Nimbus 4 BUV experiment. From Heath et al. (1977). (Copyright by the American Association for the Advancement of Science).

thus contributing to the global budgets of the radicals. The latter cannot be important for HO_x, because the lifetime of odd hydrogen is never more than a few hours below 85 km so that it is always in a photochemical equilibrium determined by its instantaneous sources and sinks, but could be significant for NO_x, which is sufficiently long lived that repeated events could influence its source strength on a global basis. In the following section, we first discuss the importance of instantaneous perturbations during particle precipitation events, and we then examine the cumulative effect of typical events in the photochemistry of NO_x.

Perhaps the most important aspect of the interactive photochemistry of ozone and the HO_x and NO_x produced by particle precipitation events is the fact that they can provide a naturally occurring test of the theory of ozone photochemistry and its control by these free radicals. The ozone changes produced by these events have therefore been commonly referred to as the "smoking gun", i. e., the tangible proof that ozone is indeed sensitive to perturbations in these constituents (natural or anthropogenic).

The possibility that odd nitrogen would be produced by energetic particle precipitation seems to have first been presented by Dalgarno (1967), but this was before the importance of odd nitrogen chemistry in the ozone balance (Crutzen, 1970; Johnston, 1971) was recognized. Crutzen et al. (1975) later suggested that solar proton events could have important effects on atmospheric ozone. A few years later, Heath et al. (1977) presented the first observations of this effect: they found that the very intense and energetic solar proton event of August, 1972 was accompanied by dramatic decreases in the ozone abundances at high latitudes, as shown, for example, in Figure 7.10, and that the observed decreases were in reasonable agreement with theoretical predictions. This event was widely studied in later papers, including those of Fabian et al. (1979), Reagan et al. (1981), Solomon and Crutzen (1981) and Rusch et al. (1981). The analysis by Solomon and Crutzen pointed out that the ozone-temperature

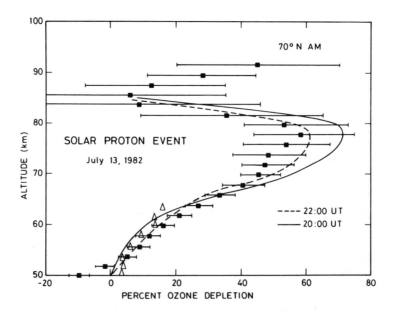

Fig. 7.11. Observed and calculated ozone depletion during the solar proton event of July 13, 1982. From Solomon et al. (1983). (Copyright by the American Geophysical Union.)

feedback, as discussed in §7.2, was demonstrated during this event. In particular, it was found that temperature changes of about 6-10K should have occurred near the stratopause as a result of the ozone depletion, and that consideration of the temperature changes therefore damped the calculated ozone changes when compared to calculations neglecting this effect.

In contrast to NO_x, the depletion of ozone related to HO_x production during particle precipitation events was first observed, and only later understood in

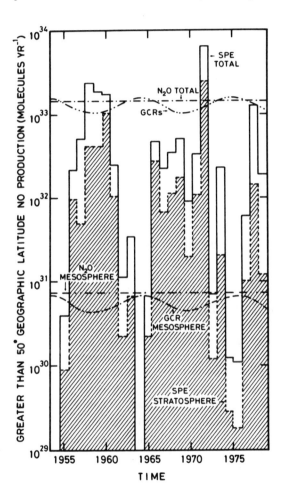

Fig. 7.12. Annual nitric oxide production rates for geographic latitudes above 50°. The solid and dashed lines represent NO_x production by SPE's in the stratosphere and mesosphere, respectively. From Jackman et al. (1980). (Copyright by the American Geophysical Union.)

terms of theory. The first observations of mesospheric ozone changes induced by HO_x production during solar proton events were presented by Weeks et al. (1972). These observations were based only on a few rocket flights, so that the exact magnitude of the observed ozone change was somewhat difficult to determine. Swider and Keneshea (1973) later suggested that these changes were probably due to production of odd hydrogen as a result of the ion chemistry, as discussed in Chapter 6. Observations by the Solar Mesosphere Explorer (SME) satellite provided a much more detailed picture of the ozone perturbation in the mesosphere during a solar proton event (Thomas et al., 1983). These observations were shown to be in very good agreement with theoretical model predictions by Solomon et al. (1983) and Jackman and McPeters (1984). Figure 7.11 presents the ozone depletion observed by SME compared to model predictions. Thorne (1980) suggested that similar changes in mesospheric ozone could occur during REP events.

These studies have shown that the response of ozone to perturbations in odd hydrogen in the mesosphere, and similarly to odd nitrogen in the upper stratosphere, is consistent with our present understanding of the chemistry of ozone destruction by these free radical species. The natural laboratory provided by the atmosphere during solar proton events represents a unique and very useful experiment.

We now turn to the question of long term effects of particle precipitation on the budget of atmospheric NO_x. As was previously mentioned, Crutzen et al. (1976) suggested that the production of NO_x in particle precipitation events could be an important part of the budget of odd nitrogen in the middle atmosphere, since the odd nitrogen input provided by a single SPE can be as large as that produced by oxidation of atmospheric N_2O in an entire year for latitudes poleward of about 50 °. These concepts were explored in a budget analysis by Jackman et al. (1980). Jackman et al. (1980) also examined the role of odd nitrogen production by galactic cosmic rays (see Figure 7.12). Orsini and Frederick (1982) presented a one-dimensional model study of the impact of the SPE related NO_x production on NO densities in the mesosphere, and found it to be an important source. Thorne (1980) suggested that REP events were also important in mesospheric NO_x production. Solomon et al. (1982) found that the production of NO_x in aurorae near 100 to 120 km could also provide an important source of NO_x in the mesosphere and stratosphere as a result of large scale transport. Such a connection had long been suspected based on an observed correlation between aurorae and winter anomaly events (e. g. Schwentek, 1971, see also §6.6).

7.5 Volcanic emissions

Volcanic eruptions occur sporadically at locations scattered around the globe. These events inject large amounts of dust and trace gases into the atmosphere. The composition of the injected gas is dependent on the chemistry of the crust and upper mantle of the Earth, and it is believed that these processes

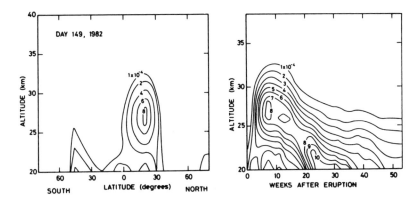

Fig. 7.13. Left: Contour plot of the aerosol extinction coefficient versus height and latitude for the two week period centered on May 29, 1982. Right: Contour plot of aerosol extinction coefficient for the first year following the El Chichon April 4, 1982 eruption. The latitude is that for which the integrated vertical depth above 20 km is a maximum during each two week period. From Thomas et al. (1983). (Copyright by the American Geophysical Union.)

had a significant influence on the evolution of the Earth's atmosphere (Chapter 1). Volcanic eruptions continue to provide a perturbation to the chemical, optical, and thermal structure of the contemporary atmosphere, and probably have signficant global climatic and local meteorological effects.

In some cases, the impact of volcanic eruptions has been particularly noticeable. The explosion of Tombora (Indonesia) in 1815 apparently decreased the global mean surface temperature by as much as a few degrees, yielding snow in June in North America (Landsberg and Albert, 1974), and famine in Europe. Particles injected into the atmosphere by the eruption of Krakatoa (Indonesia) in 1883 resulted in brilliant sunsets throughout the world.

The study of volcanic perturbations helps us to understand some fundamental aspects of atmospheric behavior, such as large scale transport processes, the role of aerosols in radiative transfer in the Earth-atmosphere system, and the importance of chemical reactions in the volcanic plume. Since the eruption of Fuego in 1974, volcanic activity seems to have increased, and this has been accompanied by a corresponding increase in research activity related to the behavior of the injected aerosols and possible climate effects.

Analysis has shown that gaseous volcanic emissions are composed largely of water vapor and carbon dioxide, as well as CO, SO_2, H_2S, H_2, NH_3, Cl, F, N_2, and CH_4. The sulfur compounds are of particular interest, because (as indicated in §5.8) they provide the main source of aerosols in the middle atmosphere, and the amount of sulfur released by volcanoes is not negligible compared to other more continuous sources.

H_2S is probably rapidly oxidized in the atmosphere, producing SO_2 (see, e. g., McKeen et al., 1983). SO_2 is then converted to sulfuric acid in the presence of water vapor. Observations by Hofmann and Rosen (1983) show that aerosols concentrated in a layer near 25 km after the eruption of El Chichon (Mexico) in April, 1982 were composed of an 80% sulfuric acid solution. These aerosols appear to have been produced mostly by chemical conversion of gaseous sulfur compounds, occuring on a time scale of weeks (Heath, 1983). The chemical conversion process was discussed in more detail in §5.8. Since their sedimentation rate is slow, the lifetime of these aerosols in the stratosphere is long, of the order of years. The aerosol cloud will begin to move and disperse as a result of dynamical effects, and can be observed at latitudes far from their point of entry into the stratosphere. For example, Figure 7.13 shows the aerosol extinction coefficient in the volcanic cloud released by El Chicon as measured at the end of May, 1982. Figure 7.13 also shows the transport of the cloud. These observations were obtained by an infrared radiometer onboard the SME satellite. The maximum aerosol density is obtained 8 weeks after the eruption, at about 27 km altitude, while 16 weeks later, the maximum density is found at only 20 km.

The increase in the aerosol loading after major volcanoes leads to both an increase in the attenuation of solar radiation by scattering and to an increase in the absorption of terrestrial and solar radiation. As a result of the former effect, a simultaneous reduction in the direct solar flux received and an increase in the scattered flux are observed. The net result of all of these processes is a heating at stratospheric altitudes and a cooling at ground level. The cooling is weak above the oceans, because of their large heat capacity, but may be observable over land, particularly in continental regions. A radiative-convective model study by Hansen et al. (1978) indicates that after the eruption of Mount Agung in 1963, the temperature of the middle stratosphere should have increased by 4-8 °C, while the surface temperature should have decreased by about 0.4 °C. Figure 7.14 shows that this model calculation seems to be in good agreement

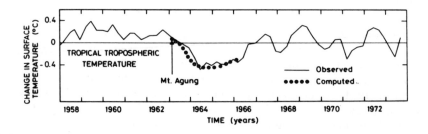

Fig. 7.14. Observed averaged tropospheric temperature in the latitude belt from 30N-30S and calculated temperature after the Mount Agung eruption in 1963. From Hansen et al. (1978). (Copyright by the American Association for the Advancement of Science).

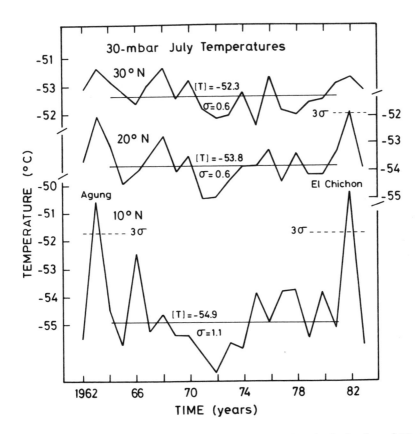

Fig. 7.15. Zonal mean temperatures ($^{\circ}$C) at 30 mb for the latitudes of 10, 20 and 30°N from 1962 to 1983. The 18 year average [T] corresponds to the period 1964-1981. From Labitzke and Naujokat (1984). (Copyright by Friedr. Vieweg and Sohn).

with observations. A similar model study by Pollack et al. (1983) for the case of El Chichon predicts an increase in the planetary albedo of about 10%, an increase in the solar radiation (compared to clear skies) of about 2-3%, and an increase in the temperature at 24 km of 3.5°C. This theoretical estimate is in good agreement with data presented by Labitzke and Naujokat (1984) and Quiroz (1983), as indicated in Figure 7.15.

The tropospheric cooling and associated mean surface temperature decrease (perhaps first suggested by Humphreys, 1940) has been characterized as a "dust veil index", describing the magnitude of such eruptions (Lamb, 1970). Statistical analysis of average temperature changes have been presented by Mass and Schneider (1977), and Miles and Gildersleeves (1978). Mass and Schneider (1977) found that the global mean surface temperature decreased by about 0.3°C during the year following major eruptions. Robock (1979) presented data indicating a clear relationship between surface temperature and volcanic activity. Turco et al. (1982) found a correlation between volcanic emissions and the ice cover in polar regions.

It thus seems that major eruptions can modify the atmospheric temperature structure, and perhaps the global circulation. Variations in the global mean surface temperature have been observed, and are probably due to changes in aerosol loading. Pollack et al. (1976) calculated the heating which would have occurred if no volcanic eruptions happened after the end of the 19th century. The increase in temperature from 1890 to the period between 1920 and 1960 would be about 0.4 ° C, i. e., more than that predicted from the observed increase in CO_2 over the same time period.

7.6 Anthropogenic emissions

Industrial processes emit a great variety of gases. Some of these have relatively short tropospheric lifetimes, and so contribute only to local pollution problems; e. g., oxides of nitrogen and sulfur, whose concentrations are very high near large cities and industrial complexes. These gases are rapidly absorbed into liquid particles and contribute to the acidification of the plumes emanating from industrial areas. These in turn increase the mean pH of rainfall, polluting lakes and streams, and posing a serious threat to certain types of wildlife, as well as dissolving the stone of some buildings.

Other industrial effluents have a much longer lifetime, and are very stable in the troposphere. These species progressively penetrate into the middle atmosphere, where they may influence global chemical and thermal processes. Two important examples of such gases are carbon dioxide and the chlorofluorocarbons, as we have already indicated in previous chapters.

7.6.1 Carbon dioxide

Anthropogenic carbon dioxide emissions come principally from combustion of fossil fuels, particularly coal, oil, and natural gas. Large scale coal gasification is presently seen as an important future source of natural gas, but will also lead to the production of large amounts of CO_2. Other artificial sources must also be considered, particularly the production and the emission of CO_2 as a by product in gasoline and oil production. Another potential source of anthropogenic release of CO_2 into the atmosphere is the destruction of wooded lands (see, e.g., Seiler and Crutzen, 1980), but the rates of global deforestation and forestation are not easy to specify. Since the global combustion rate is not well known, it is difficult to estimate with certainty the rate of CO_2 emission into the atmosphere. As a first approximation, one may assume that the amount of CO_2 emitted is directly proportional to the total global energy production, which has been regularly increasing by 5% per year over the last century. This suggests a doubling every 14 years. Figure 7.16 shows the emission of CO_2 during the 20th

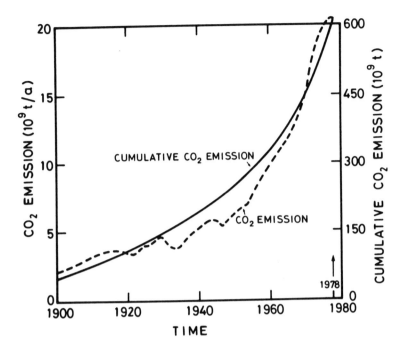

Fig. 7.16. Annual and cumulative emission of carbon dioxide in the 20th century. From Niehaus (1981). (Copyright by Reidel Publishing Company).

century, as well as the cumulative emission since 1860.

The observed increase of this gas in the atmosphere (see Fig. 5.19) reflects the fraction of "new CO_2" which is not absorbed by the terrestrial geochemical system. As shown by systematic observations since 1958, the carbon dioxide content must have been about 270 ppmv in 1850; it was near 340 ppmv in 1980, having increased by 5% per year over the last 20 years, and should reach 365 to 395 ppmv in the year 2000. Different scenarios are projected for the future but it is generally agreed that the amount of CO_2 will double by the end of the 21st century. The doubling of CO_2 is the standard condition used in perturbation studies. Climate models generally predict a mean increase in surface temperature of about 2 to 3° C with a probable error of ± 1.5° C (NRC/CRB, 1979) for a CO_2 doubling. This increase is due to a trapping effect: infrared radiation emitted by the Earth will be more readily absorbed in the lower layers of the atmosphere. Two and three-dimensional model studies show that the effect of CO_2 will be greater near the poles than at low latitudes. The model of Manabe and Wetherald (1975) predicts an increase in surface temperature of 2 to 3° between zero and 50 degrees latitude. The temperature change is predicted to be greater than 6° C above 60 degrees latitude, due to the snow albedo feedback. Important changes in the polar ice caps and in the level of the oceans are

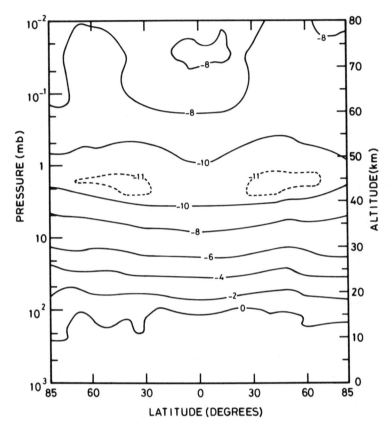

Fig. 7.17. Variation in the temperatures in the middle atmosphere for a doubling in carbon dioxide, calculated with a general circulation model. From Fels et al. (1980). (Copyright by the American Meteorological Society.)

expected to result.

In the middle atmosphere, carbon dioxide produces most of the local infrared cooling. A doubling of its concentration will decrease the stratopause temperature by about 10K according to current radiative models. The resulting coupling between the thermal and chemical structure as discussed in §7.2 has been studied with one dimensional models (Callis and Natarajan, 1981; Wuebbles et al., 1983; etc.), which suggest that the ozone density near 50 km will increase by about 20%. The associated increase in the total ozone column is about 3%, implying an important decrease in the ultraviolet radiation reaching the surface, but it must be emphasized that these studies are based only on one-dimensional models and do not include a complete treatment of dynamical feedbacks, which are likely to be important. The effects of an increase in CO_2 have also been studied with multidimensional models, in particular by Manabe

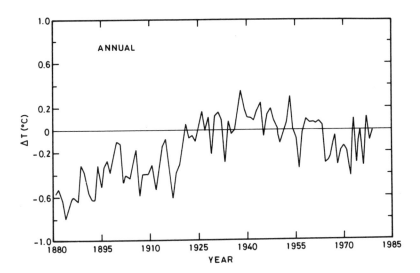

Fig. 7.18. Observed average surface temperature changes during the 20th century. From Jones et al. (1982). (Copyright by the American Meteorological Society).

and Wetherald (1975; 1980), Manabe and Stouffer (1979; 1980), Fels et al. (1980),
and Groves and Tuck (1980). Figure 7.17 shows that the stratospheric temperature perturbation calculated by the general circulation model of Fels et al.
(1980) varies little with latitude and is in good agreement with one-dimensional
model studies.

Long term observations of surface temperatures are available at several
stations around the world over the last century. The observed long term variation shown in Figure 7.18 displays a fairly regular increase of about 0.5K
between 1880 and 1940, which may be due to increasing CO_2. Since 1940, however, in spite of the continued increase in CO_2, the mean temperature has
decreased in a continuous manner, and the reasons for this behavior are
presently not well understood. The incidence of volcanic activity may play a
role in these global temperature trends, making it difficult to unambiguously
identify a CO_2 effect.

The study of the CO_2 problem requires an understanding not only of the
emission sources, but also of the transfer of CO_2 between the atmosphere, ocean
and biosphere. It is necessary to determine the processes affecting CO_2 in each
part of the biogeochemical system, their capacities for storage and the rate of
exchange from one reservoir to another. The calculation of the future increase
in the atmospheric CO_2 content is only possible if all of these parameters are
understood. Simple models of the carbon cycle (see, e. g. Laurmann and
Spreiter, 1983) include four CO_2 reservoirs: the biosphere, the atmosphere, the

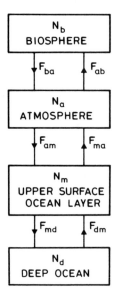

Fig. 7.19. Representation of a simplified model of CO_2 in the Earth's environment assuming exchange between 4 reservoirs. From Laurmann and Spreiter (1983).

surface ocean and the deep ocean (see Fig. 7.19). The amount of CO_2 in each reservoir i is given by the continuity equation (Bacastow and Keeling, 1973):

$$\frac{dN_i}{dt} = \sum_j (F_{ji} - F_{ij}) + F \qquad (7.8)$$

where F_{ij} is the flux from reservoir i to j and F is the rate of artificial production of carbon (combustion, etc.). This last term therefore appears only in the equation for the atmosphere. The determination of the fluxes F_{ij} must include a large number of physical and chemical considerations which will not be described here. The source strength F depends on the assumed type of energy supply for the future (e. g., coal, nuclear, etc.), and it is instructive to consider several possible scenarios. Figure 7.20 shows a model of future energy consumption by Voss (1977), described in detail by Niehaus and Williams (1979). This study is presented here for the purpose of illustration, and it should be noted that the scenarios presented by these authors may be somewhat pessimistic (see, e.g., NAS, 1983). The global energy crisis has resulted in somewhat slower energy consumption than previously anticipated. In this analysis, the assumed increase in energy consumption continues up to about 1990, but then decreases regularly after the end of the 20th century and stops completely before the year 2100. At this point, the consumption is 65 TW or 70×10^9 equivalent tons of

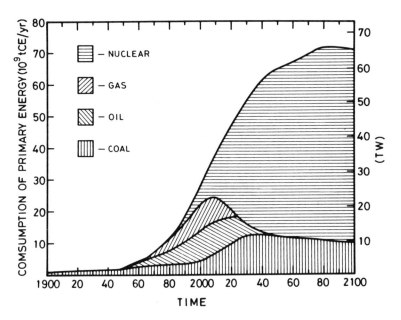

Fig. 7.20. Model of primary energy consumption through 2100 assuming a composite of energy sources. From Niehaus and Williams (1979). (Copyright by the American Geophysical Union.)

carbon per year of which only 10 TW is in the form of fossil fuels. A greater and greater part of the energy is provided by nuclear sources, implying widespread use of fast breeder reactors. The corresponding model calculations (Fig. 7.21) assume increasing emission of CO_2 through 2010, followed by a decrease due to a lower consumption of fossil fuels. The atmospheric CO_2 content never doubles its 1983 value. A regular increase in temperature up to a maximum of 2.2° C in 2100 is predicted.

Other scenarios have been presented by Niehaus and Williams. One of these assumes no nuclear energy and a return to extensive use of coal. The total energy consumption increases in 2100 to 50 TW (or 50 x 10^9 equivalent tons of carbon per year). In this case, the 1983 CO_2 content doubles in the year 2025, and the increase in temperature is 2.5° C in 2030, and 5° C in 2100, when the mixing ratio of CO_2 exceeds 1500 ppmv. Such a prediction demonstrates the danger of the widespread use of coal. These effects are global in extent.

A more favorable scenario in terms of ecological consequences consists of a dramatic reduction in the use of fossil fuels by the year 2000, with replacement by both nuclear and solar energy. The energy consumed in 2100 is 50 TW but the fossil fuel portion is only 2%. CO_2 emissions will then be considerably

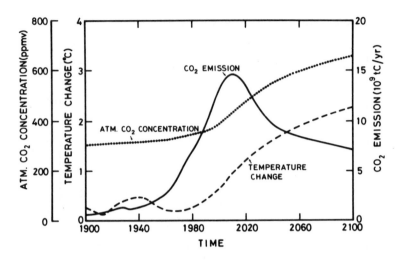

Fig. 7.21. CO_2 emission rate, atmospheric mixing ratio and surface temperature change through the year 2100 corresponding to the energy consumption scenario depicted in Fig. 7.20. (Copyright by the American Geophysical Union).

reduced after the year 2000 and the atmospheric burden will begin to decrease slightly after 2050. The predicted temperature increase will not exceed 0.8 ° C from 2040 to 2060, and will then decrease along with the CO_2 abundance.

In conclusion, the exact determination of the climatic impact of CO_2 requires detailed knowledge of the carbon cycle in the earth-atmosphere system, as well as a precise inventory of the CO_2 sources and a detailed study of all the parameters (often interactive) which play a role in global climate. Present models include too many simplifications and assumptions to consider the results as definitive, but they do indicate the serious dangers which humanity invites unless emissions of CO_2 are reduced in the future.

7.6.2. Methane

How rapidly has the amount of methane present in the atmosphere increased over the last century? This question is controversial because the apparent increases which have been reported are based on measurements made with different methods, and may simply reflect calibration errors. Rasmussen and Khalil (1981) have presented continuous measurements from Cap Meares in Oregon from the years 1979 to 1980 which suggest a mean increase of about 2.0 ± 0.5 percent per year. Blake et al. (1982) deduced an increase of about 1 percent per year based on their measurements made between 1977 and 1980.

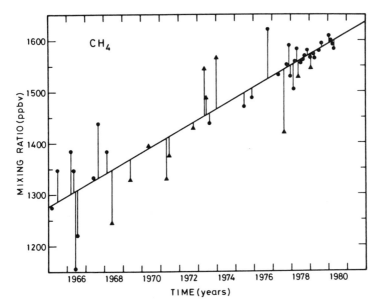

Fig. 7.22. Evolution of the tropospheric methane mixing ratio from 1965 to 1981 measured by Rasmussen and Khalil (1981), based on measurements over oceans (triangles), and over land (circles). (Copyright by the American Geophysical Union.)

The present mole fraction of methane in the troposphere is about 1.6 ppmv, while that present in the pre-industrial era may have been as low as 0.56 ppmv (Robbins et al., 1973) as suggested by measurements of air trapped in ice cores in Greenland and Antarctica, dating back 100 to 2500 years. It is not certain at present, however, that the methane content in these air bubbles is stable over time.

Most investigators agree that an increase in methane of about 1-2% per year occurred from 1977 to 1982, but the longer term variation of this gas is not as well established. Rasmussen and Khalil (1981) deduced an increase of $1.7 \pm 0.3\%$ per year since 1965 (Fig. 7.22.). However, Ehhalt et al. (1983) presented another series of observations dating back to about 1950 to which careful calibration corrections were applied, and deduced an increase of only 0.5% per year. Therefore, the amplitude of a possible variation over long time scales (100 years or more) is difficult to determine, because these data suggest that the yearly trend may have changed.

The effect of a continuous increase in methane on the chemical and thermal structure of the middle atmosphere has been studied with numerical models. A modification of the infrared cooling to space is also expected to occur for this gas, with a corresponding increase in surface temperatures. For a doubling of the present methane abundance, the model by Wang et al. (1976)

predicts a temperature increase of 0.3K, while Donner and Ramanathan (1980) suggest 0.2K and Owens et al. (1983) derive 0.34K. A methane increase will also affect the ozone and temperature profiles in the middle atmosphere. In the troposphere and lower stratosphere, the oxidation of methane leads to ozone production (see Chapter 5). In the mesosphere, an increase in methane will reduce the ozone abundance because this gas contributes to the production of odd hydrogen radicals which participate in odd oxygen destruction. In the stratosphere, several partially compensating effects will both occur: an increase in methane will produce an *increase* in OH (which destroys odd oxygen) but a related *decrease* in NO_2 and an *increase* in ClO (two other species which catalytically destroy ozone) will also occur. Finally, an increase in methane will tend to decrease the concentration of chlorine atoms and increase HCl through reaction of Cl with CH_4.

The sign and amplitude of the associated change in total ozone depend on the relative importance of each of these processes, but according to Owens et al. (1982, 1983), the net effect is to increase the ozone column. Figure 7.30 (see below) presents the calculated change in ozone concentration obtained from a one-dimensional model.

7.6.3 Nitrous oxide

In Chapter 5 we mentioned that N_2O is primarily produced by bacteria as a result of nitrification and denitrification processes in soils. The rate of production of N_2O depends on the amount of nitrogen fixed annually as well as on the ratio of the amounts of N_2O and N_2 produced by these processes. In spite of a great deal of research, the global rate of N_2O production is not presently well known. According to McElroy (1980), the global fixation of nitrogen is about 220 MT of nitrogen per year. If 5% of this nitrogen is in the form of N_2O, then the annual production of this gas is about 11 MT (N), approximately equal to the annual rate of destruction by photodissociation in the middle atmosphere. McElroy (1980) estimated that the oceanic source is probably small (less than 4 MT per year), and noted that the source related to human activities is likely to increase considerably in future years. This artificial source is due to agriculture, primarily the intensive use of nitrogen fertilizers (Crutzen, 1976; McElroy et al., 1977; Liu et al., 1977) and to waste disposal (McElroy, 1980), as well as combustion of coal and oil (Pierotti and Rasmussen, 1976; Weiss and Craig, 1976); anthropogenic sources probably already represent as much as 25-40% of the natural production. Breitenbeck et al. (1980) showed that the emission of N_2O from soils depends on the type and the amount of fertilizers used. Weiss and Craig estimate that the production of N_2O from combustion is about 1.6 MT (N) per year.

Observations of N_2O have shown that this gas is well mixed in the troposphere, with a mixing ratio of 300± 5 ppbv (Weiss, 1981; Khalil and Rasmussen,

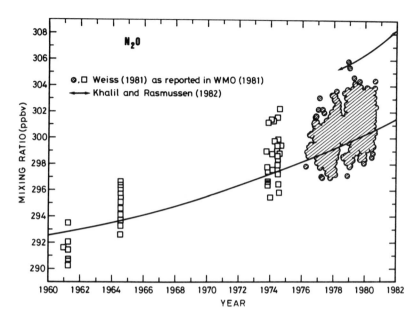

Fig. 7.23. Variation in the tropospheric mixing ratio of nitrous oxide between 1960 and 1980. From WMO (1982).

1983). A small difference between the two hemispheres (about 0.8 ppbv according to Weiss, 1981) suggests that the anthropogenic sources, which are more intense in the northern hemisphere, are beginning to influence the N_2O budget. Measurements also show that the concentration of N_2O has increased over the past twenty years (Fig. 7.23) by about 0.2± 0.1 percent per year (Golden et al., 1980; Rasmussen et al., 1981; Khalil and Rasmussen, 1981, 1983; Weiss, 1981). If this exponential increase in N_2O continues, the tropospheric abundance of N_2O will double in less than 300 years.

The lifetime of N_2O is about 100 years in the middle atmosphere (McElroy, 1980), so that this gas is completely well mixed in the troposphere, and is only destroyed in the stratosphere, where it produces nitric oxide (a species capable of catalytically destroying odd oxygen, as was discussed in Chapter 5). The effects of an increase of N_2O on ozone have been examined in several studies (see WMO, 1982). A calculated response of the ozone profile for a doubling of N_2O is shown in Figure 7.30 (below). The magnitude of the reduction of the ozone column due to such a perturbation depends on the adopted chemical scheme; in this particular calculation it is about 10%.

The increase in N_2O can also modify the thermal structure of the atmosphere. Particularly in the stratosphere, a decrease in the concentration of ozone results in atmospheric cooling, while at low altitudes, the increase in N_2O and NO_2 can lead to a small net heating; this gas is optically active in the

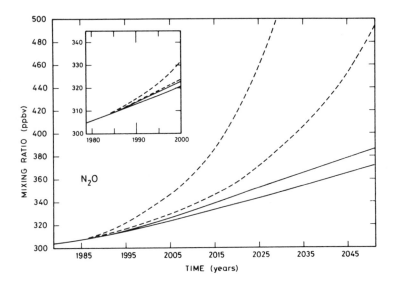

Fig. 7.24. Scenarios suggested by Weiss (1981) (dotted lines) and by Khalil and Rasmussen (1983) (solid lines) for the future nitrous oxide mixing ratios in the troposphere.

infrared and contributes to the radiation trapping in the troposphere, as does carbon dioxide, but to a much lesser extent (Donner and Ramanathan, 1980; Lacis et al., 1981). The doubling of N_2O can increase the surface temperature by 0.3K according to Donner and Ramanathan (1980). A similar study by Wang et al. (1976) suggests a temperature change of 0.44K.

Predictions of the behavior of the middle atmosphere in response to such a perturbation depend upon the scenario adopted to describe the probable evolution of industrial and agricultural emissions. Such scenarios are based on some hypothesis of future economic development. Several scenarios have been suggested by Weiss (1981) and Khalil and Rasmussen (1983). The former suggests an exponentially increasing source of 3.5-6% per year. Khalil and Rasmussen (1983) adopted a similar scenario but impose a saturation emission rate based on the expected increase in world population. The evolution of the N_2O density can thus be described by

$$\frac{d(N)}{dt} = S - \frac{N}{\tau} \tag{7.9}$$

where τ is the lifetime of N_2O (about 100 years) and where S is the total rate of N_2O production, composed of a natural part assumed to be 22.4 MT per year, and an anthropogenic part whose initial (1980) value is 6.6 MT per year. The corresponding change in the mixing ratio with these two models as a function of time is shown in Figure 7.24. Both of these models suggest an increase of about

30 ppbv by the year 2000, but the projections beyond that time are quite uncertain.

7.6.4. *Aircraft in the troposphere and lower stratosphere*

Combustion in aircraft engines produces a number of gaseous effluents which are injected into the atmosphere at altitudes between about 6 and 13 km. Most of these have no effect on the atmosphere because their production rate is negligible compared to natural processes. For example, carbon monoxide and dioxide fall into this category, as well as water vapor, which is predicted to have no global effect on ozone or on cloud formation (CIAP, 1974). As first indicated by Johnston in 1971, however, the oxides of nitrogen produced in this manner can produce important changes in the atmospheric ozone layer. Using a simple chemical model along with estimates of the number of supersonic aircraft then expected to fly in the lower stratosphere (near 20 km) in 1985, Johnston (1971) showed that the atmospheric ozone layer could be reduced by a factor as large as two. The importance of this catalytic cycle was also presented by Crutzen (1970). Because of the dynamical stability of the lower stratosphere, the nitric oxide and particulate matter injected at flight altitude would be expected to accumulate in this region before being slowly transported to the troposphere.

Concern over the possible effects of the predicted vast stratospheric fleets of SST aircraft led to an intensive study of the properties of the stratosphere. It was thus discovered that in addition to the catalytic destruction sequence

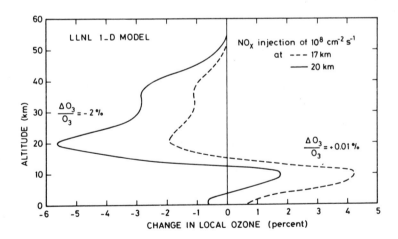

Fig. 7.25. Relative variation in the ozone concentration as a function of altitude for a NO_x injection of $1 \times 10^8 \, \text{molec cm}^{-2}\text{s}^{-1}$ at 17 and 20 km altitude, respectively. The corresponding percentage change in the ozone column is indicated. (After Wuebbles and Chang, 1981, from WMO, 1982).

involving NO and NO_2, HNO_3 and other nitrogen compounds such as NO_3 and N_2O_5 must also be considered. In this case, as in the case of the chlorofluorocarbons, model predictions have appreciably evolved as the chemical kinetics became better understood (see Fig. 7.28, §7.6.5).

The measurement of the rate of the reaction between NO and HO_2 by Howard and Evenson (1977) showed that the methane oxidation cycle (see Chapter 5) would provide an efficient source of odd oxygen in the lower stratosphere, and that the injection of nitric oxide could actually produce ozone rather than destroy it. The effect of NO_x emission on ozone thus depends directly on the altitude of aircraft flight; high altitude emissions could reduce the ozone column but low altitude emissions (upper troposphere) could increase the total ozone abundance (see, e. g., Liu et al., 1980). Figure 7.25 presents a calculation of the variation of the ozone density between 0 and 60 km in response to a continuous injection of NO of $10^8 cm^{-2}$ into a one kilometer layer at 17 and at 20 km, respectively (Wuebbles and Chang, 1981). These results depends strongly on the calculated densities of several radical species, i. e., on the adopted reaction rate constants. Low OH densities, for example, increase the concentration of NO_2 in comparison to HNO_3 and that of HCl in comparison to ClO. In this case, ozone is more sensitive to NO_x than to chlorine oxide perturbations. The theoretical evaluation of the probable influence of any chemical perturbation must include all the appropriate reactions, as well as the coupling between the various chemical families.

7.6.5 The chlorofluorocarbons (CFC's)

The importance of chlorine catalyzed destruction of odd oxygen was described in detail in Chapter 5. The chlorine atoms which are present in the atmosphere are produced by dissociation of complex molecules whose origin, either natural or artificial, is found at ground level. CH_3Cl (methyl chloride) is apparently of natural origin, and its tropospheric abundance is about 615 pptv (Singh et al., 1979). It is produced by the conversion of methyl iodide (CH_3I) originating from marine algae in the presence of chloride ions in the ocean, as well as by forest fire burning in tropical regions (Lovelock, 1975; Singh et al., 1979; Rasmussen et al., 1980). Hydrochloric acid (HCl) is emitted both by volcanoes and by industrial processes; other inorganic halides are also produced but have no effect on the middle atmosphere because these gases are rapidly "washed out" in precipitation, and therefore do not reach the stratosphere in significant quantity.

Industry produces a considerable number of halocarbons for many different purposes. Some are only intermediates in certain fabrication processes and thus are not, in principle, emitted in quantity into the atmosphere. Others are soluble, like HCl, and do not reach the middle atmosphere. A limited number of species remain which are capable of altering the protective layer of atmospheric ozone. Some of these are indicated in Table 7.1. Dichloro 1,2 ethane

(CH_2Cl-CH_2Cl) and vinyl chloride $(CH_2=CHCl)$ are produced in the largest amounts, but are not widely dispersed in the atmosphere, and are relatively rapidly destroyed in the troposphere, so that their concentration is not very great in the middle atmosphere. On the other hand, carbon tetrachloride CCl_4 is predominantly an industrial intermediate, but is found in the remote troposphere in relatively large amounts (about 130 pptv, see Rasmussen et al., 1981), indicating that this species is very stable at low altitudes. Other compounds which appear in the table (such as CHF_2Cl, $CHCl_3$, C_2H_5Cl), are emitted in small amounts, and play only a minor role in the atmospheric chlorine budget.

The production of odd chlorine in the stratosphere is primarily due to dissociation of trichloroflouromethane $(CFCl_3)$, also called F-11, dichlorodifluoromethane (CF_2Cl_2), or F-12, and to a lesser extent at present, methylchloroform (CH_3CCl_3). The importance of the first two species was pointed out by Molina and Rowland in 1974, while the role of methyl chloroform was not recognized until about 1980.

The chlorofluorocarbons F-11 and F-12 possess certain physical characteristics (particularly their high vapor pressures and low solubilities in water) which make them very useful for some industrial applications. They have been widely used as propellants in aerosol cans, as refrigerants, and as solvents and

Table 7.1. Amounts of principal halocarbons produced and released in 1973 based on several estimates (10^3 metric tons per year). Adapted from Jesson, (1982). Average mixing ratios of these species in 1981 (pptv) from WMO, 1982.

Compound	World production	World release	Tropospheric mixing ratio	
			NH	SH
CCl_4 (F-10)	950	42 to 90	135	130
$CFCl_3$ (F-11)	354	290	190	170
CF_2Cl_2 (F-12)	447	386	300	270
CHF_2Cl (F-22)	77	36	52	42
CH_3CCl_3	420	324 to 423	165 to 180	120
$CHCl_3$	225	12.4	10 to 15	3
C_2H_5Cl	550	14.6		
CH_2ClCH_2Cl	12000	565	30 to 40	20
CH_3Cl	400	7.9	650	650
CH_2Cl_2	425	346	35 to 50	20
$CCl_2=CCl_2$	740	609	40 to 50	20
$CCl_2=CHCl$	700	648	10 to 15	< 3
$CH_2=CHCl$	7100	352		

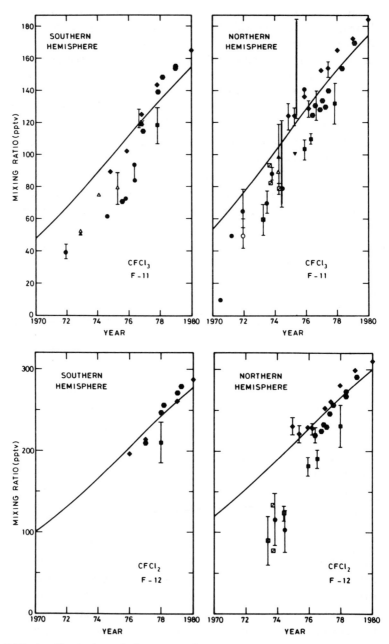

Fig. 7.26a,b. Evolution of the tropospheric mixing ratios of CFCl$_3$ and CF$_2$Cl$_2$ for the southern and northern hemispheres during the 1970's. From WMO (1982) and Logan et al. (1981).

foam blowing agents.

The total amount of F-11 and F-12 released to the atmosphere through 1980 was 4270 and 6153 tons, respectively, while the corresponding values twenty years earlier were only 250 and 640 tons, according to the Chemical Manufacturers Association (1981). The rapid increase in the emission rates of these gases increased their atmospheric mixing ratios (see Figs. 7.26a and 7.26b), but it must be noted that the rate of this increase has reduced dramatically since the mid-1970's (see Fig. 7.27, and Prinn et al., 1983a) both because of public and private environmental protection measures and because of a decrease

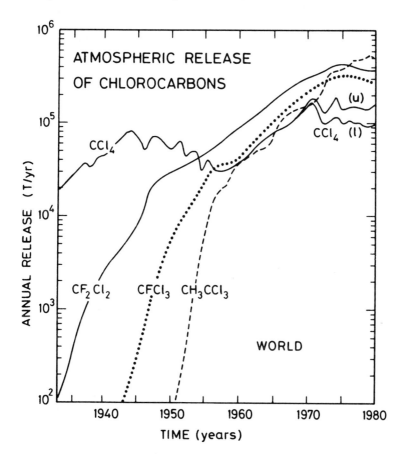

Fig. 7.27. Atmospheric release of the halocarbons CCl_4, $CFCl_3$, CF_2Cl_2, and CH_3CCl_3. Data for F-11 and F-12 were provided by the Chemical Manufacturers Association (1983). Releases for CH_3CCl_3 are from Neely and Plonka (1978) and Prinn et al. (1983b), while those for CCl_4 are from Simmonds et al. (1983) with an upper and lower limit in the later years.

in the world demand. The atmospheric lifetimes of these species are not known with complete accuracy. According to calculations done with several models (WMO, 1982), the lifetime of $CFCl_3$ is about 75 ± 15 years and that of CF_2Cl_2 is about 140 ± 30 years. Observational studies have deduced a lifetime between 15 years and infinity for $CFCl_3$ and between 20 years and infinity for CF_2Cl_2 (Lovelock, 1975; Singh et al., 1979; Rasmussen and Khalil, 1980). An intensive measurement campaign was carried out to examine the lifetimes of these species more accurately (Prinn et al., 1983a). This experiment deduced lifetimes of about 78 years for $CFCl_3$ and 69 years for CF_2Cl_2 (Cunnold et al., 1983a,b). In any case, both compounds are clearly very stable in the troposphere and are only destroyed at high altitudes through photolysis (see Chapter 5). Their tropospheric mole fractions were about 200 pptv for F-11 and 300 pptv for F-12 in 1982.

Methylchloroform is destroyed in the troposphere by reaction with the OH radical, and its lifetime is therefore only 7 ± 4 years. This species can also be

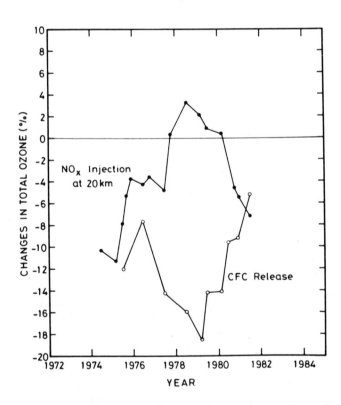

Fig. 7.28. History of *calculations* of the expected change in total ozone from CFC release and from an NO_x injection of $2\times10^8 \text{xm}^{-2}\text{s}^{-1}$ at 20 km during the 1970's. From Wuebbles et al. (1983), and WMO (1982).

photodissociated in the stratosphere.

The destruction of the CFC's ultimately produces chlorine atoms, which attack stratospheric ozone in a catalytic cycle. The appropriate reactions, involving Cl, ClO, $ClONO_2$, HOCl and HCl were discussed in Chapter 5. Several chemical cycles must be considered, and their relative importance in ozone destruction is a function of altitude. When HCl is formed, odd chlorine is sequestered in an inert reservoir, and can eventually removed from the atmosphere by rainout and washout.

The study of the effects of CFC's has often been performed with one-dimensional models (see the summary by WMO, 1982). These models can evaluate the time dependent ozone changes resulting from a continuous CFC emission, which is generally assumed to remain a constant at the 1976 level. Some models include temperature feedback, which has an important effect on the calculated ozone reductions. Figure 7.30 (see below) shows a typical calculated variation in the ozone profile including the feedback effect, and other coupled perturbations. It should be noted that the maximum decrease is found near 40 km, where it is of the order of 40%. Such a reduction changes the depth of penetration of solar ultraviolet radiation and therefore slightly increases the rate of ozone production in the lower stratosphere, so that an increase in the ozone concentration near 20 km is predicted. This increase is

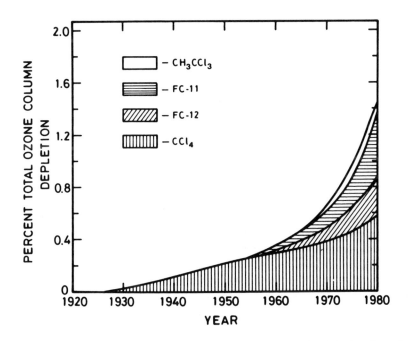

Fig. 7.29. Contribution of the 4 chlorocarbons in the calculated reduction of the ozone column as a function of time (Derwent, 1981; WMO, 1982).

also partly due to the coupling between chlorine and nitrogen compounds due to the formation of $ClONO_2$, reducing the amount of odd nitrogen in the stratosphere (another ozone destroying catalyst).

The resulting decrease in the ozone column is relatively small (a few percent), although it may be locally larger at certain latitudes. This numerical value depends strongly on the photochemical scheme which is adopted and can vary from one model to another. Because the chemical kinetic rate constants provided by laboratory studies have changed in recent years, the magnitude of the predicted changes has also undergone substantial fluctuations, and are likely to continue to do so in the future. Figure 7.28 shows, for example, that the predicted equilibrium ozone reduction was about 18 to 20% in 1979, but only 5% in 1981 due to newer values of the kinetic rates (WMO, 1982).

A calculation of the theoretical ozone reduction due to the known emission of the 4 principal chlorocarbons was presented by Derwent (1981) (see WMO, 1982, and Fig. 7.29). This study suggests that the impact of carbon tetrachloride on ozone dates back to 1930, but its relative contribution has progressively decreased since 1950 while those of F-11 and F-12 increased. The effect of this species may again become important if its emission rate continues at present values.

7.6.6 Simultaneous perturbations

The perturbations which have been discussed in the preceding sections are likely to produce simultaneous effects whose amplitude will not be equal to the sum of the individual effects. We have already seen how this may occur due to the coupling between families. As an example, the sensitivity of stratospheric ozone to a CO_2 increase will decrease if the amount of atmospheric chlorine increases. The mechanism of ozone destruction which is most sensitive to temperature, and therefore to CO_2, is the reaction $O + O_3 \rightarrow 2O_2$, but this process will become less important relative to the chlorine catalyzed ozone destruction process (which is only slightly temperature sensitive), as the atmospheric chlorine content increases.

It is thus necessary to study the simultaneous effects of several perturbations using models which include the appropriate chemical and thermal feedbacks. Such simulations have been presented by Wuebbles et al. (1983), Owens et al. (1983) and Brasseur et al. (1984), using historical records of the emissions of perturbing gases and adopting various projections for the future. Figure 7.30 shows the calculated variations in ozone densities in response to a doubling in CH_4, in CO_2, in N_2O, and for a steady injection of chlorocarbons, as well as the simultaneous response to these perturbations. The calculated variation of the ozone column compared to the pre-industrial era is presented in Figure 7.31 for the period 1970-2080. The scenario adopted for each species is described by Wuebbles et al. (1983). In this particular case, consideration of all the simultaneous perturbations suggests that the total ozone column is maintained at

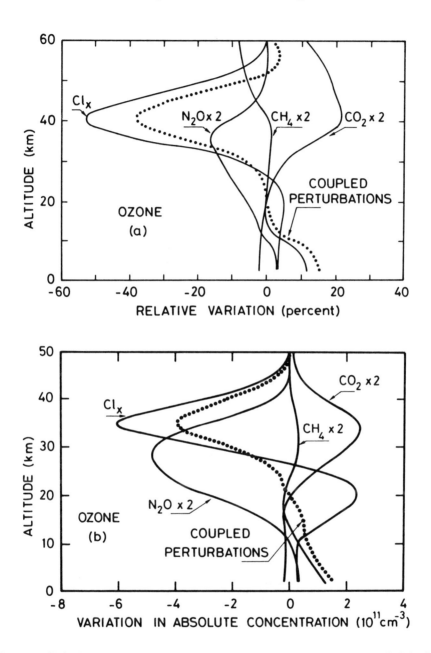

Fig. 7.30. Relative variation of the ozone concentration as a function of altitude calculated for a doubling in CH_4, in CO_2, in N_2O, and for a steady injection of chlorocarbons corresponding to the 1978 level. The ozone response to simultaneous perturbations is also indicated. From Brasseur et al. (1984)

nearly constant levels. This apparent stability, however, results from a compensating effect of locally significant perturbations when the entire column is integrated with altitude. The variation of gradients in the concentration, and thus in the temperature, may affect the general circulation and eventually the climate, but the amplitude of such an effect is difficult to estimate, even with the most sophisticated models.

The models indicate that the variation of the total column compared to the pre-industrial era will be very small, even imperceptible. Statistical analyses of ozone data (St. John et al., 1981; Reinsel et al., 1981; 1984) have not detected any change in the ozone column since 1970. Such analyses are complex because they are based on a limited number of unevenly distributed ground stations. The detection of such a signal must be extracted from data exhibiting a great degree of seasonal and dynamical variability (as we have seen in Chapters 3 and 5) as well as a variation with solar activity.

Satellite observations provide a much improved spatial coverage of the ozone layer, but their short duration and the possibility of instrument drift introduce considerable obstacles to the determination of ozone changes. The continued pursuit of observational and theoretical understanding of the fundamental processes controlling the ozone layer, its natural variability, and its

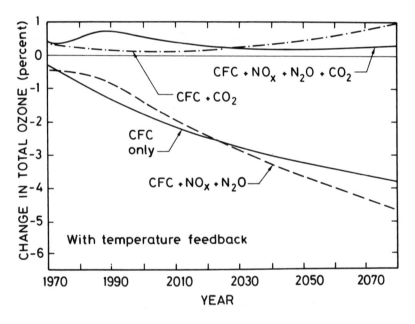

Fig. 7.31. Calculated changes in total ozone as a function of time relative to the 1911 background atmosphere for various combinations of anthropogenic perturbation scenarios. From Wuebbles et al. (1983). (Copyright by the American Geophysical Union).

response to anthropogenic perturbations remain a central focus in the study of the middle atmosphere.

References

Allen, M., J. I. Lunine and Y. L. Yung, The vertical distribution of ozone in the mesosphere and lower thermosphere, in press, J. Geophys. Res., 1984.

Angell, J.K. and J. Korshover, Global analysis of recent total ozone fluctuations, Mon. Weather Rev., 104, 63, 1976.

Angell, J.K. and J. Korshover, Global ozone variations : An update into 1976, Mon. Weather Rev., 106, 725, 1978.

Bacastow, R.A. and C.D. Keeling, Atmospheric CO2 and radiocarbon in the natural carbon cycle. II : Changes from AD 1700 to 2070 as deduced from a geochemical model, in carbon and and the Biosphere ; 86-135, US Atomic Energy Commission, Washington, DC, 1973.

Barnett, J.J., J.T. Houghton and J.A. Pyle, The temperature dependence of ozone at the stratopause, Quart. J. Roy. Met. Soc., 101, 245, 1975.

Barnett, J.J., quoted in Heath, D.F., Spatial and temporal variability of ozone as seen from space, in Proceedings of the NATO Advanced Institute on Atmospheric Ozone, A. Aikin (ed.), US Dept. of Transportation, Report FAA-EE-80-20, 1980.

Benzi, R., G. Parisi, A. Sutera and A. Vulpiani, Stochastic resonance in climatic change, Tellus, 34, 10, 1982.

Blake, D.R., E.W. Mayer, S.C. Tyler, Y. Makide, D.C. Montague and F.S. Rowland, Global increase in atmospheric methane concentrations between 1978 and 1980, Geophys. Res. Lett., 9, 477, 1982.

Brasseur, G. and P.C. Simon, Stratospheric chemical and thermal response to long-term variability in solar UV irradiance, J. Geophys. Res., 86, 7343, 1981.

Brasseur, G., P. De Baets and A. De Rudder, Solar variability and minor constituents in the lower thermosphere and in the mesosphere, Space Sci. Rev., 34, 377, 1983.

Brasseur, G. and P. De Baets, Minor constituents in the mesosphere and lower thermosphere, to be submitted to J. Geophys. Res., 1984.

Brasseur, G., A. De Rudder and C. Tricot, Stratospheric response to chemical perturbations, to be submitted to J. Atmos. Chem., 1984.

Breitenbeck, G.A., A.M. Blackmer and J.M. Bremner, Effects of different nitrogen fertilizers on emission of nitrous oxide from soil, J. Geophys. Res., 7, 85, 1980.

Brusa, R.W. and C. Frohlich, Recent solar constant determinations from high altitude ballonns, Paper presented at the symposium on the solar constant and the spectral distribution of solar irradiance, IAMAP third scientific Assembly, Published by the Radiation Commission, Boulder, CO, USA, 1982.

Budyko, M.I., The effect of solar radiation variations on the climate of the earth, Tellus, 21, 611, 1969.

Callis, L.B. and M. Natarajan, Atmospheric carbon dioxide and chloro-fluoromethanes : combined effects on stratospheric ozone, temperature and surface temperature, Geophys. Res. Lett., 8, 587, 1981.

Chemical Manufacturers Association, Private Communication, 1981.

Chemical Manufacturers Association, Private communication, 1983.

CIAP : The natural stratosphere of 1974. Climatic Impact Assessment Program, Monograph 1, Report No DOT-TST-75-51, Dept. of Transportation, Washington, D.C., 1974.

Crutzen, P.J., The influence of nitrogen oxides on the atmospheric ozone content, Quart. J. Roy. Met. Soc., 96, 320, 1970.

Crutzen, P.J., I.S.A. Isaksen and G.C. Reid, Solar proton events : stratospheric sources of nitric oxide, Science, 189, 457, 1975.

Crutzen, P.J., Upper limits on atmospheric ozone reduction following increased applications of fixed nitrogen to the soil, Geophys. Res. Lett., 3, 169, 1976.

Cunnold, D.M., R. G. Prinn, R. A. Rasmussen, P. G. Simmonds, F. N. Alyea, C. A. Cardelino, and A. J. Crawford, The atmospheric lifetime experiment 4. Results for CF_2Cl_2 based on three years of data, J. Geophys. Res., 88, 8401, 1983a.

Cunnold, D. M., R. G. Prinn, R. A. Rasmussen, P. G. Simmonds, F. N. Alyea, C. A. Cardelino, A. J. Crawford, P. J. Fraser, and R. D. Rosen, The atmospheric lifetime experiment 3. Lifetime methodology and application to three years of $CFCl_3$ data, J. Geophys. Res., 88, 8379, 1983b.

Dalgarno, A., Atmospheric reactions with energetic particles, Space Research, 7, 849, 1971.

De Baets, P., G. Brasseur and P.C. Simon, Chemical response of the middle atmosphere to solar variations, Solar Physics, 74, 349, 1981.

Derwent, R.G. (1981), Results presented in WMO (1982).

Dickinson, R.E., Method of parametrization for infrared cooling between the altitude of 30 and 70 km, J. Geophys. Res., 78, 4451, 1973.

Donner, L. and V. Ramanathan, Methane and nitrous oxide : Their effects on the terrestrial climate, J. Atmos. Sci., 37, 119, 1980.

Duetsch, H.U., The search for solar cycle-ozone relationships, J. Atmos. Terr. Physics, 41, 771, 1979.

Ebel, A. and W. Baetz, Response of stratospheric circulation at 10 mb to solar activity oscillations resulting from the sun's rotation, Tellus, 29, 41, 1977.

Ehhalt, D.H., R.J. Zander and R.A. Lamontagne, On the temporal increase of tropospheric CH_4, J. Geophys. Res., 88, 8442, 1983.

Eliassen, A., Slow thermally and frictionally controlled meridional circulations in a circular vortex, Astrophys. Norv., 5, 19, 1952.

Fabian, P., J.A. Pyle and R.J. Wells, The August 1972 solar proton event and the atmospheric ozone layer, Nature, 277, 458, 1979.

Fels, S.B., J.D. Mahlman, M.D. Schwarzkopf and R. W. Sinclair, Stratospheric sensitivity of perturbations in ozone and carbon dioxide : Radiative and dynamical response, J. Atm. Sci., 37, 2265, 1980.

Fraedrich, K., Structural and stochastic analysis of a zero-dimensional climate system, Quart. J. Roy. Met. Soc., 104, 461, 1979.

Gage, K.S. and G.C. Reid, Coherent annual and interannual variations in temperature and height fields in the tropical troposphere and lower stratosphere, Geophys. Res. Lett., 9, 1199, 1982.

Garcia, R., S. Solomon, R.G. Roble and D.W. Rusch, A numerical study of the response of the middle atmosphere to the 11 year solar cycle, Planet. Space Sci., in press, 1984.

Geller, M.A. and J.C. Alpert, Planetary wave coupling between the troposphere and the middle atmosphere as a possible sun-weather mechanism, J. Atmos. Sci., 37, 1197, 1980.

Ghazi, A. and J.J. Barnett, Ozone behavior and stratospheric thermal structure during southern hemisphere spring, Beitr. Phys. Atm, 53, 1, 1980.

Ghil, M., Climatic stability for a Sellers-type model, J. Atmos. Sci., 36, 260, 1976.

Ghil, M., Energy-balance models : an introduction, in Climatic Variations and Variability : Facts and Theories (A. Berger, ed.), Reidel Publishing Co, p. 461, 1981.

Groves, K.S. and A.F. Tuck, Stratospheric O_3-CO_2 coupling in a photochemical - radiative column model. I : Without chlorine chemistry, Quart. J. Roy. Met. Soc., 106, 125, 1980.

Haigh, J.D. and J.A. Pyle, Ozone perturbation experiments in a two-dimensional circulation model, Quart. J. Roy. Met. Soc., 108, 551, 1982.

Hampson, R.F. and D. Garvin, Chemical kinetic and photochemical data for modelling atmospheric chemistry, US Dept. of Commerce, NBS Technical note 866, 1975.

Hansen, J.E., W.C. Wang and A.A. Lacis, Mount agung eruption provides test of a global climatic perturbation, Science, 199, 1065, 1978.

Heath, D.F., A.J. Krueger and P.J. Crutzen, Solar proton event : Influence on stratospheric ozone, Science, 197, 886, 1977.

Heath, D.F., Spatial and temporal variability of ozone as seen from space, Proceedings NATO Advanced Study Inst. on atmospheric ozone (A. Aikin ed)., Report No FAA-EE-80;20, 45, 1980.

Heath, D.F., B.M. Schlesinger and H. Park, Spectral changes in the ultraviolet absorption and scattering properties of the atmosphere associated with the eruption of El Chichon : Stratospheric SO_2 budget and decay (Abstract) EOS, 64, 197, 1983.

Hines, C.O., A possible mechanism for the production of sun-weather correlations, J. Atmos. Sci., 31, 589, 1974.

Hofmann, D. J., and J. M. Rosen, Stratospheric sulfur acid fraction and mass estimate for the 1982 volcanic eruption of El Chichon, Geophys. Res. Lett., 10, 313, 1983.

Howard, C.J. and K.M. Evenson, Kinetics of the reaction of HO_2 with NO, Geophys. Res. Lett., 4, 437, 1977.

Humphreys, W.J., *Physics of the Air,* Mc Graw-Hill, (New York, N.Y.), 1940.

Jackman, C.H., J.E. Frederick and R.S. Stolarski, Production of odd nitrogen in the stratosphere and mesosphere: an intercomparison of source strengths, J. Geophys. Res., 85, 7495, 1980.

Jackman, C. H. and R. D. McPeters, The response of ozone to solar proton events during solar cycle 21: a theoretical study, submitted to J. Geophys. Res., 1984.

Jesson, J.P., Halocarbons, in Stratospheric ozone and man, CRC Press Boca Raton, Florida, Volume II, p. 29, 1982.

Johnston, H.S., Reduction of stratospheric ozone by nitrogen oxide catalysts from SST exhaust, Science, 173, 517, 1971.

Jones, P.D., T.M.L. Wigley and P.M. Kelly, Variations in surface air temperature: Part I. Northern Hemisphere, 1881-1980, Mon. Wea. Rev., 110, 59, 1982.

Khalil, M.A.K. and R.A. Rasmussen, Increases in atmospheric concentrations of halocarbons and N_2O, Geophys. Monit. Clim. Change, 9, 134, 1981.

Khalil, M.A.K. and R.A. Rasmussen, Increase and seasonal cycles of nitrous oxide in the earth's atmosphere, Tellus, 35B, 161, 1983.

Krueger, A.J., B.W. Guether, A.J. Fleig, D.F. Heath, E. Hilsenrath, R. Mc Peters and C Prabakhara, Satellite ozone measurements, Phil. Trans. R. Soc. Lon., A, 296, 191, 1980.

Labitzke, K. and B. Naujokat, On the variability and on trends of the temperature in the middle stratosphere, Beit. Phys. Atmosf., in press, 1984.

Lacis, A., J. Hansen, P. Lee, T. Mitchell and S. Lebedeff, Greenhouse effect of trace gases, 1970-1980, Geophys. Res. Lett., 8, 1035, 1981.

Lamb, H.H., Volcanic dust in the atmosphere, with a chronology and assessment of its meteorological significance, Philos. Trans. R. Soc. London, 266, 425, 1970.

Landsberg, H.E. and J.M. Albert, The summer of 1816 and volcanism, Weatherwise, 27, 63, 1974 .

Laurmann, J.A. and J.R. Spreiter, The effects of carbon cycle model error in calculating future atmospheric carbon dioxide levels, Climatic Change, 5, 145, 1983.

Liu, S.C., R.J. Cicerone and T.M. Donahue, Sources and sinks of atmospheric N_2O and the possible ozone reduction due to industrial fixed nitrogen fertilizers, Tellus, 29,251, 1977.

Liu, S.C., D. Kley, M. McFarland, J.D. Mahlman and H. Levy II, On the origin of tropospheric ozone, J. Geophys. Res., 85, 7546, 1980.

Logan, J.A., M.J. Prather, S.C. Wofsy and M.B. McElroy, Tropospheric chemistry: a global perspective, J. Geophys. Res., 96, 7210, 1981.

London, J. and M.W. Haurwitz, Ozone and sunspots, J. Geophys. Res., 68, 795, 1963.

London, J. and C.A. Reber, Solar activity and total atmospheric ozone, Geophys. Res. Lett., 6, 869, 1979.

London, J. and T. Sasamori, in *Man's Impact on the Climate,* W.H. Kellogg and G.D. Robinson (eds.) pp. 141-155, MIT, Press, Cambridge, MA, 1971.

Lovelock, J.E., Natural halocarbons in the air and in the sea, Nature, 256, 193, 1975.

Luther, F. M., D. J. Wuebbles and J. S. Chang, Temperature feedback in a stratospheric model, J. Geophys. Res., 82, 4935, 1977.

Mahlman, J.D., H. Levy II and W.J. Moxim, Three-dimensional tracer structure and behavior as simulated in two ozone precursor experiments, J. Atmos. Sci., 37, 655, 1980.

Manabe, S. and R.T. Wetherald, Thermal equilibrium of the atmosphere with a given distribution of relative humidity, J. Atmos. Sci., 24, 241, 1967.

Manabe, S. and R.T. Wetherald, The effects of doubling the CO_2 concentration on the climate of a general circulation model, J. Atmos. Sci., 32, 3, 1975.

Manabe, S. and R.J. Stouffer, A CO_2 - climate sensitivity study with a mathematical model of the global climate, Nature, 282, 491, 1979.

Manabe, S. and R.J. Stouffer, Sensitivity of a global climate model to an increase of CO_2 concentration in the atmosphere, J. Geophys. Res., 85, 5529, 1980.

Manabe, S. and R.T. Wetherald, On the distribution of climate change resulting from an increase in CO_2 content of the atmosphere, J. Atmos. Sci., 37, 99, 1980.

Mass, C. and H.H. Schneider, Statistical evidence on the influence of sunspots and volcanic dust on long-term temperature records, J. Atmos. Sci., 34, 1995, 1977.

Mastenbrook, H. J., and S. J. Oltmans, Stratospheric water vapor variability over Washington DC/Boulder CO: 1964-82, J. Atmos. Sci., 40, 2157, 1983.

McElroy, M.B., S.C. Wofsy and Y.L. Yung, The nitrogen cycle : Perturbations due to man and their impact on atmospheric N_2O and O_3, Phil. Trans. R. Soc. Lond., B277, 159, 1977.

McElroy, M.B., Sources and sinks for nitrous oxide, in Proceedings of the NATO Advanced Study Institute on Atmospheric ozone (ed. A.C. Aikin), US Dept. of Transportation Report No FAA-EE-80-20, 1980.

McKeen, S.A., S.C. Liu and C.S. Kiang, On the chemistry of stratospheric SO_2 from volcanic eruptions, in press, J. Geophys. Res., 1984.

Miles, M.K. and P.B. Gildersleeves, Volcanic dust and changes in northern hemisphere temperature, Nature, 271, 735, 1978.

Molina, M., and F. S. Rowland, Stratospheric sink for chlorofluoromethanes: Chlorine atom catalyzed destruction of ozone, Nature, 249, 810, 1974.

NAS (National Academy of Sciences), *Changing climate: report of the carbon dioxide assessment committee,* National Academy Press, (Washington, D. C.), 1983.

Nastrom, G.D. and A.D. Belmont, Preliminary results on 27-day solar rotation variation in stratospheric zonal winds, Geophys. Res. Lett., 5, 665, 1978.

Natarajan, M., L.B. Callis and J.E. Nealy, Solar UV variability : Effects on stratospheric ozone, trace constituents and thermal structure, Pure Appl. Geophys., 19, 750, 1981.

Neely, W.B. and J.H. Plonka, Estimation of the time-averaged hydroxyl radical concentration in the troposphere, Env. Sci. and Tech., 12, 317, 1978.

Niehaus, F. and H. Williams, Studies of different energy strategies in terms of their effects on the atmospheric CO_2 concentration, J. Geophys. Res., 84, 3123, 1979.

Niehaus, F., The impact of energy production on atmospheric CO_2 concentrations, in *Climatic variations and variability: facts and theories,* A. Berger, ed., (Reidel, Dordrecht), 1981.

Nicolis, C. and G. Nicolis, Environmental fluctuation effects on the global energy balance, Nature, 281, 132, 1979.

Nicolis, C., Stochastic aspects of climatic transitions - response to a periodic forcing, Tellus, 34, 1, 1982.

North, G.R., R.F. Cahalan and J.A. Coakley Jr., Energy balance climate models, Rev. Geophys. Space Phys., 19, 91, 1981.

NRC/CRB : carbon dioxide and climate. A scientific assessment, Report of the ad hoc study group on carbon dioxide and climate, Woods Hole, Mass. (Chairman J.G., Charney), 1979.

Orsini, N. and J.E. Frederick, Solar disturbances and mesospheric odd nitrogen, J. Atmos. Terr. Phys., 44, 489, 1982.

Owens, A.J., J.M. Steed, D.L. Filkin, C. Miller and J.P. Jesson, The potential effects of increased methane on atmospheric ozone, Geophys. Res. Lett., 9,

1105, 1982.

Owens, A.J., C.H. Hales, D.L. Filkin, C. Miller, A. Yokozeki, J.M. Steed and J.P. Jesson, A coupled-one-dimensional radiative convective chemistry - transport model of the atmosphere. I. Model structure and steady-state perturbation calculations, Manuscript, 1983.

Paetzold, H.K., Variation of the vertical ozone profile over middle Europe from 1951 to 1968, Ann. Geophys., 25, 347, 1969.

Penner, J.E. and J.S. Chang, The relation between atmospheric trace species variabilities and solar UV variability, J. Geophys. Res., 85, 5523, 1980.

Pierotti, D. and R.A. Rasmussen, Combustion as a source of nitrous oxide in the atmosphere, Geophys. Res. Lett., 3, 265, 1976.

Pittock, A.B., A critical look at long-term sun-weather relationships, Rev. Geophys. Space Phys., 16, 400, 1978.

Pollack, J.B., O.B. Toon, C. Sagan, A. Summers, B. Baldwin and W. Van Camp, Volcanic explosions and climatic change : A theoretical assessment, J. Geophys. Res., 81, 1071, 1976.

Pollack, J.B., O.B. Toon, E.F. Danielsen, D.J. Hofmann and J.M. Rosen, The El Clichon volcanic cloud : An introduction, Geophys. Res. Lett., 10, 989, 1983.

Prinn, R. G., P. G. Simmonds, R. A. Rasmussen, R. D. Rosen, F. N. Alyea, C. A. Cardelino, A. J. Crawford, D. M. Cunnold, P. J. Fraser, and J. E. Lovelock, The atmospheric lifetime experiment, 1. Introduction, instrumentation and overview, J. Geophys. Res., 88, 8353, 1983a.

Prinn, R.G., R.A. Rasmussen, P.G. Simmonds, F.N. Alyea, D.M. Cunnold, B.C. Lane, C.A. Cardelino and A.J. Crawford, The Atmospheric lifetime experiment. 5 : Results for CH_3CCl_3 based on three years of data, J. Geophys. Res., 88, 8415, 1983b.

Quiroz, R.S., Stratospheric temperatures during solar cycle 20, J. Geophys. Res., 84, 2415, 1979.

Quiroz, R.S., The isolation of stratospheric temperature change due to El Chichon volcanic eruption from non-volcanic signals, J. Geophys. Res., 88, 6773, 1983.

Ramanathan, V., L.B. Callis and R.E. Boughner, Sensitivity of surface temperature and atmospheric temperature to perturbations in the stratospheric concentrations of ozone and nitrogen dioxide, J. Atmos. Sci., 33, 1092, 1976.

Ramanathan, V. and R.E. Dickinson, The role of stratospheric ozone in the zonal and seasonal radiative energy balance of the earth - troposphere system, J. Atmos. Sci., 36, 1084, 1979.

Rasmussen, R.A. and M.A.K. Khalil, Atmospheric halocarbons : measurements of selected trace gases in Proceedings of the NATO Advanced Study Institute on Atmospheric Ozone (A.C. Aikin ed.), US Dept. of Transportation

Report FAA, 1980.

Rasmussen, R.A., L.E. Rasmussen, M.A.K. Khalil and R.W. Dalluge, Concentration distribution of methyl chloride in the atmosphere, J. Geophys. Res., 85, 7350, 1980.

Rasmussen, R.A. and M.A.K. Khalil, Atmospheric methane (CH_4): trends and seasonal cycles, J. Geophys. Res., 86, 9826, 1981.

Rasmussen, R.A., M.A.K. Khalil and R.W. Dalluge, Atmospheric trace gases in Antartica, Science, 211, 285, 1981.

Reagan, J.B., R.E. Meyerott, R.W. Nightingale, R.C. Gunton, R.G. Johnson, J.E. Evans and W.L. Imhof, Effects of the August 1972 solar particle events on stratospheric ozone, J. Geophys. Res., 86, 1473, 1981.

Reinsel, G.C., G.C. Tiao, M.N. Wang, R. Lews and D. Nychka, Statistical analysis of stratospheric ozone data for the detection of trends, Atmos. Environ., 15, 1569, 1981.

Reinsel, G.C., G.C. Tiao, A.J. Miller, C.L. Mateer, J. Deluisi and J.E. Frederick, Analysis of upper stratospheric umkehr ozone profile data for trends and the effect of stratospheric aerosols, in press, J. Geophys. Res., 1984.

Robbins, R.C., L.A. Cavanagh, L.J. Salas and E. Robinson, Analysis of ancient atmospheres, J. Geophys. Res., 78, 5341, 1973.

Robock, A., The "Little Ice Age" : Northern hemisphere average observations and model calculations, Science, 206, 1402, 1979.

Rottman, G.J., 27-day variations observed in solar ultraviolet (120-300 nm), Planet. Space Sci., 31, 1001, 1983.

Rundel, R.D., Action spectra and estimation of biologically effective UV radiation, Physiol. Plant., 58, 360, 1983.

Rusch, D.W., J.C. Gerard, S. Solomon, P.J. Crutzen and G.C. Reid, The effect of particle precipitation on the neutral and ion chemistry of the middle atmosphere, I. Odd nitrogen, Planet. Space Sci., 29, 767, 1981.

Schneider, S.H. and R.E. Dickinson, Climate modelling, Rev. Geophys. Space Phys., 12, 447, 1974.

Schoeberl, M.R. and D.F. Strobel, The response of the zonally averaged circulation to stratospheric ozone reductions, J. Atmos. Sci., 35, 1751, 1978.

Schwentek, H., Regular and irregular behavior of the winter anomaly in ionospheric absorption, J. Atm. Terr. Phys., 33, 1647, 1971.

Seiler, W., and P. J. Crutzen, Estimates of gross and net fluxes of carbon between the biosphere and the atmosphere from biomass burning, Climatic Change, 2, 207, 1980.

Sellers, W.D., A climate model based on the energy balance of the earth-atmosphere system, J. Appl. Met., 8, 392, 1969.

Singh, H.B., L.J. Salas, H. Shigeishi and E. Scribner, Atmospheric halocarbons and sulphur hexafluoride : Global distributions, sources and sinks, Science,

1203, 899, 1979.

Simmonds, P.G., F.N. Alyea, C.A. Cardelino, A.J. Crawford, D.M. Cunnold, B.C. Lane, J.E. Lovelock, R.G. Prinn and R.A. Rasmuusen, The atmospheric lifetime experiment. 6 : Results for carbon tetrachloride based on three years of data, J. Geophys. Res., 88, 8427, 1983.

Solomon, S. and P.J. Crutzen, Analysis of the August 1972 solar proton including chlorine chemistry, J. Geophys. Res., 86, 1140, 1981.

Solomon, S., D.W. Rusch, J.C. Gerard, G.C. Reid and P.J. Crutzen, The effect of particle precipitation on the neutral and ion chemistry of the middle atmosphere. II. Odd hydrogen, Planet. Space Sci., 29, 885, 1981.

Solomon, S., P.J. Crutzen and R.G. Roble, Photochemical coupling between the thermosphere and the lower atmosphere. I. Odd nitrogen from 50 to 120 km, J. Geophys. Res., 87, 7206, 1982.

Solomon, S., G.C. Reid, D.W. Rusch, R.J. Thomas, Mesospheric ozone depletion during the solar proton event of July 13, 1982, Geophys. Res. Lett., 10, 257, 1983.

St.John, D.S., S.P. Bailey, W.H. Fellner, J.M. Minor and R.D. Snee, Time series search for trend in total ozone measurements, J. Geophys. Res., 86, 7299, 1981.

Swider, W. and T.J. Keneshea, Decrease of ozone and atomic oxygen in the lower mesosphere during a PCA event, Planet. Space Sci., 21, 1969, 1973.

Thomas, G.E., B.M. Jakosky, R.A. West and R.W. Sanders, Satellite limb-scanning thermal infrared observations of the El Chichon stratospheric aerosol : first results, Geophys. Res. Lett., 10, 997, 1983.

Thorne, R.M., The importance of energetic particle precipitation on the chemical composition of the middle atmosphere, Pure Appl. Geophys., 118, 128, 1980.

Turco, R.P., R.C. Whitten and O.B. Toon, Stratospheric aerosols : observation and theory, Rev. Geophys. Space Phys., 20, 233, 1982.

Von der Haar, T.H. and V.E. Suomi, Measurements of the Earth's radiation budget from satellites during a five - year period. Part I: Extended time and space means, J. Atm. Sci., 28, 305, 1971.

Voss, A., Anstze zur Gesamtanalyse des Systems Mensch - Energie - Umwelt, ISR 30, Birkhuser Verlag, Basel, 1977.

Vupputuri, R.K.R., A reexamination of 11-year solar cycle effects on stratospheric thermal and chemical structure in a 2-D model, in *Weather and climate responses to solar variations* (B.M. McCormac, ed.), Colorado Associated University Press, 1982.

Wang, W.C., Y.L. Yung, A.A. Lacis, T. Mo and J.E. Hansen, Greenhouse effects due to man-made perturbations of trace gases, Science, 194, 685, 1976.

Weeks, C.H., R.S. Cuikay and J.R. Corbin, Ozone measurements in the mesosphere during the solar proton event of 2 November, 1969, J. Atmos. Sci., 29, 1138, 1972.

Weiss, R. and H. Craig, Production of atmospheric nitrous oxide by combustion, Geophys. Res. Lett., 3, 751, 1976.

Weiss, R.W., The temporal and spatial distribution of tropospheric nitrous oxide, J. Geophys. Res., 86, 7185, 1981.

Willett, H.C., The relationship of total atmospheric ozone to the sunspot cycle, J. Geophys. Res., 67, 661, 1962.

WMO: World Meteorological Organization, The stratosphere 1981, Theory and measurements, WMO Global ozone research and monitoring project, Report 11, 1982.

Wuebbles, D.J. and J.S. Chang, Calculations performed in 1981 for the WMO report - see WMO, 1982.

Wuebbles, D.J., F.M. Luther and J.E. Penner, Effect of coupled anthropogenic perturbations on stratospheric ozone, J. Geophys. Res., 88, 1444, 1983.

Zerefos, C.S. and P.J. Crutzen, Stratospheric thickness variations over the northern hemisphere and their possible relation to solar activity, J. Geophys. Res., 80, 5041, 1975.

Appendix A

Numerical values of physical constants and other data

Table 1. Physical constants and other data

Fundamental Constants

Speed of light	$c = 2.9979 \times 10^8$ m s^{-1}
Planck's constant	$h = 6.6256 \times 10^{-34}$ J s
Mass of electron	$m_e = 9.1096 \times 10^{-31}$ kg
Charge of electron	$e = 1.061 \times 10^{-19}$ C
Permittivity of free space	$\epsilon_o = 8.85 \times 10^{-12}$ F m^{-1}
First radiation constant	$c_1 = 3.7415 \times 10^8$ W m^{-2} μm^4
Second radiation constant	$c_2 = 1.4388 \times 10^4$ μmK
Photon radiation constant	$c_1{}' = 1.8836 \times 10^{19}$ s^{-1} m^{-2}
Stefan-Boltzmann constant	$\sigma = 5.6697 \times 10^{-8}$ W m^{-2} K^{-4}
Wien displacement constant	$\lambda_{max} T = 2.8978 \times 10^3$ μmK
Avogadro's number	$N_A = 6.0225 \times 10^{23}$ mole^{-1}
Loschmidt's number	$N_o = 2.6871 \times 10^{19}$ molec cm^{-3}
Boltzmann's constant	$k = 1.3806 \times 10^{-23}$ J K^{-1}
Gas constant	$R = 8.31436$ J mole^{-1} K^{-1}

The Earth

Radius (mean)	$a = 6.371 \times 10^6$ m
Surface area	$S_E = 5.10 \times 10^{14}$ m^2
Acceleration of gravity	$g = 9.81$ m s^{-2}
Distance from the sun (mean)	$R = 1$ AU $= 1.496 \times 10^{11}$ m
Rotation rate of the Earth	$\Omega = 7.292 \times 10^{-5}$ s^{-1}

Data on air and water vapor

Gas constant for water vapor	$R_v = 4.6191 \times 10^2$ J kg^{-1} K^{-1}
Gas constant for dry air	$R = 287.04$ J kg^{-1} K^{-1}
Latent heat of vaporization (288K)	$L_v = 2.26 \times 10^6$ J kg^{-1} = 540 cal g^{-1}
Specific heat of water vapor	$c_p = 4.1855$ J g^{-1} K^{-1}
Specific heat of dry air	$c_p = 1.005$ J g^{-1} K^{-1}

Saturation vapor pressure

over ice (183K)	9.33×10^{-5} mb
(193K)	5.332×10^{-4} mb
(203K)	2.58×10^{-3} mb
over liquid water (243K)	0.5088 mb
(263K)	2.8627 mb
(273K)	6.1078 mb
(283K)	12.272 mb
(303K)	42.430 mb

Appendix B

Conversion factors

Table 2. Equivalent of one atmosphere (atm) in various units of pressure

760 mm Hg	760 torr	1.01325×10^{6} dyn cm^{-2}
1013.25 mb	1.01325×10^{5} N m^{-2}	1.01325×10^{5} Pa

Pressure-Altitude Conversion Chart

PRESSURE-ALTITUDE

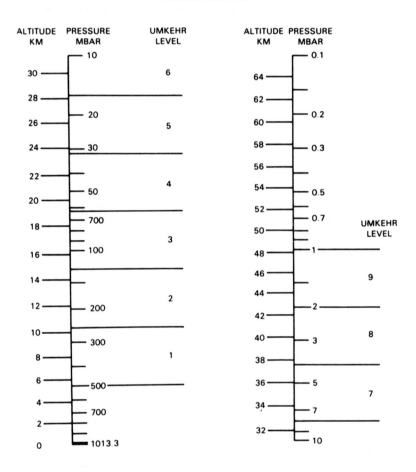

Altitudes are based on the U. S. Standard Atmosphere, 1976. The actual altitude for a given pressure may differ by as much as 5 km, depending on season, latitude, and short term variations. From World Meteorological Organization (WMO), The stratosphere 1981, theory and measurements, WMO global ozone research and monitoring project, Report no. 11, Geneva, Switzerland, 1982. See that document for a discussion of the Umkehr technique for measuring ozone.

Table 3. Conversion factors for energy units. From McCartney (1983).

Multiply by → / To obtain ↓	Joules	Ergs	eV	cm^{-1} molec^{-1}	Cal
Joules	1	1(-7)	1.6021(-19)	1.9863(-23)	4.1855
Ergs	1(7)	1	1.6021(-12)	1.9863(-16)	4.1865(7)
Electron volts	6.2418(18)	6.2418(11)	1	1.2397(-4)	2.6123(19)
cm^{-1} molec^{-1}	5.0345(22)	5.0345(15)	8.0665(3)	1	2.1072(23)
Calories	0.2389	2.3890(-6)	3.8280(-20)	4.7457(-24)	1

McCartney, E. J., *Absorption and emission by atmospheric gases*, Wiley-Interscience, (New York), 1983.

Appendix C

Reaction rate constants

As discussed in Chapters 2, 5, and 6, chemical reaction kinetics determine the rates of interaction of neutral and ionized atoms and molecules of importance in the atmosphere. Improved laboratory techniques often lead to more accurate estimates of the reaction rate constants needed for atmospheric studies (see, e.g., Fig. 7.28). In the following tables, we present estimates of the reaction rate constants appropriate to the processes indicated in this volume. We emphasize that these numbers should be checked against the most current kinetic panel evaluations for any quantitative work.

Oxygen (O_x) reactions

k_1	$O + O + M \rightarrow O_2 + M$	$4.7(-33)(300/T)^{2.0}[M]$
k_2	$O + O_2 + M \rightarrow O_3 + M$	$6.0(-34)(300/T)^{2.3}[M]$
k_3	$O + O_3 \rightarrow 2O_2$	$8.0(-12)\exp(-2060/T)$
k_{4a}	$O(^1D) + N_2 \rightarrow O(^3P) + N_2$	$1.8(-11)\exp(107/T)$
k_{4b}	$O(^1D) + O_2 \rightarrow O(^3P) + O_2$	$3.2(-11)\exp(67/T)$
k_6	$O_2(^1\Delta_g) + O_2(^3\Sigma_g^-) \rightarrow 2O_2(^3\Sigma_g^-)$	$2.22(-18)(T/300)^{0.78}$
k_7	$O(^1D) + O_3 \rightarrow 2O_2$	$1.2(-10)$
k_8	$O(^1D) + O_3 \rightarrow O_2 + 2O$	$1.2(-10)$

Carbon reactions

c_{1a}	$CH_4 + O(^1D) \rightarrow CH_3 + OH$	$1.4(-10)$
c_{1b}	$CH_4 + O(^1D) \rightarrow CH_2O + H_2$	$1.4(-11)$
c_2	$CH_4 + OH \rightarrow CH_3 + H_2O$	$2.4(-12)\exp(-1710/T)$
c_3	$CH_3 + O \rightarrow CH_2O + H$	$1.1(-10)$

c_4	$CH_3 + O_2 + M \rightarrow CH_3O_2 + M$	$4.5(-31)(300/T)^{2.0}$ [M]
c_{4a}	$CH_3 + O_2 \rightarrow CH_2O + OH$	$< 3(-16)$
c_5	$CH_3O_2 + NO \rightarrow CH_3O + NO_2$	$4.2(-12)\exp(180/T)$
c_7	$CH_3O_2 + HO_2 \rightarrow CH_3OOH + O_2$	$7.7(-14)\exp(1300/T)$
c_8	$CH_2O + OH \rightarrow CHO + H_2O$	$1.0(-11)$
c_9	$CH_2O + O \rightarrow CHO + OH$	$3.0(-11)\exp(-1550/T)$
c_{10}	$CH_2O + Cl \rightarrow HCl + CHO$	$8.2(-11)\exp(-34/T)$
c_{12}	$CHO + O_2 \rightarrow CO + HO_2$	$3.5(-12)\exp(140/T)$
c_{14}	$CH_3O_2 + CH_3O_2 \rightarrow 2CH_3O + O_2$	$1.6(-13)\exp(220/T)$
c_{15}	$CH_3O + O_2 \rightarrow CH_2O + HO_2$	$8.4(-14)\exp(-1200/T)$
c_{17}	$CH_3OOH + OH \rightarrow CH_3O_2 + H_2O$	$1.0(-11)$

Odd hydrogen (HO_x) reactions

a^*_1	$H_2O + O(^1D) \rightarrow 2OH$	$2.2(-10)$
a^*_3	$H_2 + O(^1D) \rightarrow H + OH$	$1.0(-10)$
a_1	$H + O_2 + M \rightarrow HO_2 + M$	$5.5(-32)(300/T)^{1.6}$ [M]
a_2	$H + O_3 \rightarrow OH + O_2$	$1.4(-10)\exp(-470/T)$
a_5	$OH + O \rightarrow H + O_2$	$2.2(-11)\exp(117/T)$
a_6	$OH + O_3 \rightarrow O_2 + HO_2$	$1.6(-12)\exp(-940/T)$
a_{6b}	$HO_2 + O_3 \rightarrow OH + 2O_2$	$1.4(-14)\exp(-580/T)$
a_7	$HO_2 + O \rightarrow OH + O_2$	$3.0(-11)\exp(200/T)$
a_{16}	$OH + OH \rightarrow H_2O + O$	$4.2(-12)\exp(-242/T)$
a_{17}	$OH + HO_2 \rightarrow H_2O + O_2$	$1.7(-11)\exp(416/T)$
a_{19}	$OH + H_2 \rightarrow H_2O + H$	$6.1(-12)\exp(-2030/T)$
a_{23a}	$H + HO_2 \rightarrow 2OH$	$4.2(-10)\exp(-950/T)$
a_{23b}	$H + HO_2 \rightarrow H_2 + O_2$	$4.2(-11)\exp(-350/T)$
a_{23c}	$H + HO_2 \rightarrow H_2O + O$	$8.3(-11)\exp(-500/T)$
a_{26}	$HO_2 + NO \rightarrow NO_2 + OH$	$3.7(-12)\exp(240/T)$
a_{27}	$HO_2 + HO_2 \rightarrow H_2O_2 + O_2$	$2.3(-13)\exp(590/T)$
a_{30}	$H_2O_2 + OH \rightarrow H_2O + HO_2$	$3.1(-12)\exp(-187/T)$
a_{31}	$H_2O_2 + O \rightarrow OH + HO_2$	$1.4(-12)\exp(-2000/T)$
a_{36}	$OH + CO \rightarrow H + CO_2$	$1.5(-13)(1+0.6(p/1014 \text{ mb}))$

Odd nitrogen (NO_x) reactions

b_3	$NO_2 + O \rightarrow NO + O_2$	$9.3(-12)$
b_4	$NO + O_3 \rightarrow NO_2 + O_2$	$1.8(-12)\exp(-1370/T)$
b_5	$N + NO_2 \rightarrow N_2O + O$	$3.0(-12)$
b_6	$N + NO \rightarrow N_2 + O$	$3.4(-11)$
b_7	$N + O_2 \rightarrow NO + O$	$4.4(-12)\exp(-3220/T)$
b^*_7	$N(^2D) + O_2 \rightarrow NO + O$	$5.0(-12)$
b_8	$N(^2D) + O \rightarrow N(^4S) + O$	$4.5(-13)$
b_9	$NO_2 + O_3 \rightarrow NO_3 + O_2$	$1.2(-13)\exp(-2450/T)$
b_{11}	$NO_3 + NO \rightarrow 2NO_2$	$1.3(-11)\exp(250/T)$
b_{12}	$NO_3 + NO_2 + M \rightarrow N_2O_5 + M$	See below
b_{15}	$NO_3 + O \rightarrow NO_2 + O_2$	$1.0(-11)$
b_{22}	$NO_2 + OH + M \rightarrow HNO_3 + M$	See below
b_{23a}	$NO_2 + HO_2 + M \rightarrow HO_2NO_2 + M$	See below
b_{27}	$HNO_3 + OH \rightarrow NO_3 + H_2O$	$7.2(-15)\exp(785/T)$
b_{28}	$HO_2NO_2 + OH \rightarrow H_2O + O_2 + NO_2$	$1.3(-12)\exp(380/T)$
b_{29}	$HO_2NO_2 + O \rightarrow OH + O_2 + NO_2$	$7.0(-11)\exp(-3370/T)$
b_{38}	$N_2O + O(^1D) \rightarrow N_2 + O_2$	$4.9(-11)$
b_{39}	$N_2O + O(^1D) \rightarrow 2NO$	$6.7(-11)$

Chlorofluorocarbons

$CCl_4 + O(^1D) \rightarrow$ products	$3.3(-10)$
$CFCl_3 + O(^1D) \rightarrow$ products	$2.3(-10)$
$CF_2CCl_2 + O(^1D) \rightarrow$ products	$1.4(-10)$
$CFCl_2 - CF_2Cl + O(^1D) \rightarrow$ products	$2.75(-10)$
$CClF_2 - CClF_2 + O(^1D) \rightarrow$ products	$1.62(-10)$
$CH_3CCl_3 + OH \rightarrow$ products	$5.4(-12)\exp(-1820/T)$
$CHFCl_2 + OH \rightarrow$ products	$8.9(-13)\exp(-1013/T)$
$CHF_2Cl + OH \rightarrow$ products	$7.8(-13)\exp(-1530/T)$

Chlorine reactions

d_0	$CH_3Cl+OH \rightarrow CH_2Cl+H_2O$	$1.8(-12)\exp(-1112/T)$
d_{0a}	$CH_3Cl+Cl \rightarrow CH_2Cl+HCl$	$3.4(-11)\exp(-1260/T)$
d_1	$CH_3CCl+OH \rightarrow CH_2CCl_3+H_2O$	$5.4(-12)\exp(-1820/T)$
d_2	$Cl+O_3 \rightarrow ClO+O_2$	$2.8(-11)\exp(-257/T)$
d_3	$ClO+O \rightarrow Cl+O_2$	$4.7(-11)\exp(-50/T)$
d_4	$ClO+NO \rightarrow Cl+NO_2$	$6.2(-12)\exp(+294/T)$
d_{4b}	$ClO+OH \rightarrow products$	$1.0(-11)\exp(+120/T)$
d_5	$Cl+CH_4 \rightarrow CH_3+HCl$	$9.6(-12)\exp(-1350/T)$
d_6	$Cl+H_2 \rightarrow O_2+HCl$	$3.7(-11)\exp(-2300/T)$
d_7	$Cl+HO_2 \rightarrow O_2+HCl$	$1.8(-11)\exp(170/T)$
	$Cl+HO_2 \rightarrow OH+ClO$	$4.1(-11)\exp(-450/T)$
d_8	$Cl+H_2O_2 \rightarrow HO_2+HCl$	$1.1(-11)\exp(-980/T)$
d_9	$Cl+HNO_3 \rightarrow NO_3+HCl$	$< 1.7(-14)$
d_{10}	$Cl+CH_2O \rightarrow HCO+HCl$	$8.2(-11)\exp(-34/T)$
d_{11}	$HCl+OH \rightarrow H_2O+Cl$	$2.6(-12)\exp(-350/T)$
d_{12}	$HCl+O \rightarrow OH+Cl$	$1.0(-11)\exp(-3340/T)$
d_{31}	$ClO+NO_2+M \rightarrow ClONO_2+M$	See below
d_{32}	$ClONO_2+O \rightarrow products$	$3.2(-12)\exp(-808/T)$
d_{32a}	$ClONO_2+OH \rightarrow products$	$1.2(-12)\exp(-333/T)$
d_{32b}	$ClONO_2+Cl \rightarrow products$	$6.8(-12)\exp(+160/T)$
d_{33}	$ClO+HO_2 \rightarrow HOCl+O_2$	$4.6(-13)\exp(710/T)$
d_{34}	$HOCl+OH \rightarrow H_2O+ClO$	$3.0(-12)\exp(-150/T)$
d_{34a}	$HOCl+O \rightarrow OH+ClO$	$1.0(-11)\exp(-2200/T)$
d_{34b}	$HOCl+Cl \rightarrow OH+2Cl$	$3.0(-12)\exp(-130/T)$
d_{35}	$Cl+O_2+M \rightarrow ClOO+M$	$2.0(-33)(\frac{300}{T})^{1.4}[M]$

Bromine reactions

e_0	$CH_3Br+OH \rightarrow CH_2Br+H_2O$	$6.1(-13)\exp(-825/T)$
e_1	$C_2H_4Br_2+OH \rightarrow products+2Br$	$7.3(-12)\exp(-1000/T)$
e_2	$Br+O_3 \rightarrow BrO+O_2$	$1.4(-11)\exp(-755/T)$
e_3	$BrO+O \rightarrow Br+O_2$	$3.0(-11)$
e_4	$BrO+NO \rightarrow Br+NO_2$	$8.7(-12)\exp(265/T)$
e_{4b}	$BrO+OH \rightarrow Br+HO_2$	$1.0(-11)$

e_5	$BrO + ClO \rightarrow Br + OClO$	$6.7(-12)$
	$BrO + ClO \rightarrow Br + Cl + O_2$	$6.7(-12)$
e_6	$BrO + BrO \rightarrow 2Br + O_2$	$1.4(-12)\exp(150/T)$
	$BrO + BrO \rightarrow Br_2 + O_2$	$6.0(-14)\exp(600/T)$
e_7	$Br + HO_2 \rightarrow HBr + O_2$	$8.0(-13)$
e_8	$Br + H_2O_2 \rightarrow HBr + HO_2$	$< 2.0(-15)$
e_9	$Br + CH_2O \rightarrow HBr + HCO$	$1.7(-11)\exp(-800/T)$
e_{11}	$HBr + OH \rightarrow Br + H_2O$	$1.1(-11)$
e_{12}	$HBr + O \rightarrow Br + OH$	$6.6(-12)\exp(-1540/T)$
e_{13}	$BrO + NO_2 + M \rightarrow BrONO_2 + M$	$5.0(-31)[M](300/T)^{2.0}$
e_{15}	$BrO + HO_2 \rightarrow HOBr + O_2$	$5.0(-12)$

Fluorine reactions

f_2	$F + O_3 \rightarrow FO + O_2$	$2.8(-11)\exp(-226/T)$
f_3	$FO + O \rightarrow F + O_2$	$5.0(-11)$
f_4	$FO + NO \rightarrow F + NO_2$	$2.6(-11)$
f_5	$F + CH_4 \rightarrow HF + CH_3$	$3.0(-10)\exp(-400/T)$
f_6	$F + H_2 \rightarrow HF + H$	$1.6(-10)\exp(-525/T)$
f_7	$F + H_2O \rightarrow HF + OH$	$4.2(-11)\exp(-400/T)$

Sulfur reactions

$S + O_2 \rightarrow SO + O$	$2.3(-12)$
$COS + O \rightarrow CO + SO$	$2.1(-11)\exp(-2200/T)$
$COS + OH \rightarrow products$	$3.9(-13)\exp(-1780/T)$
$SO + O_2 \rightarrow SO_2 + O$	$2.4(-13)\exp(-2370/T)$
$SO + O_3 \rightarrow SO_2 + O_2$	$3.6(-12)\exp(-1100/T)$
$SO + NO_2 \rightarrow SO_2 + NO$	$1.4(-11)$
$SO_2 + OH + M \rightarrow HSO_3 + M$	$3.0(-31)\exp(300/T)^{2.9}$
$HSO_3 + O_2 \rightarrow HO_2 + SO_3$	$4.0(-13)$
$HSO_3 + OH \rightarrow H_2O + SO_3$	$1.0(-11)$
$SO_2 + HO_2 \rightarrow SO_3 + OH$	$< 1.0(-18)$
$SO_2 + O + M \rightarrow SO_3 + M$	$3.4(-32)\exp(-1130/T)$
$SO_3 + H_2O \rightarrow H_2SO_4$	$9.0(-13)$

Equivalent bimolecular rates for termolecular reactions

z(km)	$b_{12}(N_2O_5)$	$b_{22}(HNO_3)$	$b_{23a}(HO_2NO_2)$	$d_{31}(ClONO_2)$
10	1.47 (-12)	1.27 (-11)	1.59 (-12)	2.67 (-12)
20	1.22 (-12)	6.43 (-12)	6.12 (-13)	8.19 (-13)
30	6.54 (-13)	1.83 (-12)	1.39 (-13)	1.61 (-13)
40	1.91 (-13)	2.77 (-13)	2.02 (-14)	2.12 (-14)
50	5.80 (-14)	7.18 (-14)	5.34 (-15)	5.24 (-15)

Equilibrium constants (cm^3 molecule^{-1})

$HO_2 + NO_2 \rightarrow HO_2NO_2$	$2.33(-27)\exp(10870/T)$
$NO_2 + NO_3 \rightarrow N_2O_5$	$1.52(-27)\exp(11153/T)$
$Cl + O_2 \rightarrow ClOO$	$2.43(-25)\exp(2979/T)$
$CH_3O_2 + NO_2 \rightarrow CH_3O_2NO_2$	$1.30(-28)\exp(11192/T)$

Reactions of positive ions with neutrals

$O^+ + N_2 \rightarrow NO^+ + N$	$1.2(-12)$
$N^+ + O_2 \rightarrow NO^+ + O$	$2.6(-10)$
$N^+ + O_2 \rightarrow O_2^+ N$	$3.1(-10)$
$N_2^+ + O_2 \rightarrow O_2^+ + N_2$	$5.1(-11)$
$N_2^+ + O \rightarrow NO^+ + N$	$1.4(-10)$
$O_2^+ + O_2 + M \rightarrow O_4^+ + M$	$2.6(-30)(\frac{300}{T})^{3.2}[M]$
$O_2^+ + NO \rightarrow NO^+ + O_2$	$4.4(-10)$
$O_2^+ + N \rightarrow NO^+ + O$	$1.2(-10)$
$O_4^+ + O \rightarrow O_2^+ + O_3$	$3.0(-10)$
$O_4^+ + H_2O \rightarrow O_2^+ \cdot H_2O + O_2$	$1.5(-9)$
$O_4^+ + O_3 \rightarrow O_5^+ + O_2$	$1.0(-10)$
$O_5^+ + H_2O \rightarrow O_2^+ \cdot H_2O + O_3$	$1.2(-9)$
$O_2^+ \cdot H_2O + H_2O \rightarrow H_3O^+ \cdot OH + O_2$	$1.0(-9)$
$O_2^+ \cdot H_2O + H_2O \rightarrow H_3O^+ + OH + O_2$	$2(-10)$
$H^+(H_2O) + H_2O + M \rightarrow H^+(H_2O)_2 + M$	$3.4(-27)(\frac{300}{T})^{4.0}[M]$
$H^+(H_2O)_2 + M \rightarrow H^+(H_2O) + H_2O + M$	$9.6(11)T^{-5}e^{-17100/T}$

$H^+(H_2O)_2+H_2O+M \rightarrow H^+(H_2O)_3+M$	$2.3(-27)(\frac{300}{T})^{4.0}[M]$
$H^+(H_2O)_3+M \rightarrow H^+(H_2O)_2+H_2O+M$	$1.95(11)T^{-5}e^{-11000/T}$
$H^+ \cdot (H_2O)_3+H_2O+M \rightarrow H^+ \cdot (H_2O)_4+M$	$2.4(-27)(\frac{300}{T})^{4.0}[M]$
$H^+ \cdot (H_2O)_4+M \rightarrow H^+ \cdot (H_2O)_3+H_2O+M$	$1.36(11)T^{-5}e^{-8360/T}$
$H_3O^+ \cdot OH+H_2O \rightarrow H^+(H_2O)_2+OH$	$1.4(-9)$
$NO^++H_2O+M \rightarrow NO^+ \cdot H_2O+M$	$1.8(-28)(\frac{308}{T})^{4.7}[M]$
$NO^++CO_2+M \rightarrow NO^+ \cdot CO_2+M$	$7.0(-30)(\frac{300}{T})^{3.0}[M]$
$NO^++N_2+M \rightarrow NO^+ \cdot N_2+M$	$2.0(-31)(\frac{300}{T})^{4.4}[M]$
$NO^+ \cdot CO_2+M \rightarrow NO^++CO_2+M$	$3.1(4)T^{-4}e^{-4590/T}$
$NO^+ \cdot CO_2+H_2O \rightarrow NO^+ \cdot H_2O+CO_2$	$1.0(-9)$
$NO^+ \cdot N_2+M \rightarrow NO^++N_2+M$	$1.5(6)T^{-5.4}e^{-2450/T}$
$NO^+ \cdot N_2+CO_2 \rightarrow NO^+ \cdot CO_2+N_2$	$1.0(-9)$
$H^+(H_2O)+N_2O_5 \rightarrow NO_2^+ \cdot H_2O+HNO_3$	$1.3(-9)$
$H^+(H_2O)+HNO_3 \rightarrow NO_2^+ \cdot H_2O+H_2O$	$1.6(-9)$
$NO_2^+ \cdot H_2O+H_2O+M \rightarrow NO_2^+(H_2O)_2+M$	$2.0(-27)[M]$
$NO_2^+ \cdot (H_2O)_2+H_2O \rightarrow H^+(H_2O)_2+HNO_3$	$2.0(-10)$
$NO^+ \cdot H_2O+H_2O+M \rightarrow NO^+ \cdot (H_2O)_2+M$	$1.0(-27)(\frac{308}{T})^{4.7}[M]$
$NO^+ \cdot H_2O+CO_2+M \rightarrow NO^+ \cdot H_2O \cdot CO_2+M$	$7.0(-30)(\frac{300}{T})^{3.0}[M]$
$NO^+ \cdot H_2O+N_2+M \rightarrow NO^+ \cdot H_2O \cdot N_2+M$	$2.0(-31)(\frac{300}{T})^{4.4}[M]$
$NO^+ \cdot (H_2O)_2+H_2O+M \rightarrow NO^+ \cdot (H_2O)_3+M$	$1.0(-27)(\frac{308}{T})^{4.7}[M]$
$NO^+ \cdot (H_2O)_2+CO_2+M \rightarrow NO^+ \cdot (H_2O)_2 \cdot CO_2+M$	$7.0(-30)(\frac{300}{T})^{3.0}[M]$
$NO^+ \cdot (H_2O)_2+N_2+M \rightarrow NO^+ \cdot (H_2O)_2 \cdot N_2+M$	$2.0(-31)(\frac{300}{T})^{4.4}[M]$
$NO^+ \cdot (H_2O)_3+H_2O \rightarrow H^+ \cdot (H_2O)_3+HNO$	$7.0(-11)$
$NO^+ \cdot H_2O \cdot CO_2+M \rightarrow NO^+ \cdot H_2O+CO_2+M$	$3.1(4)T^{-4}e^{-4025/T}$
$NO^+ \cdot H_2O \cdot CO_2+H_2O \rightarrow NO^+ \cdot (H_2O)_2+CO_2$	$1.0(-9)$
$NO^+ \cdot H_2O \cdot N_2+M \rightarrow NO^+ \cdot H_2O+N_2+M$	$1.5(6)T^{-5.4}e^{-2150/T}$
$NO^+ \cdot (H_2O)_2 \cdot CO_2+M \rightarrow NO^+ \cdot (H_2O)_2+CO_2+M$	$3.1(4)T^{-4}e^{-3335/T}$
$NO^+ \cdot (H_2O)_2 \cdot CO_2+H_2O \rightarrow NO^+(H_2O)_3+CO_2$	$1.0(-9)$
$NO^+ \cdot H_2O \cdot N_2+CO_2 \rightarrow NO^+ \cdot H_2O \cdot CO_2+N_2$	$1.0(-9)$
$NO^+ \cdot (H_2O)_2 \cdot N_2+M \rightarrow NO^+ \cdot (H_2O)_2+N_2+M$	$1.5(6)T^{-5.4}e^{-1800/T}$
$NO^+ \cdot (H_2O)_2 \cdot N_2+CO_2 \rightarrow NO^+ \cdot (H_2O)_2 \cdot CO_2+N_2$	$1.0(-9)$

$H_3O^+ + CFCl_3 \rightarrow$ products $\qquad\qquad$ 4(−10)

$H_3O^+ + CF_2Cl_2 \rightarrow$ products $\qquad\qquad \leqslant 1(−11)$

$H^+(H_2O) + CH_3CN \rightarrow H^+(CH_3CN) + H_2O \qquad$ 4.5(−9)

$H^+(H_2O)_2 + CH_3CN \rightarrow H^+(CH_3CN)\cdot H_2O + H_2O \qquad$ 4.0(−9)

$H^+(H_2O)_3 + CH_3CN \rightarrow H^+(CH_3CN)\cdot(H_2O)_2 + H_2O \qquad$ 3.6(−9)

$H^+(H_2O)_4 + CH_3CN \rightarrow H^+(CH_3CN)(H_2O)_3 + H_2O \qquad$ 3.3(−9)

$H^+(CH_3CN) + CH_3CN + M \rightarrow H^+(CH_3CN)_2 + M \qquad$ 1(−25) [M]

$H^+(H_2O)(CH_3CN)_3 + H_2O \rightarrow H^+(H_2O)_2(CH_3CN)_2 + CH_3CN < 10^{-12}$

Acetonitrile cluster ion reactions (X=CH₃CN)

(1) k_f = forward reaction rate $(cm^6 s^{-1})$ [3 body]

(2) ΔH = kcal/mole

(3) ΔS = cal/mole K

Reaction	$k_f^{(1)}$	$-\Delta H^{(2)}$	$-\Delta S^{(3)}$
$H^+\cdot H_2O + H_2O + M \Leftrightarrow H^+(H_2O)_2 + M$	$2.55(-17)T^{-4}$	31.6	24.3
$H^+(H_2O)_2 + H_2O + M \Leftrightarrow H^+(H_2O)_3 + M$	$2.25(-8)T^{-7.5}$	19.5	21.7
$H^+(H_2O)_3 + H_2O + M \Leftrightarrow H^+(H_2O)_4 + M$	$2.9(-7)T^{-8.1}$	17.9	28.4
$H^+(H_2O)_4 + H_2O + M \Leftrightarrow H^+(H_2O)_5 + M$	$1.52(-7)T^{-4}$	12.7	23.4
$H^+(H_2O)_5 + H_2O + M \Leftrightarrow H^+(H_2O)_6 + M$	$3.2(9)T^{-15.3}$	11.6	25
$H^+(H_2O)_6 + H_2O + M \Leftrightarrow H^+(H_2O)_7 + M$	$3.2(9)T^{-15.3}$	10.7	26.1
$H^+X + H_2O + M \Leftrightarrow H^+XH_2O + M$	$2.18(-17)T^{-4}$	24.8	28.4
$H^+XH_2O + H_2O + M \Leftrightarrow H^+X(H_2O)_2 + M$	$2.1(-8)T^{-7.5}$	17.5	25.1
$H^+X(H_2O)_2 + H_2O + M \Leftrightarrow H^+X(H_2O)_3 + M$	$2.79(-7)T^{-8.1}$	15.6	24.8
$H^+X(H_2O)_3 + H_2O + M \Leftrightarrow H^+X(H_2O)_4 + M$	$1.48(7)T^{-14}$	11.2	21.8
$H^+X(H_2O)_4 + H_2O + M \Leftrightarrow H^+X(H_2O)_5 + M$	$3.15(9)T^{-15.3}$	10.4	23.4
$H^+X(H_2O)_5 + H_2O + M \Leftrightarrow H^+X(H_2O)_6 + M$	$3.16(9)T^{15.3}$	10.1	25.1
$H^+X_2 + H_2O + M \Leftrightarrow H^+X_2H_2O + M$	$2.04(-8)T^{-7.5}$	15.9	24.6
$H^+X_2H_2O + H_2O + M \Leftrightarrow H^+X_2(H_2O)_2 + M$	$2.73(-7)T^{-8.1}$	15.3	25.2
$H^+X_2(H_2O)_2 + H_2O + M \Leftrightarrow H^+X_2(H_2O)_3 + M$	$1.14(7)T^{-14}$	10.3	22.3
$H^+X_2(H_2O)_3 + H_2O + M \Leftrightarrow H^+X_2(H_2O)_4 + M$	$3.11(9)T^{-15.3}$	9.7	21.5

$H^+X_3H_2O+M \Leftrightarrow H^+X_3H_2O+M$	$2.69(-7)T^{-8.1}$	27.2	32.9
$H^+X_3H_2O+H_2O+M \Leftrightarrow H^+X_3(H_2O)_2+M$	$1.43(7)T^{-14}$	9.7	22
$H^+(H_2O)+X+M \Leftrightarrow H^+XH_2O+M$	$2.15(-17)T^{-4}$	46.7	29.3
$H^+(H_2O)_2+X+M \Leftrightarrow H^+X(H_2O)_2+M$	$1.77(-8)T^{-7.5}$	32.6	30.1
$H^+(H_2O)_3+X+M \Leftrightarrow H^+X(H_2O)_3+M$	$2.2(-7)T^{-8.1}$	28.7	33.2
$H^+(H_2O)_4+X+M \Leftrightarrow H^+X(H_2O)_4+M$	$1.13(7)T^{-14}$	22	26.6
$H^+(H_2O)_5+X+M \Leftrightarrow H^+X(H_2O)_5+M$	$2.33(9)T^{-15.3}$	19.7	26.6
$H^+(H_2O)_6+X+M \Leftrightarrow H^+X(H_2O)_6+M$	$2.3(9)T^{-15.3}$	18.2	27.1
$H^+X+X+M \Leftrightarrow H^+X_2+M$	$1.66(-17)T^{-4}$	30.2	29
$H^+XH_2O+X+M \Leftrightarrow H^+X_2H_2O+M$	$1.59(-8)T^{-7.5}$	23.4	24.7
$H^+X(H_2O)_2+X+M \Leftrightarrow H^+X_2(H_2O)_2+M$	$2.06(-7)T^{-8.1}$	21.2	24.8
$H^+X(H_2O)_3+X+M \Leftrightarrow H^+X_2(H_2O)_3+M$	$1.08(7)T^{-14}$	15.9	22.3
$H^+X(H_2O)_4+X+M \Leftrightarrow H^+X_2(H_2O)_4+M$	$2.26(9)T^{-15.3}$	14.4	22
$H^+X_2+X+M \Leftrightarrow H^+X_3+M$	$1.51(-8)T^{-7.5}$	9.3	19
$H^+X_2H_2O+X+M \Leftrightarrow H^+X_3H_2O+M$	$1.98(-7)T^{-8.1}$	20.6	27.3
$H^+X_2(H_2O)_2+X+M \Leftrightarrow H^+X_3(H_2O)_2+M$	$1.05(7)T^{-14}$	15	24.1

Reactions between negative ions and neutrals

$e^-+O_2+M \rightarrow O_2^-+M$	$3.1(-31)(300/T)^{2.5}$
$O^-+O_3 \rightarrow O_3^-+O$	$8.0(-10)$
$O^-+CO_2+M \rightarrow CO_3^-+M$	$3.1(-28)[M]$
$O^-+NO_2 \rightarrow NO_2^-+O$	$1.0(-9)$
$O^-+H_2 \rightarrow OH^-+H$	$3.2(-11)$
$O^-+CH_4 \rightarrow OH^-+CH_3$	$1.0(-10)$
$O^-+O_2(^1\Delta g) \rightarrow e+O_3$	$3.0(-10)$
$O^-+H_2 \rightarrow e+H_2O$	$6.0(-10)$
$O^-+O \rightarrow e+O_2$	$1.9(-10)$
$O^-+NO \rightarrow e+NO_2$	$2.8(-10)$
$O^-+HCl \rightarrow Cl^-+OH$	$2.0(-9)$
$O^-+HNO_3 \rightarrow NO_3^-+OH$	$3.0(-9)$
$O_2^-+O \rightarrow O^-+O_2$	$1.5(-10)$
$O_2^-+O_3 \rightarrow O_3^-+O_2$	$7.8(-10)$
$O_2^-+NO_2 \rightarrow NO_2^-+O_2$	$7.0(-10)$

$O_2^- + CO_2 + M \rightarrow CO_4^- + M$	$4.7(-29)[M]$
$O_2^- + O_2 + M \rightarrow O_4^- + M$	$3.4(-31)[M]$
$O_2^- + O_2(^1\Delta g) \rightarrow e + 2O_2$	$2.0(-10)$
$O_2^- + O \rightarrow e + O_3$	$1.5(-10)$
$O_2^- + H \rightarrow e + HO_2$	$1.4(-9)$
$O_2^- + H_2O + M \rightarrow O_2^- \cdot H_2O + M$	$2.2(-28)[M]$
$O_2^- \cdot H_2O + CO_2 \rightarrow CO_4^- + H_2O$	$5.8(-10)$
$O_2^- \cdot H_2O + NO \rightarrow NO_3^- + H_2O + M$	$2.0(-10)$
$O_2^- + HCl \rightarrow Cl^- + HO_2$	$1.6(-9)$
$O_2^- + HNO_3 \rightarrow NO_3^- + HO_2$	$2.8(-9)$
$O_3^- + O \rightarrow O_2^- + O_2$	$2.5(-10)$
$O_3^- + H \rightarrow OH^- + O_2$	$8.4(-10)$
$O_3^- + CO_2 \rightarrow CO_3^- + O_2$	$5.5(-10)$
$O_3^- + NO_2 \rightarrow NO_3^- + O_2$	$2.8(-10)$
$O_3^- + NO \rightarrow NO_3^- + O$	$4.5(-12)$
$CO_3^- + N_2O_5 \rightarrow NO_3^- + CO_2 + NO_3$	$2.8(-10)$
$CO_3^- + O \rightarrow O_2^- + CO_2$	$1.1(-10)$
$CO_3^- + H \rightarrow OH^- + CO_2$	$1.7(-10)$
$CO_3^- + NO \rightarrow NO_2^- + CO_2$	$1.0(-11)$
$CO_3^- + NO_2 \rightarrow NO_3^- + CO_2$	$2.0(-10)$
$CO_3^- + HNO_3 \rightarrow NO_3^- + HCO_3$	$8.0(-10)$
$NO_2^- + O_3 \rightarrow NO_3^- + O_2$	$1.2(-10)$
$NO_2^- + NO_2 \rightarrow NO_3^- + NO$	$2.0(-13)$
$NO_2^- + H \rightarrow OH^- + NO$	$3.0(-10)$
$NO_2^- + HCl \rightarrow Cl^- + HNO_2$	$1.4(-9)$
$NO_2^- + HNO_3 \rightarrow NO_3^- + HNO_2$	$1.6(-9)$
$NO_2^- + N_2O_5 \rightarrow NO_3^- + 2NO_2$	$7.0(-10)$
$OH^- + NO_2 \rightarrow NO_2^- + OH$	$1.1(-9)$
$OH^- + O_3 \rightarrow O_3^- + OH$	$9.0(-10)$
$OH^- + CO_2 + M \rightarrow HCO_3^- + M$	$7.6(-28)[M]$
$OH^- + H \rightarrow e + H_2O$	$1.4(-9)$
$OH^- + O \rightarrow e + HO_2$	$2.0(-10)$
$O_4^- + O \rightarrow O_3^- + O_2$	$4.0(-10)$
$O_4^- + CO_2 \rightarrow CO_4^- + O_2$	$4.3(-10)$
$O_4^- + NO \rightarrow NO_3^- + O_2$	$2.5(-10)$

$O_4^- + H_2O \rightarrow O_2^- (H_2O) + O_2$ $1(-10)$

$CO_4^- + O_3 \rightarrow O_3^- + CO_2 + O_2$ $1.3(-10)$

$CO_4^- + O \rightarrow CO_3^- + O_2$ $1.4(-10)$

$CO_4^- + H \rightarrow CO_3^- + OH$ $2.2(-10)$

$CO_4^- + NO \rightarrow NO_3^- + CO_2$ $4.8(-11)$

$CO_4^- + H_2O \rightarrow O_2^- \cdot H_2O + CO_2$ $2.5(-10)$

$Cl^- + H \rightarrow HCl + e$ $9.3(-10)$

$Cl^- + HNO_3 \rightarrow NO_3^- + HCl$ $1.6(-9)$

$Cl^- + N_2O_5 \rightarrow NO_3^- + ClNO_2$ $9.4(-10)$

Photodetachment and photodissociation rates (s^{-1})

$O^- + hv \rightarrow e + O$ 1.4

$O_2^- + hv \rightarrow e + O_2$ 0.38

$OH^- + hv \rightarrow e + OH$ 1.1

$NO_2^- + hv \rightarrow e + NO_2$ $8.0(-4)$

$CO_3^- + hv \rightarrow O^- + CO_2$ 0.15

$O_3^- + hv \rightarrow O^- + O_2$ 0.47

$CO_4^- + hv \rightarrow O_2^- + CO_2$ $6.2(-3)$

$O_4^- + hv \rightarrow O_2^- + O_2$ 0.24

Recombination rates $(cm^3 s^{-1})$

X^- represents negative ions, while Y^+ represents positive cluster ions

$X^- + Y^+ \rightarrow$ neutrals $6.0(-8)$

$e + O_2^+ \rightarrow$ neutrals $1.9(-7)(\frac{300}{T})^{0.5}$

$e + NO^+ \rightarrow$ neutrals $2.3(-7)(\frac{300}{T})^{0.5}$

$e + Y^+ \rightarrow$ neutrals $4.0(-6)$

Appendix D

Estimated mixing ratio profiles

This appendix presents calculated mixing ratio, temperature, air density, and oxygen and ozone column content profiles obtained by a one-dimensional radiative convective photochemical model (see Chapter 7) for mid-latitudes at equinox. The mixing ratios indicated here are consistent with the photochemical rate expressions given in Appendix C. These values should only be considered estimates based on presently accepted photochemistry.

Atmospheric Parameters

Altitude (km)	Temperature (K)	H (km)	[M] (cm^{-3})	Pressure (mb)	Typical K$_{zz}$ (cm^2 s^{-1})
0	288.8	8.6	2.5(19)	1013.0	1.0(5)
5	259.3	7.7	1.5(19)	542.0	1.0(5)
10	229.7	6.8	8.5(18)	269.0	2.0(4)
15	212.6	6.3	4.2(18)	122.0	4.7(3)
20	215.5	6.4	1.8(18)	55.0	3.1(3)
25	218.6	6.5	8.3(17)	25.0	6.8(3)
30	223.7	6.6	3.7(17)	11.5	1.4(4)
35	235.1	7.0	1.7(17)	5.4	2.6(4)
40	249.9	7.4	7.7(16)	2.7	4.8(4)
45	266.1	7.9	3.8(16)	1.4	8.8(4)
50	271.0	8.0	2.0(16)	0.73	1.6(5)
55	265.3	7.9	1.1(16)	0.38	2.8(5)
60	253.7	7.5	5.7(15)	0.20	4.3(5)
65	237.0	7.0	3.0(15)	0.10	5.0(5)
70	220.2	6.5	1.5(15)	0.046	4.0(5)
75	203.4	6.0	7.4(14)	0.021	2.5(5)
80	186.7	5.5	3.4(14)	0.0089	2.0(5)
85	170.0	5.0	2.3(14)	0.0055	2.0(5)

90	177.5	5.3	1.1(14)	0.0027	5.0(5)
95	193.0	5.7	4.9(13)	0.0013	1.0(6)
100	209.2	6.2	2.2(13)	0.00064	1.5(6)

Mixing Ratios

Alt (km)	N_2O	CH_4	H_2O	CH_3Cl	CCl_4
0	3.0(-7)	1.6(-6)	1.4(-2)	7.0(-10)	1.0(-10)
5	3.0(-7)	1.6(-6)	1.0(-3)	6.4(-10)	1.0(-10)
10	3.0(-7)	1.6(-6)	3.6(-5)	6.2(-10)	1.0(-10)
15	2.9(-7)	1.5(-6)	2.9(-6)	5.7(-10)	9.0(-11)
20	2.3(-7)	1.3(-6)	3.3(-6)	4.0(-10)	4.8(-11)
25	1.5(-7)	9.4(-7)	3.9(-6)	2.1(-10)	9.4(-12)
30	9.0(-8)	6.7(-7)	4.4(-6)	1.1(-10)	6.2(-13)
35	4.4(-8)	4.5(-7)	4.8(-6)	4.5(-11)	8.0(-15)
40	2.1(-8)	3.0(-7)	5.2(-6)	1.8(-11)	4.5(-17)
45	1.1(-8)	2.1(-7)	5.4(-6)	7.6(-12)	
50	6.7(-9)	1.6(-7)	5.5(-6)	4.2(-12)	
55	4.6(-9)	1.4(-7)	5.6(-6)	2.9(-12)	
60	3.4(-9)	1.3(-7)	5.6(-6)	2.3(-12)	
65	2.7(-9)	1.2(-7)	5.7(-6)	1.9(-12)	
70	2.3(-9)	1.0(-7)	5.7(-6)	1.8(-12)	
75	2.0(-9)	8.5(-8)	5.7(-6)	1.5(-12)	
80	1.8(-9)	7.0(-8)	5.6(-6)	1.4(-12)	
85	1.6(-9)	5.5(-8)	3.4(-6)	1.3(-12)	
90	1.5(-9)	4.2(-8)	2.1(-6)	1.2(-12)	
95	1.3(-9)	3.3(-8)	1.2(-6)	1.1(-12)	
100	1.2(-9)	2.9(-8)	7.2(-7)	1.1(-12)	

Alt (km)	$CFCl_3$	CF_2Cl_2	O_x	O_3	Ozone column
0	1.7(-10)	2.9(-10)	3.0(-8)	3.0(-8)	9.5(18)
5	1.7(-10)	2.8(-10)	5.1(-8)	5.1(-8)	9.1(18)
10	1.7(-10)	2.8(-10)	8.9(-8)	8.9(-8)	8.7(18)
15	1.6(-10)	2.7(-10)	5.2(-7)	5.2(-7)	8.1(18)
20	9.7(-11)	2.2(-10)	2.7(-6)	2.7(-6)	6.3(18)
25	3.1(-11)	1.4(-10)	5.9(-6)	5.9(-6)	3.7(18)
30	5.5(-12)	7.4(-11)	8.3(-6)	8.3(-6)	1.7(18)
35	3.2(-13)	3.0(-11)	8.0(-6)	8.0(-6)	5.6(17)
40	1.2(-14)	1.0(-11)	5.1(-6)	5.1(-6)	1.7(17)
45	5.9(-16)	4.1(-12)	3.3(-6)	3.3(-6)	5.8(16)

50	1.8(-12)	2.2(-6)	2.1(-6)	2.0(16)
55	9.1(-13)	1.6(-6)	1.4(-6)	7.6(15)
60	5.2(-13)	1.6(-6)	1.0(-6)	2.8(15)
65	3.2(-13)	1.9(-6)	8.6(-7)	9.5(14)
70	2.0(-13)	2.6(-6)	4.0(-7)	3.1(14)
75	1.5(-13)	3.8(-6)	1.8(-7)	1.6(14)
80		3.0(-5)	2.7(-7)	1.1(14)
85		4.6(-4)	7.4(-7)	6.0(13)
90		4.1(-3)	3.6(-7)	1.2(13)
95		1.8(-2)	1.1(-7)	2.6(12)
100		5.3(-2)	1.1(-7)	1.0(12)

Alt (km)	$O(^3P)$	$O(^1D)$	CO	H	OH
0	4.0(-17)	1.1(-22)	1.0(-7)		6.0(-14)
5	3.1(-16)	6.0(-22)	6.8(-8)		1.0(-13)
10	1.4(-15)	2.1(-21)	5.6(-8)		4.1(-14)
15	3.1(-14)	2.9(-20)	3.7(-8)		3.5(-14)
20	8.8(-13)	4.9(-19)	1.8(-8)		1.5(-13)
25	1.0(-11)	5.5(-18)	1.2(-8)		9.5(-13)
30	8.9(-11)	5.4(-17)	1.1(-8)		6.2(-12)
35	7.1(-10)	3.9(-16)	1.2(-8)	3.2(-16)	3.3(-11)
40	4.9(-9)	1.8(-15)	1.3(-8)	1.6(-14)	1.2(-10)
45	2.9(-8)	5.2(-15)	1.8(-8)	4.2(-13)	2.8(-10)
50	1.0(-7)	9.5(-15)	3.6(-8)	4.1(-12)	4.2(-10)
55	1.2(-7)	1.4(-14)	8.9(-8)	2.2(-11)	5.1(-10)
60	6.0(-7)	1.9(-14)	2.0(-7)	9.7(-11)	6.2(-10)
65	1.0(-6)	2.5(-14)	3.9(-7)	4.5(-10)	7.9(-10)
70	2.2(-6)	2.7(-14)	7.0(-7)	2.5(-9)	1.4(-9)
75	3.6(-6)	2.6(-14)	1.1(-6)	2.0(-8)	3.1(-9)
80	3.0(-5)	8.8(-14)	1.6(-6)	2.0(-7)	4.3(-9)
85	4.6(-4)	5.2(-13)	2.1(-6)	4.6(-6)	2.1(-9)
90	4.1(-3)	9.1(-13)	2.6(-6)	7.2(-6)	7.1(-10)
95	1.8(-2)	4.1(-12)	3.1(-6)	9.0(-6)	2.6(-10)
100	5.3(-2)	4.5(-11)	3.4(-6)	1.0(-5)	2.0(-10)

Alt (km)	HO_2	H_2O_2	active nitrogen NO_y	NO	NO_2
0	4.0(-11)	1.8(-8)			
5	7.5(-12)	4.7(-10)			
10	5.4(-13)	1.7(-12)	3.7(-10)	8.2(-11)	3.6(-11)
15	1.7(-13)	8.8(-14)	1.6(-9)	2.5(-10)	1.9(-10)
20	9.6(-13)	1.1(-12)	7.2(-9)	5.4(-10)	9.9(-10)

Estimated mixing ratio profiles

25	6.7(-12)	2.1(-11)	1.4(-8)	1.1(-9)	2.2(-9)
30	2.9(-11)	1.3(-10)	1.7(-8)	2.7(-9)	4.2(-9)
35	6.8(-11)	1.7(-10)	1.7(-8)	6.5(-9)	6.2(-9)
40	9.8(-11)	8.3(-11)	1.5(-8)	1.0(-8)	3.7(-9)
45	1.5(-10)	5.0(-11)	1.3(-8)	1.2(-8)	1.0(-9)
50	2.0(-10)	3.4(-11)	1.1(-8)	1.1(-8)	2.3(-10)
55	2.3(-10)	2.4(-11)	9.5(-9)	9.4(-8)	5.3(-11)
60	2.7(-10)	1.9(-11)	8.3(-9)	8.3(-8)	1.4(-11)
65	3.4(-10)	1.8(-11)	7.7(-9)	7.7(-9)	4.1(-12)
70	5.8(-10)	3.3(-11)	8.6(-9)	8.6(-9)	2.6(-12)
75	1.4(-9)	1.1(-10)	1.4(-8)	1.4(-8)	5.1(-12)
80	1.5(-9)	8.2(-11)	3.1(-8)	3.1(-8)	2.2(-12)
85	4.1(-9)	5.2(-12)	8.4(-8)	8.4(-8)	1.3(-13)
90	9.8(-11)	1.2(-13)	3.0(-7)	3.0(-7)	1.3(-14)
95	3.3(-11)	4.7(-15)	1.2(-6)	1.2(-6)	4.6(-15)
100	1.0(-11)	2.0(-16)	5.0(-6)	5.0(-6)	4.0(-15)

Alt (km)	HNO_3	HO_2NO_2	N_2O_5	active chlorine Cl_y	Cl
0					
5					
10	2.0(-10)	3.4(-11)	1.1(-11)	8.3(-11)	4.6(-17)
15	9.4(-10)	3.5(-11)	6.4(-11)	6.5(-11)	4.4(-17)
20	4.2(-9)	1.9(-10)	5.3(-10)	8.5(-10)	2.1(-15)
25	6.9(-9)	3.4(-10)	1.1(-9)	1.7(-9)	2.1(-14)
30	6.0(-9)	2.4(-10)	1.3(-9)	2.1(-9)	1.7(-13)
35	2.1(-9)	6.1(-11)	7.7(-10)	2.3(-9)	1.4(-12)
40	4.2(-10)	4.9(-12)	2.2(-10)	2.3(-9)	6.8(-12)
45	3.8(-11)	2.8(-13)	5.6(-11)	2.4(-9)	2.1(-11)
50	2.6(-12)	1.8(-14)	1.7(-11)	2.4(-9)	3.8(-11)
55	2.1(-13)	1.5(-15)	5.3(-12)	2.4(-9)	5.2(-11)
60	2.1(-14)	1.5(-16)	1.3(-12)	2.4(-9)	6.6(-11)
65	2.8(-15)		2.3(-13)	2.4(-9)	8.1(-11)
70	9.5(-16)		3.2(-14)	2.4(-9)	8.9(-11)
75			4.4(-15)	2.4(-9)	8.2(-11)
80			2.3(-15)	2.4(-9)	1.0(-10)
85			3.0(-15)	2.4(-9)	3.3(-10)
90			4.8(-15)	2.4(-9)	1.4(-9)
95			8.6(-15)	2.4(-9)	2.2(-9)
100			4.0(-14)	2.4(-9)	2.4(-9)

Alt (km)	ClO	HCl	ClONO$_2$	HOCl	O$_2$ column (cm^{-2})
0					4.8(24)
5					2.5(24)
10	2.0(-14)	8.3(-11)	5.5(-13)	1.4(-15)	1.2(24)
15	3.0(-14)	6.3(-11)	1.5(-12)	4.0(-16)	5.5(23)
20	3.6(-12)	6.6(-10)	1.8(-10)	1.1(-13)	2.5(23)
25	4.1(-11)	8.1(-10)	8.3(-10)	3.7(-12)	1.2(23)
30	1.9(-10)	8.2(-10)	1.0(-9)	2.9(-11)	5.6(22)
35	6.0(-10)	1.2(-9)	4.3(-10)	8.1(-11)	2.8(22)
40	8.6(-10)	1.4(-9)	2.9(-11)	6.1(-11)	1.4(22)
45	5.2(-10)	1.8(-9)	4.6(-13)	2.1(-11)	7.5(21)
50	2.0(-10)	2.1(-9)	6.4(-15)	5.0(-12)	4.1(21)
55	7.4(-11)	2.3(-9)	1.4(-16)	1.2(-12)	2.1(21)
60	3.0(-11)	2.3(-9)	4.8(-18)	3.5(-13)	1.0(21)
65	1.1(-11)	2.3(-9)		1.1(-13)	4.7(20)
70	3.6(-12)	2.3(-9)		3.8(-14)	2.2(20)
75	8.3(-13)	2.3(-9)		1.3(-14)	1.0(20)
80	1.9(-13)	2.3(-9)		2.1(-15)	4.2(19)
85	9.0(-14)	2.1(-9)			2.4(19)
90	2.3(-14)	1.0(-9)			1.2(19)
95	2.9(-15)	2.3(-10)			5.7(18)
100	1.2(-15)	6.0(-11)			2.9(18)

Index